Mathematical and Numerical Modelling
of Heterostructure Semiconductor Devices:
From Theory to Programming

T0214385

*To Pam*

E.A.B. Cole

# Mathematical and Numerical Modelling of Heterostructure Semiconductor Devices: From Theory to Programming

 Springer

E.A.B. Cole
Department of Applied Mathematics
University of Leeds
Leeds LS2 9JT
UK
amt6eac@maths.leeds.ac.uk

ISBN 978-1-84882-936-7        e-ISBN 978-1-84882-937-4
DOI 10.1007/978-1-84882-937-4
Springer London Dordrecht Heidelberg New York

British Library Cataloguing in Publication Data
A catalogue record for this book is available from the British Library

Library of Congress Control Number: 2009941547

Mathematics Subject Classification (2000): 15A06, 15A18, 35J05, 35J10, 35J35, 35Q20, 35Q40, 65F15, 65N06, 65N22, 65N25, 65N50, 65N55, 65N75, 65Z05, 80A10, 81Q05, 81Q15, 81Q20, 82B05, 82B10, 82B30, 82D37, 82D80

*Cover design*: deblik

Printed on acid-free paper

Springer is part of Springer Science+Business Media (www.springer.com)

# Preface

Part of my lecturing work in the School of Mathematics at the University of Leeds involved teaching quantum mechanics and statistical mechanics to mathematics undergraduates, and also mathematical methods to undergraduate students in the School of Electronic and Electrical Engineering at the University. The subject of this book has arisen as a result of research collaboration on device modelling with members of the School of Electronic and Electrical Engineering.

I wanted to write a book which would be of practical help to those wishing to learn more about the mathematical and numerical methods involved in heterojunction device modelling. I have introduced only a comparatively small number of topics, and the reader may think that other important topics should have been included. But of the topics which I have introduced, I hope that I have given the reader some practical advice concerning the implementation of the methods which are discussed. This practical advice includes demonstrating how the implementation of the methods may be tailored to the specific device being modelled, and also includes some sections of computer code to illustrate this implementation. I have also included some background theory regarding the origins of the routines.

I have sought to produce an appropriate blend of theory and practical tips. I have not sought to obtain complete solutions—for example, $I-V$ characteristic results have not been included. The mathematical treatment may not be rigourous enough for the likes of the hardened mathematician, and the device treatment may not be detailed enough for the hardened device modeller. But I hope that I have given a balanced blend of both aspects, so that the reader will be confident enough to tackle subjects which are not covered in this book, particularly in relation to heterojunction device modelling.

When applying the mathematical and numerical routines, I have concentrated mainly on the heterojunction devices MESFET and the HEMT. In particular, the modelling of the HEMT involves problems which are encountered in many situations—coupled highly nonlinear sets of equations, and the explicit solution of the Schrödinger equation. An understanding of the methods involved in the modelling of these devices should give the reader confidence to apply them to other devices. The discretisation of the modelling equations will be described mainly in terms of finite differences.

Part I of the book introduces the basic physical theory which forms the backbone of semiconductor device modelling. The subjects of quantum mechanics and statistical mechanics are introduced in detail, and targeted towards device applications. This Part also includes discussions of the Effective Mass Approximation, the density of states in the quantum wells which form in HEMTs, the Boltzmann Transport Equation from which the transport equation are derived, and the Wigner Transport Equation.

In my collaborative work on device modelling, I have used all of the methods described in Part II to obtain results in device simulations. Part II presents the discussion of some of the main mathematical and numerical methods which are to be applied to the solution of the coupled modelling equations. Chapters include introductions to the Newton method in its specific application to device modelling, upwinding, a phaseplane method which can be used to obtain solutions of the device equation in a way that enables these equations to be easily modified when the device model changes, the multigrid method, a chapter on the numerical solution of the Schrödinger equation, and a chapter on rectangular grid generation. I have also included a chapter on Genetic Algorithms, although this is considered by some to be a method more closely associated with "softer" disciplines. This method cannot be used solely for a full device simulation, but I have found the method to be very useful in solving some subsidiary optimisation problems. Part II also contains short sections of code, written in simple C++, to illustrate how some of the methods can be implemented.

I have written this book for anyone who is interested in learning about, or refreshing their knowledge of, some of the basic mathematical and computational aspects of device modelling. This could include upper-level undergraduates, both mathematicians who would like to enter the world of device modelling, and electronic engineering students who would like an introduction to some of the mathematical and numerical methods applied to device modelling. Also included would be those wishing to learn about, or refresh their knowledge of, the basic subjects of quantum mechanics and statistical mechanics. The work could also be of interest to researchers already working on device modelling, and who would perhaps like to get a different perspective on the subject. I have included a section on simple coding in C++; this should be of use to programmers using other languages, and to undergraduates who have not had much exposure to programming at all.

There is a website associated with this book. It contains a list of errata, together with complete working listings of some of the programmes described in the book. Visit the Springer website at www.springer.com.

Finally, I would like to acknowledge the influence of two people. To Chris Snowden who, as Professor of Microwave Engineering at the University of Leeds, communicated his endless enthusiasm to all who came into contact with him. Also to Peter Landsberg who, as my research supervisor many years ago at Cardiff, taught me that I should not always trust the equation

$$A = B + \text{wishful thinking.}$$

Leeds                                                                    *Eric Cole*

# Contents

# Acronyms

| | |
|---|---|
| ADI | Alternating direction implicit |
| AlGaAs | Aluminium Gallium Arsenide |
| BE | Bose-Einstein |
| BTE | Boltzmann transport equation |
| CAD | Computer aided design |
| EMA | Effective mass approximation |
| FAS | Full Approximation Storage |
| FD | Fermi-Dirac |
| FET | Field Effect Transistor |
| GA | Genetic algorithms |
| GaAs | Gallium Arsenide |
| HEMT | High electron mobility transistor |
| InGaAs | Indium Gallium Arsenide |
| MESFET | Metal Semiconductor Field Effect Transistor |
| MC | Monte Carlo |
| SA | Simulated annealing |
| SOR | Successive over-relaxation |
| SUR | Successive under-relaxation |
| TS | Tsallis |

# Part I
# Overview and physical equations

# Chapter 1
# Overview of device modelling

Mathematical and numerical modelling plays an integral part in the development and simulation of new semiconductor devices. Generally, there are three main strands to the Computer Aided Design (CAD) approach to physical simulations. First, there is the problem of deciding which physical microscopic processes should be included in the physical description. Second, this description must be fed into the simulation of individual devices. Third, these device simulations must be fed into the simulations of the circuits in which these devices are integral. Due to the complexity of the physical structure of these devices, it is no longer possible to model all of this with simple equations which, hopefully, have analytic solutions.

The general modelling process consists of a number of decisions and steps which the modeller must take. Having decided on the particular device to be modelled, decisions must be made in roughly the following order:

1. what physical processes must be included in the description of the device;
2. how to idealise the device to a simpler physical structure;
3. what mathematical approach to take;
4. what equations to use, and what will be the primary macroscopic variables which will quantify the system;
5. how to overlay the idealised system with a discrete grid, and how to discretise the equations on this grid;
6. what methods should be used to solve the equations;
7. how to compare the computed results with observed results.

For example, the idealisation in step 2 could involve choosing a suitably simplified geometric shape. The mathematical approach in step 3 could be a Monte Carlo approach, or one based on the moments of the Boltzmann transport equation. In step 4, a useful set of primary variables could be the electron density $n$, the electrostatic potential $\psi$, and the electron temperature $T_e$. However, this set is inconvenient when modelling devices in which quantum wells are involved, in which case a more useful set would be $\psi$, $T_e$, and the Fermi potential $\phi$. In the event that step 7 indicates that the comparison is not satisfactory, the modeller must then decide which of the previous steps to return to.

E.A.B. Cole, *Mathematical and Numerical Modelling of Heterostructure Semiconductor Devices: From Theory to Programming*, DOI 10.1007/978-1-84882-937-4_1, © Springer-Verlag London Limited 2009

Equivalent circuit modelling has long been a mainstay in device modelling. In this method, a system of circuit elements is used to model the electrical behaviour of devices. These circuit elements have to be both linear and non-linear, but there is a limit to the use of this process in the description of ever more complex devices being developed today. It is then more appropriate to go back further into the physics of device processing, and to build modelling techniques from that basis. Two of the main approaches used in the mathematical modelling are the Monte Carlo (MC) approach, and the approach whereby transport equations are based on the moments of the Boltzmann Transport Equation (BTE).

In the MC approach, a large number of charged particles (or rather, clouds of particles) are subjected to random scattering once the scattering mechanism and other physical features have been specified. This method is often used to solve self-consistently with the Poisson equation. A further use of the MC approach is in the solution of differential equations which are being used in device modelling. The MC approach is extremely powerful when used to obtain information regarding the detailed physical processes of devices. However, it can be somewhat time-consuming and unwieldy when it comes to the rapid development of new device structures.

The approach based on the transport equations derived from the BTE provides one of the most flexible and widely used method. These equations are normally the carrier concentration equation, the momentum conservation equation, and energy transport equation; these are solved self-consistently with the Poisson equation. The complexity of these equations means that no analytic solutions exist, and all solutions must be obtained numerically. Further, these equations contain terms such as carrier mobilities, effective masses, and lifetimes; these quantities can be specified by making reference to data from MC simulations which has been previously obtained. Quantum effects are generally hidden from view in this approach, having been included in the assumptions which led to the generation of the moment equations. However, in the case of the High Electron Mobility Transistor (HEMT) in which quantum wells are formed at material interfaces, the Schrödinger equation must also be solved self-consistently with the four equations already mentioned.

Device equations must be solved on grids of up to three spatial directions. Devices such as the p-n junction can often be satisfactorily modelled using one spatial direction only. Other devices often have a structure in which the dimension in one direction is very much larger than those in the other two; in these cases, a solution of the equations on a two-dimensional cross section will be sufficient. The case studies described in this book will focus on this last case, but the techniques described will be applicable to all dimensional structures.

The whole modelling process—both in the development of the governing equations and their numerical solution—is often a trade-off between two opposing needs; on the one hand the need to obtain rapid and accurate numerical solutions of the modelling equations, and on the other hand the need to be able to modify the governing equations (by perhaps adding extra terms) with the minimum of difficulty. Ideally, it is useful to have a number of solution techniques available. One such technique would enable the governing equations to be easily modified, but which produces a solution at a slower rate. Another would be a faster, more sophisticated numerical technique for all subsequent solutions after it has been established that the governing equations have been sufficiently refined. In this book, attention will be confined to the processing of these equations using hand-written programming techniques, rather than through the use of packages such as SPICE and pSPICE (Horenstein 1995).

The device modeller is faced with a hierarchy of approximations. First, the physical processes are modelled using equations, which are mainly differential equations. In forming these equations, the modeller has had to decide which physical influences to include, and which ones to leave out. Next, these equations must be discretised on a finite grid, and this discretisation is then an approximation to the original differential equations. Finally, these discretised equations can themselves be solved only approximately.

A brief introduction will be given to the main types of devices for which the mathematical and numerical methods in this book will be appropriate. This will be followed by a description of the modelling equations, and of some of the mathematical and numerical techniques used to solve these equations.

## 1.1 Devices

Only a relatively small number of devices can realistically be considered in such a work as this, and only a brief outline of the main types of common semiconductor devices will be given here. The reader should turn to other work for detailed descriptions of these devices (Horenstein 1995; Miles 1989; Streetman 1990).

In *bipolar* devices, the current is carried by both electrons and holes. These include the p-n junction, and the bipolar junction transistors (BJT) which consist of two p-n junctions close together to form either an n-p-n or a p-n-p three-terminal transistor. The FET (Field Effect Transistor) is a *unipolar* device with at least three terminals, in which the current involves only the majority carriers. The current-voltage characteristics between the main terminals (the source and drain) is controlled by the input on at least one other terminal (the gate and perhaps a fieldplate). The varying input on the gate gives rise to a larger or smaller depletion region beneath it, causing a narrowing or broadening of the carrier conduction channel. Fig. 1.1 illustrates a simplified outline of this device. The *gate length* is the distance between the edges of the source and drain which are closest to each other, and are usually of the order of 0.1 to 1.0 microns. The width $W$ can vary from several microns to several millimeters. Examples of the FET are

- The JFET (Junction Field Effect Transistor). In the n-channel JFET, $n^+$ regions are doped beneath the source and drain. A highly-doped p-type channel is introduced into the n-type channel between the source and drain; this effectively provides a p-n junction between the gate and the n-type channel.

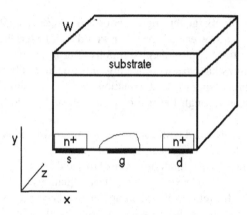

**Fig. 1.1** Simplified view of a FET showing the depletion region and the three contacts source $s$, gate $g$ and the drain $d$. $n^+$ regions are doped beneath the source and drain.

- The MOSFET (Metal Oxide Semiconductor Field Effect Transistor). In an enhancement-type n-channel MOSFET based on silicon, $n^+$ regions are diffused beneath the source and drain contacts into a lightly-doped p-type substrate. The aluminium gate is separated from the p-type silicon by a thin oxide layer. In stand-alone devices, the source can be connected directly to the substrate, but not so when these devices are combined into a multi-transistor integrated circuit.
- The MESFET (MEtal Semiconductor Field Effect Transistor). A typical MESFET device consists of an undoped Gallium Arsenide (GaAs) substrate on which is placed an n-doped channel with $n^+$ implants beneath the source and drain. The gate is usually formed with a metal layer placed on the n-doped channel. The advantage of using GaAs over silicon is that it has a much higher electron mobility—approximately up to an order of magnitude more. This means that high-speed transistors can be used in such devices as high frequency amplifiers and digital logic circuits.
- The HEMT (High Electron Mobility Transistor). On the face of it, the conductivity of the MESFET can be increased by increasing the doping of the GaAs, thereby raising the number of charge carriers and hence increasing the conductivity of the channel. However, the carriers suffer increased scattering due to the increase in the number of donor sites. One way around this is to channel the electrons into a layer of GaAs which is sandwiched between two layers of Aluminium Gallium Arsenide (AlGaAs). Electrons from the donors in the AlGaAs fall into a quantum well and flow unhindered by significant scattering along the channel. This quantum well is formed from the mismatch in the band structures of the GaAs and AlGaAs. A simple HEMT device is shown in Fig. 1.2. At its simplest, it consists of a semi-insulating substrate, followed by a layer of undoped GaAs, followed by a layer of doped AlGaAs. This is followed by caps of $n^+$ GaAs on which a source and drain are situated. A gate contact is recessed to make contact with the AlGaAs layer. HEMTs with larger bandgaps can be

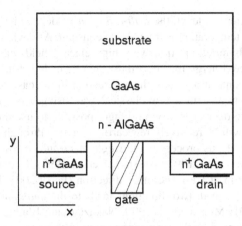

**Fig. 1.2** Simplified view of a HEMT showing the four layers which lie parallel to the $x$-axis, and the three contacts source, gate, and drain.

made by introducing extra thin layers of undoped AlGaAs and InGaAs between the n-AlGaAs and GaAs to form a *pseudomorphic* HEMT. A fourth recessed contact, called a *fieldplate*, is often introduced, and allows higher electric fields to develop in the gate region (Hussain et al. 2003; Karmalkar and Mishra 2001; Sakura et al. 2000; Wakejima et al. 2001). These additions are discussed more fully in Sect 6.7.

Photonic devices play an important role in such devices as CD and DVD players, and require semiconductor devices to transform optical signals into currents. Conversely, light-emitting diodes transform electrical signals into emitted light. In the *photodiode*, an optical lense plays the part of a third terminal alongside the emitter and base of a diode. With the incoming light signal switched off, the current in a reverse biased diode is zero. When the light signal is switched on, a current will flow due to the generation of extra electron-hole pairs. In the case of the *phototransistor*, the base terminal of an n-p-n transistor is supplemented with the input from a light source, effectively making it as a fourth terminal. Hence the device responds to both the light input and to the injected current at the base. A *light-emitting diode* can be made from a silicon p-n junction, or GaAs. Electron and hole recombination in the depletion region under reverse bias conditions causes light to be emitted. However, this light is not monochromatic. A *laser diode* can be made to produce coherent radiation; a p-n junction is sandwiched between two polished surfaces which are parallel to the diode length. The light intensity will build up in this resonant cavity, and the light will leak out of one of the mirrors which is made to be partially transmitting.

*Towards smaller devices*

The modelling of the above devices can utilise the transport equations which are obtained from the moments of the BTE—these transport equations are derived in

Chapter 6. At the simplest level, the *drift-diffusion* model can be used for the case in which the carrier temperature is treated as a constant. As devices become smaller in the move towards minuterisation, very high electric fields can form in certain regions of the device. Energy transport must then be included in the model. Apart from the explicit quantisation along the $y$-direction in the case of the HEMT, the charge carriers are treated in a semiclassical way using a continuum approach. However, as the size decreases, it is no longer possible to use this continuum approach, since there will be relatively few carriers passing through the device at any one time. The limit of this process is the *single electron transistor* (Nakazato and Blaikie 1994; Pothier et al. 1993; Wasshuber et al. 1998). Quantum confinement will then take place in at least one other direction (Ferry 2001; Ferry and Goodnick 1997). Confinement in two directions leads to the *quantum wire* (Bertoni et al. 2001; Ferry 2001; Meirav et al. 1990; Nakazato and Blaikie 1994; Pothier et al. 1993; Wasshuber et al. 1998) and confinement in three dimensions leads to the *quantum dot* (Ezaki et al. 1998; Kumar et al. 1990; Scholze et al. 1998). A major problem encountered in creating such small devices is one of reproducability.

*Devices considered in this book*

Only a small subset of the above devices will be considered in this book, with the main device to be considered being the HEMT. However, this device presents many of the challenges experienced when modelling many other devices, and many of the detailed mathematical and numerical techniques presented for the modelling of the HEMT will apply to the modelling of other devices. The methods described here will apply to the following device characteristics:

- Non-uniform doping profiles.
- multi-contacts: source, gate, drain, fieldplate.
- multi-recess.
- Thin layers—tens of nm.
- At least two spatial dimensions. We can often get away with two since devices are often appreciably longer in the third direction, hence allowing for two-dimensional cross sections.
- mostly uniplolar n-type devices.
- Energy transport.
- Solution of the Schrödinger equation required in quantum wells, with non-constant mass.

## 1.2 Physical theory and modelling equations

The basis of the physical equations lies in the twin topics of quantum mechanics and statistical mechanics. The modelling equations are often a hybrid consisting of explicitly quantum and classical methods.

## 1.2.1 Physical theory

*Quantum mechanics*

The charge carriers in devices are quantum particles, and a full understanding of quantum theory is needed in order to be able to both obtain realistic results and to interpret these results in meaningful ways. Although we shall see that the major device equations are derived by taking moments of the classical BTE, the behaviour of the charge carriers must be interpreted in the light of quantum mechanical ideas. Moreover, when considering the effect of quantum wells formed at the interfaces of different material layers in HEMTs, the Schrödinger equation must be solved directly.

The Schrödinger equation is a second order partial differential equation. This equation must be solved for the state function $\Psi(\mathbf{r}, t)$ which is a function of spatial position $\mathbf{r}$ and time $t$. The parameters which must be specified in the equation are the effective particle mass $m_e$ and the potential $V(\mathbf{r}, t)$. When the potential $V$ is explicitly a function of time $t$, that is, $\partial V / \partial t \neq 0$, then the equation is called the Time Dependent Schrödinger Equation (TDSE). The TDSE simplifies to the Time Independent Schrödinger Equation (TISE) when the function $V$ is independent of time $t$. As in the case of all partial differential equations, the solutions of the Schrödinger equation can only be found through the application of realistic boundary and continuity conditions. The Schrödinger equation can be solved exactly only a very small handful of artificial potential forms, but it is important to study these solutions in order to build an insight into how quantum solutions behave. In the next Chapter, a small number of simple one dimensional potential wells and barriers will be detailed, along with the interpretation of the reflection and transmission coefficients which ensue. The case of a three dimensional spherically symmetric potential will also be considered, with the hydrogen atom as a special case.

In real devices, however, the situation is not so simple. The effective mass is invariably not a constant, but is a function of position within the device. The potential $V$ is not a simple specified analytic function, but is a function of the electrostatic potential which must be solved from the Poisson equation whose input parameters are themselves functions of the solution of the TISE. In short, there is no way of obtaining analytical solutions of the Schrödinger equation for real devices, and numerical solutions must be sought.

Let the gradient operator with respect to the spatial coordinates be denoted by $\nabla_{\mathbf{r}}$. Then since the effective mass $m_e$ is not generally constant, it is not appropriate to use the usual form $-\hbar^2/(2m_e)\nabla_{\mathbf{r}}^2$ for the kinetic energy operator in the TISE, since the form of this operator will not be hermitian. Hence a brief study of the operator nature of observables is necessary in order to obtain an appropriate hermitian form.

The study of spin and identical particles, together with the Pauli Exclusion Principle, leads to the properties of Bose-Einstein (BE) and Fermi-Dirac (FD) statistics; these properties will be derived fully in Chapter 3.

*Statistical mechanics*

The results of the basic theory of statistical mechanics can be applied in many different fields, but in this book the basic results of the theory will be derived for, and applied to, only those topics most relevant to device modelling. The Zeroth Law of Thermodynamics gives rise to the concept of empirical temperature, the First Law gives rise to the concept of internal energy, and the Second Law gives rise to the related ideas of absolute temperature and entropy. The Second Law indicates that, in its relaxation to equilibrium, the thermodynamic entropy of a system cannot decrease. By defining a statistical entropy in terms of the probabilities for the microscopic states of the system, which has the same properties as the thermodynamic entropy, it is shown how this statistical entropy has the same form as the Shannon information content.

By maximising this statistical entropy using the assumptions that the total average energy and total average particle numbers are constant, we arrive at the Grand Canonical distribution function. This function involves summations over particle numbers. By taking upper limits for these summations specified by the Pauli Exclusion Principle (an upper limit of 1 in the case of fermions, and unlimited in the case of bosons), we arrive at the Fermi-Dirac and Bose-Einstein distribution functions.

Once the FD and BE distribution functions have been found, then average quantities such as particle density, system pressure and internal energy can be calculated from these distributions in terms of summations over discrete states. An important next step is to obtain the *continuous approximation*; this is a prescription whereby these summations can be evaluated as integrals—we are very much happier dealing with integrals and derivatives, rather than with discrete summations.

One application of this analysis is that of *Bose-Einstein Condensation*. This is the process in which a collection of bosons crowds into the energy ground state when their density is kept constant and the temperature is reduced below a characteristic condensation temperature. The device charge carriers satisfy FD statistics and, although not directly applicable to the process of modelling the devices considered in this book, the nature of BE condensation is directly analogous to the process of *Simulated Annealing*. This is the process used in the minimisation of functions using genetic algorithms, in which populations of solutions are "bred" subject to certain rules. An artificial temperature is introduced into the process, and this temperature is lowered slowly until a population is bred in which the average function value, taken over the resulting population, reaches its lowest value. This process is discussed more fully in Chapter 14.

The *Bloch theorem* applies to the solution of the TISE when the potential is periodic, as in the case of a crystal structure. The result of this theorem is then used in the *Effective Mass Approximation* (EMA). This approximation is used to replace the full unsolvable TISE with a simpler one for an envelope function which is assumed to be slowly varying over the span of a crystal unit cell.

Away from the quantum wells which are formed at the interfaces of different materials, electron energies and particle numbers are calculated in terms of standard Fermi integrals. The lower limit of these integrals is taken as the bottom of the conduction band. However, the situation in quantum wells is not so straightforward, because discrete energy levels are found in the wells, and these make discrete contributions to the energy and particle number values. Expressions must be found for these values in terms of the discrete solutions of the Schrödinger equation in the wells, and the forms of these expressions will depend on whether the electrons are confined in one or two dimensions.

## 1.2.2 Modelling equations

The process of device modelling begins by obtaining a set of working equations which are distilled from the underlying physical principles. This set consists of two main subsets:

- a subset consisting of equations—most usually differential equations—for the primary physical quantities;
- a subset of subsidiary relations linking the equations of the first subset. These relations will normally provide expressions for physical data based on material properties, and will also provide linking equations for any intermediate variables used in the main set of differential equations.

The modelling equations of many current semiconductor devices are based on the classical BTE, with several quantum effects bolted on to provide a *semiclassical* approach. The equations derived in the first part of this book will be based on the BTE, which is a classical integro-differential equation for a distribution function $f(\mathbf{r}, \mathbf{p}, t)$, for which the carrier's position $\mathbf{r}$ and momentum $\mathbf{p}$ are specified at time $t$. There are several ways in which the BTE can be used when modelling devices. One main method is to use a Monte Carlo approach to finding the distribution function when all of the carrier scattering processes have been specified (Kalos and Whitlock 2008; Kelsall 1998; Landau and Binder 2005; Lugli 1993; Mietzner et al. 2001). The main physical quantities can then be calculated once this distribution function has been found. Generally, this method is computationally expensive; however, it is used to provide data on such quantities as mobilities and lifetimes, and on some of the subsidiary relations described above. A second approach, and one which will be adopted here, is to obtain a set of differential equations based on the first three moments of the BTE. This method provides the time-dependent equations for the carrier concentration, momentum conservation, and energy transport. These equations will also contain quantities which have been provided by previously run Monte Carlo simulations.

Since the definition of the Boltzmann distribution function $f$ assumes that both the position and momentum of a particle are specified simultaneously and precisely, this approach is not strictly applicable for the description of the charge carriers which are quantum particles. The Wigner distribution function, which is defined in terms of the wave function satisfied by the Schrödinger equation, presents a satisfactory method of ammending the approach based on the BTE.

*The Boltzmann Transport Equation and its moments*

The derivation of the BTE and its moments will be discussed fully in Sect. 6.1. Let $f(\mathbf{r}, \mathbf{p}, t)d\mathbf{r}d\mathbf{p}$ be defined as the number of carriers at time $t$ in the region of space $d\mathbf{r}$ centred on position $\mathbf{r}$, and in the region of momentum space $d\mathbf{p}$ centred on momentum $\mathbf{p}$. Let $\nabla_{\mathbf{r}}$ and $\nabla_{\mathbf{p}}$ denote the gradient operators with respect to the position and momentum coordinates respectively. Then the BTE takes the form

$$\frac{\partial f}{\partial t} + \mathbf{F} \cdot \nabla_{\mathbf{p}} f + \mathbf{v} \cdot \nabla_{\mathbf{r}} f = \left(\frac{\partial f}{\partial t}\right)_c$$

where $(\partial f/\partial t)_c$ simply denotes the rate at which these collisions take place. The quantities $\mathbf{F}$ and $\mathbf{v}$ denote the force and particle velocity respectively. A general moment of the equation is found by multiplying it by a general function $A(\mathbf{p})$ and integrating over all $\mathbf{p}$-space. The first three moments of the equation are found by taking $A(\mathbf{p}) = 1$, $A(\mathbf{p}) = \mathbf{p}$, and $A(\mathbf{p}) = m_e v^2/2$ in turn.

Many devices being modelled can be assumed to be unipolar, and this reduces the degree of complexity. Consequently, we will only be concerned here with unipolar modelling. Discussions of bipolar modelling can be found elsewhere (Graaff and Klaassen 1990; Liou 1992; Tsai et al. 2002). Again, we will only be concerned with one level of carrier, and not consider carriers in multivalley structures (Blotekjaer 1970; Cheng and Chennupati 1995; Sandborn et al. 1989).

Let $q$ denote the magnitude of the electron charge, $n$ denote the particle density, $W$ denote the total carrier energy, $\mathbf{E}$ denote the electric field, $T_e$ denote the electron temperature, and $k_B$ denote the Boltzmann constant. Then the first three moments of the BTE will yield the following equations:

*Carrier concentration equation*

$$\frac{\partial n}{\partial t} + \nabla \cdot (n\mathbf{v}) = G$$

where $G$ is the net carrier generation and recombination rate.

*Momentum conservation equation*

$$\frac{\partial \mathbf{p}}{\partial t} + \nabla(nk_B T_e) + (\mathbf{v} \cdot \nabla)\mathbf{p} + \mathbf{p}\nabla \cdot \mathbf{v} + nq\mathbf{E} = \left(\frac{\partial \mathbf{p}}{\partial t}\right)_c.$$

*Energy transport equation*

$$\frac{\partial W}{\partial t} + qn\mathbf{E} \cdot \mathbf{v} + \nabla_{\mathbf{r}}(nW\mathbf{v}) = \left(\frac{\partial W}{\partial t}\right)_c.$$

The minaturisation of devices means that extremely high electric fields can develop at some positions in the device, and in this case the carrier temperature $T_e$ can reach very high values at certain points. Consequently, the full energy transport model is necessary. However, in cases in which the carrier temperature is considered to be constant throughout the device, the *drift-diffusion* model can be appropriate. This model is given by

$$\frac{\partial n}{\partial t} = \frac{1}{q} \nabla_\mathbf{r} \cdot \mathbf{J},$$

$$\mathbf{J} = q \mu n \mathbf{E} + q D \nabla_\mathbf{r} n$$

where $\mu$ is the electron mobility and $D \equiv \mu k_B T_e / q$ is the diffusion coefficient. These equations must be supplemented with the Poisson equation

$$\nabla_\mathbf{r} \cdot (\varepsilon_0 \varepsilon_r \nabla_\mathbf{r} \psi) = -\rho,$$

where $\varepsilon_0$ and $\varepsilon_r$ are the permittivity of free space and relative permittivity respectively, $\psi$ is the electrostatic potential, and $\rho$ is the charge density.

### The Wigner distribution function

The carriers in any semiconductor device are quantum particles. However, the BTE is a classical transport equation in which the particle position $\mathbf{r}$ and momentum $\mathbf{p}$ are defined simultaneously and precisely, and this prescription does not apply to quantum particles. The *Wigner function* and its associated transport differential equation replace the Boltzmann distribution and the BTE, to produce a full quantum description of the transport (Bordone et al. 2001; Grubin and Buggeln 2001; Wigner 1932). This function and associated equation are derived in Sect. 6.4. The Wigner function is defined as

$$f_W(\mathbf{r}, \mathbf{p}, t) \equiv \frac{1}{(\pi \hbar)^d} \int_{-\infty}^{\infty} d\mathbf{y} \overline{\Psi(\mathbf{r} + \mathbf{y}, t)} \Psi(\mathbf{r} - \mathbf{y}, t) e^{2i\mathbf{p} \cdot \mathbf{y}/\hbar}$$

where $\int_{-\infty}^{\infty} d\mathbf{y}$ denotes the integral $\int_{-\infty}^{\infty} dy_1 \cdots \int_{-\infty}^{\infty} dy_d$ for a system with spatial dimension $d$. This functional form is not unique, is not always positive, and cannot be regarded as the simultaneous probability of finding position $\mathbf{r}$ and momentum $\mathbf{p}$. It satisfies the equation

$$\frac{\partial f_W}{\partial t} + \frac{\mathbf{p}}{m_e} \cdot \nabla_\mathbf{r} f_W - \frac{1}{(\pi \hbar)^d} \int_{-\infty}^{\infty} d\mathbf{w} \, V_W(\mathbf{r}, \mathbf{w}) f_W(\mathbf{r}, \mathbf{p} - \mathbf{w}) = 0$$

where the function $V_W(\mathbf{r}, \mathbf{w})$ is the *Wigner potential*, defined in terms of the physical potential $V(\mathbf{r})$, by

$$V_W(\mathbf{r}, \mathbf{w}) \equiv \frac{1}{\hbar} \int_{-\infty}^{\infty} d\mathbf{q} \, \sin(\mathbf{w}.\mathbf{q}/\hbar) \{V(\mathbf{r} + \mathbf{q}) - V(\mathbf{r} - \mathbf{q})\}.$$

### The Schrödinger equation applied to the HEMT

The TISE must be solved explicitly in the quantum wells formed at the interfaces of the different material layers in the HEMT. It is often sufficient to solve this equation in one dimensional strips along the $y$-direction which is perpendicular to the material interfaces. This is because the potential is very often much more rapidly varying in this direction than along the $x$-direction, which is parallel to the contact edges. Consequently, when the HEMT simulation takes place on a two dimensional cross section, the equation to be solved for the energy eigenvalues $\lambda_k$ and eigenfunctions $\xi_k$ is

$$-\frac{\hbar^2}{2}\frac{\partial}{\partial y}\left(\frac{1}{m_e}\frac{\partial \xi_k}{\partial y}\right) + V(x, y)\xi_k = \lambda_k \xi_k,$$

whose solutions are subject to the normalisation condition

$$\int dy |\xi_k(y)|^2 = 1.$$

The precise form of the potential will be given in Sect. 6.8. But note that, although we are solving the equation along the $y$-direction, the potential $V$ will also depend on the $x$ coordinate. This situation arises because the potential is, in part, a function of the electrostatic potential. Hence , although the equation is being solved in one dimensional strips along the $y$ direction, the eigensolution in the case of the HEMT will be functions of both the $x$ and $y$ coordinates.

*The overall nature of the modelling equations*

The modelling equations for the HEMT are highly nonlinear differential equations. Based on the moments of the BTE, a minimum set of these equations will consist of the carrier concentration equation, momentum conservation equation, energy transport equation, the Poisson equation, and the Schrödinger equation. These are also supplemented with subsidiary equations linking the electron densities with the main dependent variables. Parameters such as mobilities and lifetimes which occur in the equations are not constants, but are nonlinear functions of the average particle energy $\xi$, which itself is a function of the electron temperature $T_e$. The electron densities in the quantum wells are functions of the primary dependent variables, and of the eigensolutions of the Schrödinger equation. These electron densities appear on the right hand side of the Poisson equation which must be solved in order to provide the potential energy of the Schrödinger equation.

In short, there is no way in which analytic solutions of these equations can be found for realistic devices. Solutions must be obtained numerically; the chapters in Part II detail a significant set of the main methods used in the solution of these equations.

## 1.3 Mathematical and numerical techniques

The equations describing the behaviour of heterostructure semiconductor devices are generally highly nonlinear. Except in a number of artificially simplified cases, the solutions of these equations must be performed numerically. Many numerical and computational techniques have been developed for this purpose, with the aim of providing accurate results in relatively inexpensive computational terms. To go back a stage in the modelling process, a yet more basic problem for the modeller is to produce the governing equations which fully include as many relevant physical processes as possible. If the computed solutions do not compare well with corresponding observed results, then the modelling equations must be modified. Very often, this modification results in considerable effort in preparation for the new numerical solution—involving the calculation of derivatives, re-writing of code, etc. But however the physical model is modified, it is important to remember that it is the physical model which should drive the mathematics, and not vice versa.

The whole modelling process—both development of the governing equations and their numerical solution—is then often a trade-off between two opposing needs; on the one hand the need to obtain rapid and accurate numerical solutions, and on the other hand the need to be able to modify the governing equations (by perhaps adding extra terms) with the minimum of difficulty. Ideally, it would be useful to have a solution technique which would enable the governing equations to be easily modified, and then to use a faster, more sophisticated numerical technique for all subsequent solutions after it has been established that the governing equations have been sufficiently refined.

Having decided on a physical model, whose specification includes the actual device structure and modelling equations, the next step is to is to decide what system of coordinates to impose on the model. This will be decided partly by the geometry of the device. The modelling equations will then be discretised on this coordinate system. The simplest device structure is one in which the device edges and material interfaces are all perpendicular straight lines. In this case, a rectangular grid is often the most appropriate coordinate system to use. The simplest of these rectangular grids is the uniform grid, in which all of the grid spacings along a direction are equal. This is certainly the easiest grid on which to discretise the equations. However, there may be regions of the device where the grid spacings should be close together due, for example, to high electric fields in those regions. It is then wasteful of computing resources to have such a high grid density in regions where quantities are varying slowly. In this case, it is more appropriate to use a non-uniform grid, with fewer grid lines in the more benign regions of the device. In this case, however, a little more work must be done to code the equations.

In order to avoid having too many grid points in regions where they are not needed, and for devices which do not have a regular shape, the use of irregular grids may be more appropriate (Barton 1989; Tsai et al. 2002). For example, triangulation and general quadrilateral shapes can be used for discretisation in two dimensions. *Finite Boxes*, in which grid lines terminate within the device, can be used (Franz et al. 1983; Selberherr 1984; Tsai et al. 2002). In the *Finite Elements method* (Asenov et al. 1993; Brenner and Scott 1994; Fish 2008; Mobbs 1989; Selberherr 1984), trial functions are chosen, and equations are assembled at grid points from those at the surrounding grid points. This method is particularly suited to the modelling of irregular domains, and grid refinement is straightforward, but error analysis in this case is not so easy to carry out as in the finite difference case. In the *Boundary Element method* (Ingham 1989; Kelmanson 2000; Morton and Suli 1990) applied to equations such as the Poisson equation, the divergence form of the equation can be integrated over a finite volume, and Gauss' theorem then used to convert the result into surface integrals. It is only the surface which then needs to be discretised, rather than the whole interior region.

Device modelling equations are generally highly nonlinear, and many methods exist for the solution of such nonlinear equations. Let $\mathbf{X} \equiv (X_0, X_1, \ldots, X_M)^T$ be an $(M + 1)$-component column vector whose components are the $M + 1$ indepen-dent variables $X_i$ (here and throughout the book, the superscript $T$ will denote the transpose). Let $\{F_i(\mathbf{X}); i = 0, 1, \ldots, M\}$ be a set of generally nonlinear functions of the variables. Then a set of nonlinear equations which has the form

$$F_i(\mathbf{X}) = 0 \quad (i = 0, 1, \ldots, M) \tag{1.1}$$

must be solved for the $M + 1$ unknown values $X_0, X_1, \ldots, X_M$. The most basic Newton iteration scheme seeks a solution $\mathbf{X}^{(k+1)} = \mathbf{X}^{(k)} + \delta^{(k)}$ at the $(k + 1)$th iteration step. The corretion term $\delta^{(k)}$ is given by the linear approximation

$$\mathbf{J}^{(k)}\delta^{(k)} = -\mathbf{F}(\mathbf{X}^{(k)}), \tag{1.2}$$

where the $(M + 1) \times (M + 1)$ matrix $\mathbf{J}$, called the *Jacobian* matrix, has elements given by

$$J_{ij}^{(k)} \equiv \frac{\partial F_i(\mathbf{X}^{(k)})}{\partial X_j^{(k)}}. \tag{1.3}$$

In its application to device modelling, the quantities $\mathbf{X}$ must be described in terms of the physical variables which appear in the device equations. It will be shown in Chapter 10 how appropriate groupings of the physical variables at a point can be made so that the quantities $X_i$ are themselves column vectors. For example, if the primary variables whose values are to be solved are the electrostatic potential $\psi$, the electron temperature $T_e$ and the Fermi potential $\phi$, then we can take

$$X_i \equiv (\psi_i, \phi_i, T_{ei})^T$$

where the label $i$ will denote a grid point. In the direct method of solving Equa-tion (1.2), the matrix $\mathbf{J}^{(k)}$ has to be inverted. In practical terms, the matix $\mathbf{J}^{(k)}$ will be a sparse matrix, and the direct method will generally use sparse matrix meth-ods to solve Equation (1.2). It will be shown in Chapter 10 how the use of sparse matrices can be avoided by using a modified Newton method, and how the form of Equation (1.2) is to be modified in order to avoid overshoot of the required solution.

As well as the direct method for solving linear equations such as Equation (1.2), the iteratves method also plays is an important role in solving these equations. This method is generally slower than the direct method, but does not rely on the user having to invert large matrices. Suppose that it is required to solve the set of linear equations

$$\mathbf{AX} = \mathbf{d} \tag{1.4}$$

where $\mathbf{A}$ is a given $(M + 1) \times (M + 1)$ matrix with constant coefficients and $\mathbf{d}$ is a given constant $(M + 1)$-component column vector. If the iterative process

$$\mathbf{X}^{(k+1)} = (\mathbf{I} - \mathbf{A})\mathbf{X}^{(k)} + \mathbf{d} \tag{1.5}$$

converges, then the converged quantity will be the solution of Equation (1.4). The convergence, and the rapidity of it, will be influenced by the values of the eigenvalues of the matrix $\mathbf{I} - \mathbf{A}$. Using the *relaxation method*, Equation (1.5) is replaced with the form

$$\mathbf{X}^{(k+1)} = (1 - w)\mathbf{X}^{(k)} + w((\mathbf{I} - \mathbf{A})\mathbf{X}^{(k)} + \mathbf{d}) \tag{1.6}$$

where $w$ is a real value lying in the range $0 < w < 2$. The scheme is called *successive under-relaxation* (SUR) if $0 < w < 1$, and *successive over-relaxation* (SOR) if $1 < w < 2$. It will be shown in Chapter 7 how (i) the quantity $w$ is allowed to change as the iterations progress: $w \equiv w^{(k)}$, (ii) how the quantity $w^{(k)}$ will be a $3 \times 3$ matrix if, as in the above example, each quantity $X_i$ is a 3-component column vector, and (iii) how Equation (1.6) can be modified using the LU decomposition of the matrix $\mathbf{A}$.

Generally, relaxation methods are much simpler to code than direct methods, although routines take longer to run. In the *Gummel method* (Gummel 1964), the equations are solved successively. For example, the iterated solution of the Poisson equation is fed into the discretised current continuity equation, whose iterated solution is then fed into the energy transport equation. In the *phase plane method* (Cole 2004), the equations are written as a set of first order equations in an extended set of variables, and the solution is followed in the *phase plane* whose coordinates consist of this extended set (Blanchard et al. 2002; King et al. 2008). The phase plane is then filled with trajectories which converge to the equilibrium points of the system of equations, and the natures of these equilibrium points (nodes, saddle points, centres, etc.) are analysed by studying the linear approximations of the equations near the equilibrium points. This method allows the governing equations to be easily modified without having to alter too much code; this method and its application will be discussed in Chapter 11.

In the iterative solution of the equations on a normal grid, the high frequency components of the errors are eliminated rapidly, but the convergence then becomes slow. In the *multigrid method* (Bodine et al. 1993; Bramble 1993; Brandt 1977; Brandt et al. 1983; Briggs and McCormick 2000; Dick et al. 1999; Grinstein et al. 1983; Press et al. 2002; Zhu and Cangellaris 2006) the equations and their partial solutions are moved up and down through a succession of finer and coarser grids. In this way, different error frequencies are eliminated on different grids. This method allows the equations to be solved on the initial fine grid with a significant decrease in computing time. The equations are solved exactly on the coarsest grid, and limited numbers of iterations are performed on the intermediate grids. The coarsest grid must be coarse enough in order for an exact solution to be obtained rapidly on this grid but, in the application of this method to device modelling, it must be ensured that the essential physical details of the device (layer and recess structures) are preserved on each grid.

The task of solving the full time-dependent problem

$$\frac{\partial \mathbf{Y}}{\partial t} = \mathbf{F}(\mathbf{Y}), \tag{1.7}$$

where $\mathbf{Y} \equiv \mathbf{Y}(\mathbf{r}, t)$, is much harder than solving for the steady state solution. Using a uniform time discretisation

$$t = t_0 + r\Delta t \quad (r = 0, 1, \ldots),$$

the simple Euler scheme

$$\mathbf{Y}^{(r+1)} = \mathbf{Y}^{(r)} + \Delta t \mathbf{F}(\mathbf{Y}^{(r)}) \tag{1.8}$$

is completely unsatisfactory because, unless the time step $\Delta t$ is excessively small, errors quickly accumulate. It will be shown in Chapter 7 that the alternative *Crank-Nicholson* method based on the solution

$$\mathbf{Y}^{(r+1)} = \mathbf{Y}^{(r)} + \Delta t . \frac{1}{2} \left( \mathbf{F}(\mathbf{Y}^{(r+1)}) + \mathbf{F}(\mathbf{Y}^{(r)}) \right) \tag{1.9}$$

puts no restriction on the size of $\Delta t$, although other restrictions apply. One drawback of this method is that it is totally implicit, making it difficult to use. An alternative semi-implicit method (Snowden 1988) uses the sequential scheme

$$Y_j^{(r+1)} = Y_j^{(r)} + \frac{1}{2} \left( F_j(Y_0^{(r)}, \ldots, Y_j^{(r+1)}, \ldots, Y_M^{(r)}) \right.$$

$$\left. + F_j(Y_0^{(r)}, \ldots, Y_j^{(r)}, \ldots, Y_M^{(r)}) \right),$$

$$(j = 0, 1, \ldots, M) \tag{1.10}$$

in which each quantity $Y_j^{(r+1)}$ is iterated using a relaxation method. The *alternating direction implicit* (ADI) method allows the solution to be found using only triangular matrices by using half time steps $\Delta t/2$.

As will be seen in Chapter 7, a von Neumann stability analysis of the time discretisation scheme shows that the magnitude and direction of the electric field $\mathbf{E}$ plays an important part on the stability of the iterative scheme. An *upwinding* method is a discretisation method which utilises the flow of an influence to a grid point from neighbouring grid points. The original scheme of Scharfetter and Gummel (1968) was used for a constant electron temperature, and the scheme has been applied in the case of a varying temperature (Cole and Snowden 2000; Tang 1984; Tang and Ieong 1995).

The Schrödinger equation must be solved explicitly in the modelling of the HEMT. This equation can be solved exactly only for a very limited class of artificial potentials, and numerical solutions are generally necessary. The potential is given in terms of the other physical variables which describe the system. and so the equation must be solved self-consistently with the other modelling equations. For example, the potential is partly a function of the electrostatic potential, which is the solution of the Poisson equation. The right hand side of the Poisson equation is a function of electron densities, which themselves must be calculated from the eigen-solutions of the Schrödinger equation using Fermi integrals. The evaluation of the Fermi integrals, and associated Fermi integrals which arise in the calculations in the quantum wells, is a time-consuming process, and approximate methods are required in this evaluation; these are described in Chapter 8. The Schrödinger equation can normally be solved as a stand-alone equation, after the other modelling equations have been grouped together and iterated simultaneously using relaxation methods. Approximation and numerical methods for the solution of the Schrödinger equation include both time-dependent and time-independent perturbation theory, variational methods, Fast Fourier Transforms (Trellakis and Ravaioli 2001), and the trial function method (Das Sarma et al. 1979; Lehmann and Jasiukiewicz 2002; Ng and Khoie 1991; Norris et al. 1985; Stern and Das Sarma 1984; Valadares and Sheard 1993). These methods will be discussed in Chapter 13.

## 1.4 What is in this book, and its limitations

The methods which will be described in this book relate mainly to the steady state solutions of the set of modelling equations using finite differences. This set will consist of the Poisson equation, the Schrödinger equation, and those based on the first few moments of the BTE. The devices which will be used to illustrate the mathematical and numerical methods wil be mainly the MESFET and HEMT.

Part I presents the basic physical theory of the subject. An introduction to quantum mechanics is given in Chapter 2, and Chapter 3 presents the theory of statistical mechanics leading to a description of the Grand Canonical distribution function, which in turn leads to Bose-Einstein and Fermi-Dirac statistics. Density of states formulae, blackbody radiation, vibrational and rotational aspects of specific heat, Schottky peaks, Bose-Einstein condensation and thermionic emission are discussed in Chapter 4. Applications relating to device modelling—the Bloch theorem, position-dependent mass, the effective mass approximation, and the density of states in quantum wells—are discussed in Chapter 5. The BTE is introduced in Chapter 6, together with the models based on the first three moments of this equation. The Wigner equation is also introduced.

Part II presents some of the mathematical and numerical methods which are used to solve the device modelling equations. Short sections of code are given in order to illustrate the implementation of some of the methods. The discretisation of the Poisson and Schrödinger equations are introduced in terms of finite differences in Chapter 7. This chapter also includes descriptions of the solution of linear equations, time discretisation, and function fitting and updating. Since sections of code written in C++ are given in some later chapters, this chapter also includes a brief description of some of the simpler elements of the language. Chapter 8 introduces the Fermi integrals and associated integrals which arise in the calculation of electron densities and energies, and describes how they may be approximated. Chapter 9 introduces the upwinding method. Chapter 10 presents a description of the Newton method for solving nonlinear equations, with a description of how this method can be directed towards device modelling by the judicious grouping of the physical variables. The phaseplane method is introduced in Chapter 11, and the multigrid method is introduced in Chapter 12. The approximate and numerical solutions of the Schrödinger equation are discussed in Chapter 13. Genetic algorithms and simulated annealing are introduced in Chapter 14. These methods have their limitations, and they certainly cannot be applied for the full solution of the device equations. But I have found them to be useful in subsidiary optimisation problems for which the standard, and more rigorous, optimisation methods are difficult to implement. The process of imposing, and refining/de-refining, a non-uniform rectangular grid over a device is discussed in Chapter 15.

# Chapter 2
# Quantum mechanics

The charge carriers in semiconductor devices—electrons and holes—are quantum particles. Any researcher in device modelling must have a thorough grounding in the theory of quantum mechanics, because there is no point in finding solutions to the quantum equations without being able to sensibly interpret these solutions in the quantum context.

## 2.1 The physical basis of quantum mechanics

The subject of quantum mechanics had a very messy beginning. The old certainties of Newtonian mechanics were swept away when new observations negated these ideas, and a traumatic period followed during which its replacement theory, in the form of quantum mechanics, developed in fits and starts.

Up to about 1900, the old ideas of classical physics were accepted. Among them were:

- The success of Newton's laws of motion.
- Variables representing observable quantities were real numbers, usually continuous over some interval. For example, the energy of simple harmonic motion $\ddot{x} = -\omega x$ is $\frac{1}{2}mA^2\omega^2$, where $A$ is the amplitude and $m$ is the mass; this energy could have any non-negative value by adjusting the value of $A$.
- Any two or more observables on a system could be measured simultaneously and precisely; for example, the position $x(t)$ and momentum $m\dot{x}(t)$ of the particle, or the $x$ and $y$ components of angular momentum.
- Light travelled in waves and could be reflected, refracted and diffracted. Young's diffraction experiments (1803) confirmed this, with further confirmation by Maxwell's electromagnetic theory (1864).
- Discovery of X-rays by Roentgen in 1895. They were diffracted as light by von Laue.
- Discovery of the electron in 1897. J.J. Thompson showed that the electron behaved as a Newtonian particle, and measured the ratio charge/mass.

E.A.B. Cole, *Mathematical and Numerical Modelling of Heterostructure Semiconductor Devices: From Theory to Programming*, DOI 10.1007/978-1-84882-937-4_2, © Springer-Verlag London Limited 2009

- Measurements could be made with unlimited accuracy, or would be eventually.

But experiments were beginning to be performed, the outcomes of which could not be explained by the ideas of classical physics:

1. A *blackbody* is one which absorbs all incident radiation with no reflection. It was found that the calculated spectrum of radiation from a blackbody at temperature $T$ did not correspond with experimental results. In fact, the total energy of the radiation calculated using the classical theory of radiation gave an infinite answer (the *ultraviolet catastrophe*).
2. The calculated temperature dependence of specific heats $dU/dT$ of bodies at low temperatures did not agree with experimental results. Experiment showed a fall-off at low temperatures, while calculation showed no fall-off.
3. Electrons are ejected from a solid which is radiated with light. Classical theory could not explain why the energies of the emitted electrons depended strongly on the frequency of the incident radiation, rather than its intensity (Hertz 1887).
4. The *Compton effect* (1923): X-rays with frequency $\nu$ strike a block of paraffin. Electrons are knocked out, together with radiation of frequency $\nu'$. The energy of the ejected electrons was $E = h\nu - h\nu'$. Hence the X-rays behaved as though the had particle momentum. This effect was also seen in problem 3.
5. In the *Davisson-Germer* experiment of 1927, a beam of electrons aimed at a crystal lattice produced a diffraction pattern on a screen placed behind the crystal; the electrons behaved as though they possessed some sort of wave motion.
6. The spectral experiments of Ritz on radiation emitted from solids showed that the wavelengths appeared as a set of discrete values.
7. The classical idea of the atom was that of a charged electron in orbit about a nucleus. However, the acceleration of the charged particle should cause it to radiate, lose energy, and spiral into the nucleus. This did not happen.

In 1901, Max Planck resolved problem (1) by suggesting that electromagnetic radiation was emitted in discrete packets, or *quanta*. The energy of each quantum is

$$E = h\nu$$

where $\nu$ is the frequency of the radiation and $h$ is a constant (now known as Planck's constant) with the value $h = 6.626 \times 10^{-34}$ Joule-secs. Using this suggestion, Einstein resolved problem (3) in 1905 and problem (2) in 1907. The quantum of light is called the *photon*. The three main conclusions from all of this were (i) energy appears in discrete quanta; (ii) electromagnetic radiation and matter both have a particle—wave duality; (iii) certain observable quantities have only discrete values.

To explain the particle—wave duality aspect, it is found that the momentum $p$ of a particle and its associated radiation wavelength $\lambda$ must be related by the *de Broglie relation*

$$p\lambda = h.$$

This explains the Compton effect and the diffraction pattern associated with electrons striking a crystal lattice. Hence

$$E = h\nu, \tag{2.1}$$

$$p\lambda = h, \tag{2.2}$$

$$\hbar \equiv h/(2\pi) = 1.05 \times 10^{-34} \text{ Joule-secs.}$$

The *Uncertainty Principle* deals with the fact that certain pairs of observables cannot be measured simultaneously and precisely. For example, consider an experiment in which light is beamed at a screen containing two slits. The light passes through the slits and a diffraction pattern is formed on a photosensitive screen placed behind the slits. This illustrates the particle—wave duality of the incoming radiation: its wave property causes the wave fronts to spread out from the two slits and to interact constructively and destructively on the screen behind the slits; its particle property causes electrons to be ejected from this screen to enable the diffraction pattern to be seen. Now reduce the intensity of the light so that only one photon at a time passes through the apparatus: over a length of time the same diffraction pattern will build up. Now put detectors behind the slits so that we can determine through which slit the individual photon passes—this destroys the diffraction pattern. So we cannot know the slit through which the photon passes (its position) while knowing its momentum (through measurement of the diffraction pattern).

Any experiment, thought or otherwise, will produce the following result: that if $\Delta x$ and $\Delta p_x$ are the uncertainties in the $x$ components of position and momentum, then

$$\Delta x \, \Delta p_x \geq \frac{1}{2}\hbar. \tag{2.3}$$

Similarly, if $\Delta E$ and $\Delta t$ are the uncertainties in the measurement of the energy of a system and the time at which it is measured, then

$$\Delta t \, \Delta E \geq \frac{1}{2}\hbar. \tag{2.4}$$

These results will be derived later using the operator formalism.

## 2.2 The Schrödinger equation

All this suggests that a particle can be thought of as a *wave packet*, which is composed of a group of plane waves of nearly the same wavelength $\lambda$ which interfere destructively except around a certain point. The de Broglie relation in Equation (2.2) gives

$$p = \frac{h}{\lambda}, \quad \text{or} \quad \Delta p = -\frac{h}{\lambda^2}\Delta\lambda.$$

This result corresponds neatly with the uncertainty relation of Equation (2.3):

- a small spread in $\lambda$ ($\Delta\lambda$ small, $\Delta p$ small) gives a large spread in $x$ ($\Delta x$ large);
- a large spread in $\lambda$ ($\Delta\lambda$ large, $\Delta p$ large) gives a small spread in $x$ ($\Delta x$ small).

A *wave function* $\Psi(\mathbf{r}, t)$ will be associated with the particle, where $\mathbf{r}$ represents a space position and $t$ represents time. Postulate that $|\Psi(\mathbf{r}, t)|^2 d^3\mathbf{r}$ is the probability of finding the particle in the element of volume $d^3\mathbf{r}$ surrounding the point with position $\mathbf{r}$ at time $t$ (postulate of Max Born 1926). We must find an evolution equation for the wave function, the form of which has universal validity; that is, its coefficients, apart from the value of the particle mass and potential, should be constants so that different solutions may be superposed.

### 2.2.1 Derivation of the time-dependent Schrödinger equation

A very crude derivation will be given of this evolution equation in a very special case. Suppose the particle has a constant mass $m$ and is moving in one dimension along the $x$ axis, in the direction of increasing $x$. It is moving in a constant potential $V$ and has energy $E$. Suppose that its momentum $p$ is known exactly ($\Delta p = 0$), so that nothing is known of its position ($\Delta x = \infty$). Then $|\Psi(\mathbf{r}, t)|^2$ is a constant for all $x$ and $t$. The particle will be represented by a plane wave with wavelength $\lambda = h/p$. The solutions of the wave equation

$$\frac{\partial^2 \Psi}{\partial x^2} = \frac{k^2}{\omega^2} \frac{\partial^2 \Psi}{\partial t^2}$$

are $\Psi = e^{\pm i(kx - \omega t)}$, where $\omega = 2\pi \nu$ and $k = 2\pi/\lambda$: the positive sign in the solution is taken for motion along the positive $x$ direction. Now

$$\omega = 2\pi \nu = \frac{2\pi E}{h} = \frac{E}{\hbar} \quad \text{and}$$

$$k = \frac{2\pi}{\lambda} = \frac{2\pi p}{h} = \frac{p}{\hbar}.$$

Hence

$$\Psi(x, t) = e^{\frac{i}{\hbar}(px - Et)} \tag{2.5}$$

It follows from this result that

$$i\hbar \frac{\partial \Psi}{\partial t} = E\Psi \tag{2.6}$$

$$-i\hbar \frac{\partial \Psi}{\partial x} = p\Psi \tag{2.7}$$

$$-\frac{\hbar^2}{2m} \frac{\partial^2 \Psi}{\partial x^2} = \frac{p^2}{2m} \Psi \tag{2.8}$$

Hence the classical energy equation $E = p^2/(2m) + V$ leads to the suggested replacement

$$E\Psi = \frac{p^2}{2m}\Psi + V\Psi \quad \Rightarrow$$

$$i\hbar\frac{\partial\Psi}{\partial t} = -\frac{\hbar^2}{2m}\frac{\partial^2\Psi}{\partial x^2} + V\Psi. \tag{2.9}$$

This suggested substitution has been made in the very special case of a particle whose momentum is known exactly, and is moving in one dimension in a constant potential. Now postulate that the equation holds for a particle moving in three dimensions in a potential $V(\mathbf{r}, t)$:

$$i\hbar\frac{\partial}{\partial t}\Psi(\mathbf{r}, t) = -\frac{\hbar^2}{2m}\nabla^2\Psi(\mathbf{r}, t) + V(\mathbf{r}, t)\Psi(\mathbf{r}, t). \tag{2.10}$$

This is the *Time-Dependent Schrödinger Equation* (TDSE).

## 2.2.2 The time-independent Schrödinger equation

The Schrödinger equation separates when the potential $V$ is independent of $t$: $V \equiv V(\mathbf{r})$. In this case, look for a solution in the form

$$\Psi(\mathbf{r}, t) = u(\mathbf{r})v(t). \tag{2.11}$$

Then Equation (2.10) becomes

$$i\hbar u\frac{dv}{dt} = -\frac{\hbar^2}{2m}v\nabla^2 u + V(\mathbf{r})uv$$

or

$$i\hbar\frac{1}{v}\frac{dv}{dt} = -\frac{\hbar^2}{2m}\frac{1}{u}\nabla^2 u + V(\mathbf{r}) \equiv C \tag{2.12}$$

where $C$ is a constant of separation. The resulting equation for $v(t)$ has the solution $v(t) = \exp(-\frac{i}{\hbar}Ct)$, and so

$$\Psi(\mathbf{r}, t) = u_C(\mathbf{r})e^{-\frac{i}{\hbar}Ct}, \tag{2.13}$$

where $u \equiv u_C$ denotes the fact that $u$ is now a function of $C$. Differentiation of Equation (2.13) with respect to $t$ shows that $i\hbar\frac{\partial\Psi}{\partial t} = C\Psi$, and comparison of this result with that in Equation (2.6) leads to the identification $C \equiv E$. Hence Equation (2.13) becomes

$$\Psi(\mathbf{r}, t) = u_E(\mathbf{r})e^{-\frac{i}{\hbar}Et}. \tag{2.14}$$

Since the Schrödinger equation is linear, the general solution is the superposition

$$\Psi(\mathbf{r}, t) = \sum_E u_E(\mathbf{r})e^{-\frac{i}{\hbar}Et}, \tag{2.15}$$

where the range over which $E$ is summed will depend on the particular system being considered. The spatial part of Equation (2.12) gives

$$\left(-\frac{\hbar^2}{2m}\nabla^2 + V(\mathbf{r})\right)u(\mathbf{r}) = Eu(\mathbf{r}).\tag{2.16}$$

This is the *Time-Independent Schrödinger Equation* (TISE).

By defining the *Hamiltonian* operator

$$H \equiv -\frac{\hbar^2}{2m}\nabla^2 + V,\tag{2.17}$$

the TISE Equation (2.16) can be written

$$Hu(\mathbf{r}) = Eu(\mathbf{r}).\tag{2.18}$$

The full Schrödinger Equation (2.10) then becomes

$$i\hbar\frac{\partial\Psi}{\partial t} = H\Psi.\tag{2.19}$$

## 2.3 Boundary and continuity conditions, and parity

The solution of the Schrödinger equation is not complete without the specification of boundary conditions and continuity conditions. Further, certain properties are held by the solution when the potential possesses some symmetry.

### 2.3.1 Boundary and continuity conditions

Since $|\Psi(\mathbf{r}, t)|^2$ is interpreted as a probability density, we should attempt to normalise the solution using the condition

$$\int |\Psi(\mathbf{r}, t)|^2 d^3\mathbf{r} = 1$$

where the integral is taken over the whole of the physical region of the system. In general, there are two main types of wave function:

1. those for which the normalisation integral converges. These correspond to well localised wave packets, restrained by the potential $V(\mathbf{r})$.
2. those for which the normalisation integral diverges. These correspond to travelling harmonic waves which are not localised or restrained. In this case, we require that the wave function is bounded at great distances in all directions; we

are not interested in those solutions which are unbounded, since these correspond to indefinitely increasing probability.

In all cases, we require that both $\Psi$ and $\nabla\Psi$ are continuous, finite and single valued in order to uniquely represent a definite physical situation. This ensures that the probability density $|\Psi|^2$ is finite and continuous. However, we must relax the condition that $\nabla\Psi$ is continuous at certain infinite discontinuities in the potential.

### 2.3.2 Parity

Suppose that the potential is an even function of $\mathbf{r}$, that is, $V(\mathbf{r}) = V(-\mathbf{r})$. Then

$$-\frac{\hbar^2}{2m}\nabla^2\Psi(\mathbf{r}) + V(\mathbf{r})\Psi(\mathbf{r}) = E\Psi(\mathbf{r}), \quad \text{and}$$

$$-\frac{\hbar^2}{2m}\nabla^2\Psi(-\mathbf{r}) + V(\mathbf{r})\Psi(-\mathbf{r}) = E\Psi(-\mathbf{r}).$$

Hence $\Psi(\mathbf{r})$ and $\Psi(-\mathbf{r})$ satisfy the same equation for a given value of $E$. Unless there are two or more linearly independent solutions corresponding to the energy $E$, then $\Psi(-\mathbf{r}) = \alpha\Psi(\mathbf{r})$ for some constant $\alpha$. Hence $\Psi(\mathbf{r}) = \alpha\Psi(-\mathbf{r})$, from which it follows that $\alpha = \pm 1$. It then follows that $\Psi(-\mathbf{r}) = \pm\Psi(\mathbf{r})$, and hence the solutions for an even function $V(\mathbf{r})$ are even or odd: they are said to have *even parity* or *odd parity* respectively.

## 2.4 The probability current density

An overline on any complex quantity will denote its complex conjugate so that, in particular, $|\Psi|^2 = \overline{\Psi}\Psi$. It follows from Equation (2.10) that

$$\frac{\partial\Psi}{\partial t} = -\frac{\hbar}{2im}\nabla^2\Psi + \frac{1}{ih}V\Psi$$

$$\frac{\partial\overline{\Psi}}{\partial t} = \frac{\hbar}{2im}\nabla^2\overline{\Psi} - \frac{1}{ih}V\overline{\Psi}$$

where the reality of the potential $V$ has been used: $\overline{V} = V$. Hence

$$\begin{aligned}
\frac{\partial}{\partial t}|\Psi|^2 &= \frac{\partial}{\partial t}(\Psi\overline{\Psi}) = \overline{\Psi}\frac{\partial\Psi}{\partial t} + \Psi\frac{\partial\overline{\Psi}}{\partial t} \\
&= -\frac{\hbar}{2im}\overline{\Psi}\nabla^2\Psi + \frac{\hbar}{2im}\Psi\nabla^2\overline{\Psi} \\
&= -\frac{\hbar}{2im}\nabla\cdot(\overline{\Psi}\nabla\Psi - \Psi\nabla\overline{\Psi}) \\
&= -\nabla\cdot\mathbf{s}
\end{aligned} \tag{2.20}$$

where the vector quantity $\mathbf{s}$ is defined as

$$\mathbf{s} \equiv \frac{\hbar}{2im}(\overline{\Psi}\nabla\Psi - \Psi\nabla\overline{\Psi}). \tag{2.21}$$

This leads to the *continuity equation*

$$\frac{\partial}{\partial t}|\Psi|^2 + \nabla \cdot \mathbf{s} = 0. \tag{2.22}$$

The probability of finding the particle at time $t$ in the fixed volume $R$ surrounded by surface $A$ is $P_t = \int_R |\Psi|^2 d^3\mathbf{r}$. Hence

$$\frac{dP_t}{dt} = \int_R \frac{\partial}{\partial t}|\Psi|^2 d^3\mathbf{r} = -\int_R (\nabla \cdot \mathbf{s})d^3\mathbf{r} = -\int_A s_n dA.$$

Hence $s_n dA$ is the probability per unit time that the particle crosses out of the area $dA$ at the point on the surface at which the unit normal vector is $\mathbf{n}$. For this reason, the quantity $\mathbf{s}$ defined in Equation (2.21) is called the *probability current density*.

## 2.5 One dimensional motion

In all of the following sections we consider cases in which the potential $V$ is independent of time, with motion confined to the $x$-direction. We will solve the TISE exactly only for a small number of special cases.

### 2.5.1 General considerations

Suppose that, without loss of generality, the potential $V(x)$ is given by

$$V(x) \to 0 \quad \text{as } x \to -\infty$$
$$\to V_0 > 0 \quad \text{as } x \to \infty$$

with $V_{min} \equiv \min\{V(x)\}$. The situation is shown generally in Fig. 2.1.

A particle with mass $m$ and energy $E$ moves in this potential. The TISE is

$$Eu = Tu + V(x)u$$

where $T$ is the kinetic operator $-\hbar^2/(2m)d^2/(dx^2)$. We will use angular brackets on any observable quantity $A$ to denote its average, or *expectation value* $\langle A \rangle$—an exact definition of expectation value will be given in Sect. 2.6. It follows that

$$E = \langle T \rangle + \langle V \rangle \geq V_{min}$$

**Fig. 2.1** The potential having the properties $V(x) \to 0$ as $x \to -\infty$ and $V(x) \to V_0 > 0$ as $x \to \infty$.

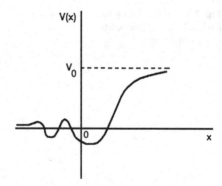

**Fig. 2.2** The points at which the potential $V(x)$ meet the energy $E$.

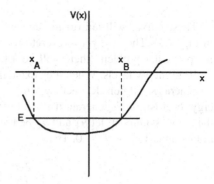

since $\langle T \rangle \geq 0$.

(i) *Bound state*

As $x \to \pm\infty$, the TISE Equation (2.16) becomes

$$u'' = \frac{2m}{\hbar^2}(V_0 - E)u \quad \text{as } x \to \infty$$

$$u'' = -\frac{2m}{\hbar^2}Eu \quad \text{as } x \to -\infty.$$

For a bound state in which the particle is not allowed to go to infinity in either direction, we must have $u \to 0$ as $x \to \pm\infty$. This means that $V_0 - E > 0$ and $E < 0$. Hence bound states occur only if $V_{min} \leq E < 0$. In this case, let $V(x) = E$ at the positions $x = x_A$ and $x = x_B$ as shown in Fig. 2.2. Since $u'' = (2m/\hbar^2(V(x)-E)u$, then for $x_A < x < x_B$ we have the two cases ($u > 0$, $u'' < 0$, $u'$ decreases), and ($u < 0$, $u'' > 0$, $u'$ increases). Hence in both cases we get an oscillating behaviour of $u$ inside the given range, and the curve of $u(x)$ against $x$ bends towards the $x$-axis. This behaviour is shown in Fig. 2.3. Immediately outside the interval $x_A < x < x_B$ we have the two possibilities ($u > 0$, $u'' > 0$, $u'$ increases) and ($u < 0$, $u'' < 0$, $u'$ decreases). Hence in both cases of $u$ the curve of $u(x)$ against $x$ bends away from the $x$-axis. Generally, the curves bend exponentially away from the $x$-axis as $x \to \pm\infty$, and the solution then becomes infinite.

**Fig. 2.3** The oscillating behaviour of $u(x)$ as it bends
towards the $x$-axis.

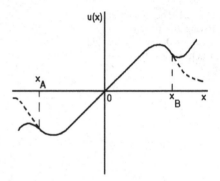

These curves will tail off to zero only for certain values of $E$ (the dotted lines in Fig. 2.3). Thus we get a discrete spectrum for the bound states. Depending on the particular problem, there will be a finite or an infinite number of, or even no, discrete energy levels. When they exist, these energy levels will be *non-degenerate*. Degeneracy and non-degeneracy will be defined properly later: we say that an energy level is non-degenerate if there corresponds to it only one solution. To prove the above statement, let $u_1$ and $u_2$ be two solutions corresponding to the energy $E$ in the range $V_{min} < E < 0$. Then

$$\frac{u_1''}{u_1} = \frac{2m}{\hbar^2}(V - E) = \frac{u_2''}{u_2}.$$

Hence

$$u_1'' u_2 - u_2'' u_1 = 0, \quad \text{or} \quad u_1' u_2 - u_1 u_2' = \text{constant}.$$

This constant is called the *Wronskian*. Since $u_1 = u_2 = 0$ at $x = \pm\infty$, the constant must be zero. Then $u_1'/u_1 = u_2'/u_2$, or $u_1 = const \times u_2$. Hence $u_1$ and $u_2$ are the same up to a multiplicative constant.

(ii) $0 < E < V_0$

If $x$ is sufficiently large and negative, then we can neglect $V(x)$. The corresponding TISE is then $u'' + (2m/\hbar^2)Eu$ which has solution $u = a\cos(kx + \delta)$ where $k \equiv \sqrt{2mE/\hbar^2}$. Then $|u|$ does not tend to zero as $x \to -\infty$, and the particle is not bounded for large negative $x$. On the other hand, as $x \to \infty$, the TISE becomes $u'' = (2m/\hbar^2)(V_0 - E)u$ which has solution $u = be^{-k_1 x}$ where $k_1 \equiv \sqrt{2m(V_0 - E)/\hbar^2}$. Now there is no term $e^{k_1 x}$ since this would cause the solution to diverge as $x \to \infty$. The energies $E$ are again non-degenerate since the quantity $u_1' u_2 - u_1 u_2'$ vanishes as $x \to \infty$.

(iii) $E > V_0$

Let us first look at the effect of the momentum operator $p$ on the quantities $e^{\pm ikx}$, where $k$ is a positive constant. From Equation (2.7) we have

$$pe^{\pm ikx} = -i\hbar \frac{d}{dx} e^{\pm ikx} = \pm k e^{\pm ikx}.$$

**Fig. 2.4** The potential of a single finite step with height $V_0$ situated at $x = 0$.

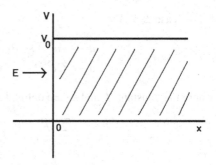

Hence the term $e^{ikx}$ corresponds to a particle moving to the right, while the term $e^{-ikx}$ corresponds to a particle moving to the left. Then as $x \to -\infty$, the solution of the TISE is

$$u \to a\cos(kx + \delta) = pe^{ikx} + qe^{-ikx}$$

and, as $x \to \infty$, the TISE $u'' = -(2m/\hbar^2)(E - V_0)$ has solution

$$u \to re^{ik_2x} + se^{-ik_2x}$$

where $k_2 \equiv \sqrt{2m(E - V_0)/\hbar^2}$. From what we have seen above, the separate terms relate to motion in the positive and negative directions. The solutions are now doubly degenerate, since both of the independent solutions $e^{\pm ikx}$ and $e^{\pm ik_2x}$ correspond to a given value of $E$.

### 2.5.2 Reflection and Transmission coefficients

Consider the piecewise constant potential step shown in Fig. 2.4. The potential is again given by

$$V(x) \to 0 \quad \text{as } x \to -\infty$$
$$\to V_0 > 0 \quad \text{as } x \to \infty$$

A particle of mass $m$ with energy $E$ is incident on the barrier from the left. If $E < V_0$, then classically the particle does not have enough energy to surmount the barrier, and will be reflected back. Quantum mechanically, however, there will be a probability that the particle will cross the barrier—this is the process of *tunneling* which will be quantified later. If $E > V_0$, then classically the carrier will carry on over the barrier, but with reduced velocity. Quantum mechanically, there is a probability that the particle will be reflected back. In order to see this, we solve the TISE

$$u'' + \frac{2m}{\hbar^2}(E - V(x))u = 0$$

for the case $E > V_0$:

$$u \to Ae^{ikx} + Be^{-ikx} \quad \text{as } x \to -\infty \tag{2.23}$$
$$u \to Ce^{ik_2x} \quad \text{as } x \to \infty \tag{2.24}$$

where the constants $k$ and $k_2$ are defined as

$$k \equiv \sqrt{\frac{2mE}{\hbar^2}}, \qquad k_2 \equiv \sqrt{\frac{2m(E - V_0)}{\hbar^2}}.$$

The first term on the right hand side of Equation (2.23) represents the incident wave, while the second represents the reflected wave. The term on the right hand side of Equation (2.24) represents the transmitted wave: note that there is no reflected wave in that region. The one-dimensional equivalent of the probability current density defined in Equation (2.21) is

$$s \equiv \frac{\hbar}{2im} \left( \bar{u}\frac{du}{dx} - u\frac{d\bar{u}}{dx} \right). \tag{2.25}$$

Evaluating and equating this quantity at both ends of the $x$-axis, it is easily shown that

$$\frac{\hbar}{m}k|A|^2 - \frac{\hbar}{m}k|B|^2 = \frac{\hbar}{m}k_2|C|^2 \tag{2.26}$$

Note that the quantity on the left hand side of this equation does not contain cross terms involving $A$ and $B$. Consequently, the probability current densities of the incident, reflected and transmitted parts separately are

$$s_{inc} = \frac{\hbar k}{m}|A|^2, \qquad s_{ref} = -\frac{\hbar k}{m}|B|^2, \qquad s_{tr} = \frac{\hbar k_2}{m}|C|^2.$$

The *reflection coefficient* $R$ and the *transmission coefficient* $T$ are defined by

$$\text{Reflection coefficient: } R \equiv \left| \frac{s_{ref}}{s_{inc}} \right| = \left| \frac{B}{A} \right|^2 \tag{2.27}$$

$$\text{Transmission coefficient: } T \equiv \left| \frac{s_{tr}}{s_{inc}} \right| = \frac{k_2}{k} \left| \frac{C}{A} \right|^2. \tag{2.28}$$

Using Equation (2.26) it is easily shown that

$$T + R = 1. \tag{2.29}$$

It is also easily shown that these coefficients are the same for incidence in either direction for the case $E > V_0$. This is easily proved by noting that, for particles moving from the right, the solution of the TISE is

$$u_1 \to C_1e^{-ikx} \quad \text{as } x \to -\infty$$
$$u_1 \to A_1e^{-ik_2x} + B_1e^{ik_2x} \quad \text{as } x \to \infty.$$

Evaluation of the Wronskian $uu'_1 - u_1u'$ at both ends gives $kAC_1 = k_2A_1C$. Hence

$$\text{Transmission coefficient (L to R)} = \frac{k_2}{k}\left|\frac{C}{A}\right|^2$$

$$= \frac{k}{k_2}\left|\frac{C_1}{A_1}\right|^2$$

$$= \text{Transmission coefficient (R to L).}$$

### 2.5.3 Single finite step for $E < V_0$

Fig. 2.4 shows a single time-independent finite step of height $V_0$ at $x = 0$, with the potential given by

$$V(x) = \begin{cases} 0 & \text{for } x \leq 0 \\ V_0 & \text{for } x > 0 \end{cases} \tag{2.30}$$

A particle of constant mass $m$ approaches the barrier from the left, now with energy $E < V_0$. The TISE is written down for the two regions:

$$-\frac{\hbar^2}{2m}\frac{d^2u}{dx^2} = Eu \quad (x \leq 0) \tag{2.31}$$

$$-\frac{\hbar^2}{2m}\frac{d^2u}{dx^2} + V_0u = Eu \quad (x > 0), \tag{2.32}$$

and these have solutions

$$u(x) = \begin{cases} A\sin\alpha x + B\cos\alpha x & \text{for } x \leq 0 \\ Ce^{-\beta x} + De^{\beta x} & \text{for } x > 0 \end{cases} \tag{2.33}$$

where

$$\alpha \equiv \frac{1}{\hbar}\sqrt{2mE}, \qquad \beta \equiv \frac{1}{\hbar}\sqrt{2m(V_0 - E)}.$$

Since $u(x)$ must be bounded everywhere, then $D = 0$. Since $u$ and $u'$ are continuous at $x = 0$, then

$$B = C, \qquad A\alpha = -\beta C. \tag{2.34}$$

Hence

$$u(x) = \begin{cases} A(\sin\alpha x - \frac{\alpha}{\beta}\cos\alpha x) & \text{for } x \leq 0 \\ -\frac{\alpha}{\beta}Ae^{-\beta x} & \text{for } x \geq 0. \end{cases} \tag{2.35}$$

This simple problem illustrates one of the important differences between the classical and quantum interpretations of the situation. Classically, since the energy $E$ of the incoming particle is less than the height of the barrier, then the particle would

**Fig. 2.5** The potential of the
infinite square well with width
$2a$ centred on $x = 0$.

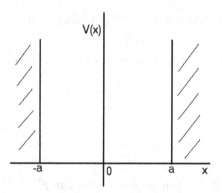

bounce back, and not appear to the right of the barrier. However, this quantum calculation shows that there is a non-zero probability of the particle being to the right
of the barrier. This tunnelling process plays an important part in the transmission of
electrons through potential barriers in semiconductors.

### 2.5.4 Infinite barrier

The particle now approaches an infinite barrier at $x = 0$. We can use the result found
in the previous section by taking the limit $V_0 \to \infty$. In this limit, $\beta \to \infty$, and so
Equation (2.34) shows that we must have $C \to 0$ in order to keep $A$ finite. Hence
$u(x) = 0$ for all $x \geq 0$: this means that there is now zero probability that the particle
will cross the barrier to the other side. The infinite potential means that the particle
encounters an impenetrable barrier.

### 2.5.5 Infinite square well

The particle is now confined in an infinite square well, shown in Fig. 2.5. The potential is given by

$$V(x) = \begin{cases} 0 & \text{for } x > -a \text{ and } x < a \\ \infty & \text{otherwise} \end{cases} \tag{2.36}$$

The TISE to be solved in the region $-a < x < a$ is

$$-\frac{\hbar^2}{2m}\frac{d^2u}{dx^2} = Eu \quad (E > 0), \tag{2.37}$$

which has the solution

**Fig. 2.6** The potential of the
finite square well with width
$2a$ and depth $V_0$.

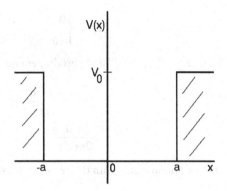

$$u(x) = A \sin kx + B \cos kx, \quad k \equiv \frac{1}{\hbar}\sqrt{2mE}. \tag{2.38}$$

Since $u = 0$ at $x = \pm a$, then

$$A \sin ka + B \cos ka = 0$$
$$-A \sin ka + B \cos ka = 0$$

giving
$$A \sin ka = 0, \qquad B \cos ka = 0.$$

Since $A$ and $B$ cannot simultaneously be zero (otherwise there would be zero probability if finding the particle in the box), then we have the two possible cases

1. $A = 0$ and $\cos ka = 0$, giving $ak = n\pi/2$ where n is an odd integer,
2. $B = 0$ and $\sin ka = 0$, giving $ak = n\pi/2$ where n is an even integer.

Hence
$$u_n(x) = \begin{cases} B \cos \frac{n\pi x}{2a} & \text{for } n \text{ odd} \\ A \sin \frac{n\pi x}{2a} & \text{for } n \text{ even.} \end{cases} \tag{2.39}$$

Note the even and odd parity in this solution, since the potential $V(x)$ is symmetric. Since $k^2 = n^2\pi^2/(4a^2)$, then the allowed energy levels are given by

$$E \equiv E_n = \frac{\pi^2\hbar^2n^2}{8ma^2}, \quad n = 1, 2, 3, \ldots. \tag{2.40}$$

Note that the value $E = 0$ (or $n = 0$) is not allowed, since this again would imply that the particle is not in the region. Normalisation gives $A = B = 1/\sqrt{a}$.

### 2.5.6 Finite square well

The particle now moves in a finite square well, shown in Fig. 2.6. The potential is given by

$$V(x) = \begin{cases} 0 & \text{for } x > -a \text{ and } x < a \\ V_0 > 0 & \text{otherwise} \end{cases} \qquad (2.41)$$

Then $E > V_{min} = 0$. The Schrödinger equation is written down for the two regions:

$$-\frac{\hbar^2}{2m}\frac{d^2u}{dx^2} = Eu \quad (|x| \le a) \qquad (2.42)$$

$$-\frac{\hbar^2}{2m}\frac{d^2u}{dx^2} + V_0 u = Eu \quad (|x| \ge a), \qquad (2.43)$$

and for bound states with $0 < E < V_0$ these equations have solutions

$$u(x) = \begin{cases} A \sin kx + B \cos kx & \text{for } |x| \le a \\ Ce^{-k_1 x} + De^{k_1 x} & \text{for } |x| \ge a \end{cases} \qquad (2.44)$$

where

$$\alpha \equiv \frac{1}{\hbar}\sqrt{2mE}, \qquad \beta \equiv \frac{1}{\hbar}\sqrt{2m(V_0 - E)}.$$

In order to keep the solution finite, we must have $C = 0$ in the region $x < a$, and $D = 0$ in the region $x > a$. Since $u$ and $u'$ are continuous at $x = \mp a$, we must also have

$$A \sin ka + B \cos ka = Ce^{-k_1 a}$$
$$-A \sin ka + B \cos ka = De^{-k_1 a}$$
$$kA \cos ka - kB \sin ka = -k_1 Ce^{-k_1 a}$$
$$kA \cos ka + kB \sin ka = k_1 De^{-k_1 a}$$

Hence it follows that

$$2A \sin ka = (C - D)e^{-k_1 a}, \qquad 2kA \cos ka = -k_1(C - D)e^{-k_1 a}$$
$$2B \cos ka = (C + D)e^{-k_1 a}, \qquad 2kB \sin ka = k_1(C + D)e^{-k_1 a}.$$

The only possible solutions are given by

1.  even parity $A = 0$, $C = D$, $k \tan ka = k_1$,
2.  odd parity $B = 0$, $C = -D$, $k \cot ka = -k_1$.

Hence the energy levels are given by

$$\sqrt{E} \tan\left(a\sqrt{2mE/\hbar^2}\right) = \sqrt{V_0 - E} \qquad (2.45)$$

$$\sqrt{E} \cot\left(a\sqrt{2mE/\hbar^2}\right) = -\sqrt{V_0 - E}. \qquad (2.46)$$

See Problems 2.1 and 2.2.

## 2.5.7 δ-function potential

Consider the δ-function potential centred at $x = 0$:

$$V(x) = V_0\delta(x).  \tag{2.47}$$

We first consider the case $V_0 > 0$, so that $E > 0$. The TISE to be solved is

$$-\frac{\hbar^2}{2m}u'' + V_0\delta(x)u = Eu,  \tag{2.48}$$

which has solution

$$u = Ae^{ikx} + Be^{-ikx}  \quad (x < 0)  \tag{2.49}$$
$$u = Ce^{ik_2x}  \quad (x > 0),  \tag{2.50}$$

where $k \equiv \sqrt{2mE/\hbar^2}$. Integrating both sides of Equation (2.48) from $-\varepsilon$ to $+\varepsilon$ where $\varepsilon > 0$, we have

$$-\frac{\hbar^2}{2m}[u'(\varepsilon) - u'(-\varepsilon)] + V_0u(0) = E\int_{-\varepsilon}^{\varepsilon} u(x)dx.$$

Hence as $\varepsilon \to 0$ and using the fact that $u$ is continuous at $x = 0$, it follows that

$$\lim_{\varepsilon \to 0}[u'(\varepsilon) - u'(-\varepsilon)] = \frac{2m}{\hbar^2}V_0u(0).$$

On using Equations (2.49) and (2.50), this becomes

$$\lim_{\varepsilon \to 0}[ikCe^{ik\varepsilon} - ikAe^{-ik\varepsilon} + ikBe^{-ik\varepsilon}] = C\frac{2m}{\hbar^2}V_0,$$

or

$$ikC - ikA + ikB = \frac{2mV_0}{\hbar^2}C.$$

This result, combined with the continuity condition $A + B = C$, eventually gives

$$B = \frac{-\frac{imV_0}{\hbar^2 k}A}{1 + \frac{mV_0}{\hbar^2 k}i},  \qquad C = \frac{A}{1 + \frac{mV_0}{\hbar^2 k}i},$$

giving the reflection coefficient

$$R = \left|\frac{B}{A}\right| = \frac{1}{\frac{\hbar^4 k^2}{m^2 V_0^2} + 1}.  \tag{2.51}$$

This result holds for both $V_0 > 0$ and $V_0 < 0$. If $V_0 > 0$ then $E > 0$. If $V_0 < 0$ then we can also have $E < 0$: in this case, the solution will be

**Fig. 2.7** The potential of the
square barrier with width $a$
and height $V_0$.

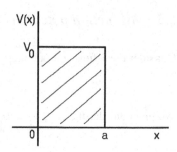

$$u(x) = \begin{cases} Pe^{-k'x} & \text{for } x > 0 \\ Qe^{k'x} & \text{for } x < 0 \end{cases}$$

where $k' \equiv \sqrt{-2mE}/\hbar$. The continuity conditions at $x = 0$ will give

$$P = Q, \quad -k'P - k'Q = \frac{2m}{\hbar^2}V_0P$$

giving $k' = mV_0/\hbar^2$. Hence there is only one allowed energy value, which is given
by

$$E \equiv E_0 = -\frac{mV_0^2}{2\hbar^2}.$$

Then for $E > 0$, the reflection coefficient may be shown to be (see Problem 2.3)

$$R = \frac{|E|}{|E| + |E_0|}. \tag{2.52}$$

### 2.5.8  Square potential barrier

The potential of the square barrier is shown in Fig. 2.7. This is given by

$$V(x) = \begin{cases} 0 & \text{for } x < 0 \text{ and } x > a \\ V_0 > 0 & \text{for } 0 < x < a. \end{cases}$$

For $E > V_0$, the solution of the TISE is

$$u = Ae^{ikx} + Be^{-ikx} \quad \text{for } x \leq 0 \tag{2.53}$$
$$u = Ce^{ikx} \quad \text{for } x \geq a \tag{2.54}$$
$$u = Fe^{ik_2x} + Ge^{-ik_2x} \quad \text{for } 0 \leq x \leq a \tag{2.55}$$

where

$$k \equiv \sqrt{\frac{2mE}{\hbar^2}}, \quad k_2 \equiv \sqrt{\frac{2m(E - V_0)}{\hbar^2}}.$$

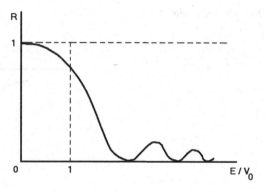

**Fig. 2.8** Perfect transmission ($R = 0, T = 1$) achieved when $k_2a = n\pi$ for $n = 1, 2, \ldots$.

The continuity conditions at $x = 0$ and $x = a$ give

$$A + B = F + G$$
$$ikA - ikB = ik_2F - ik_2G$$
$$Ce^{ika} = Fe^{ik_2a} + Ge^{-ik_2a}$$
$$ikCe^{ika} = ik_2Fe^{ik_2a} - ik_2Ge^{-ik_2a},$$

which can be solved to give

$$B = \frac{(k^2 - k_2{}^2)(1 - e^{2ik_2a})}{(k + k_2)^2 - (k - k_2)^2 e^{2ik_2a}} A,$$

$$C = \frac{4kk_2e^{i(k_2-k)a}}{(k + k_2)^2 - (k - k_2)^2 e^{2ik_2a}} A.$$

The transmission and reflection coefficients may then be calculated as

$$T = \left|\frac{C}{A}\right|^2 = \left(1 + \frac{V_0{}^2 \sin^2(k_2a)}{4E(E - V_0)}\right)^{-1}, \tag{2.56}$$

$$R = \left|\frac{B}{A}\right|^2 = \left(1 + \frac{4E(E - V_0)}{V_0{}^2 \sin^2(k_2a)}\right)^{-1}. \tag{2.57}$$

Perfect transmission ($R = 0, T = 1$) is achieved when $k_2a = n\pi$ for $n = 1, 2, \ldots$, that is, the barrier contains an integral number of half wavelengths. This is the case in the transmission of light through thin refracting layers. This situation is shown in Fig. 2.8.

In the case for which $E < V_0$, we may use the above results with $k_2 = ik_3$, together with the quantity $k_3 \equiv \sqrt{2m(V_0 - E)}/\hbar$. It then follows that

$$T = \left|\frac{C}{A}\right|^2 = \left(1 + \frac{V_0{}^2 \sinh^2(k_3a)}{4E(E - V_0)}\right)^{-1}, \tag{2.58}$$

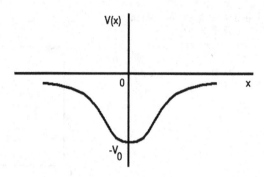

**Fig. 2.9** The potential
$V(x) = \text{sech}^2 x$.

$$R = \left|\frac{B}{A}\right|^2 = \left(1 + \frac{4E(E - V_0)}{V_0^2 \sinh^2(k_3 a)}\right)^{-1}. \tag{2.59}$$

In the situation in which $k_3 a \gg 1$, then $T \approx 16E(V_0 - E)V_0^{-2}e^{-2k_3 a}$ which is a very small quantity.

### 2.5.9 The sech² potential

We investigate the potential

$$V(x) = -V_0 \, \text{sech}^2 \alpha x \tag{2.60}$$

for $V_0 > 0$: this is plotted in Fig. 2.9. With a change of units, this potential is often used to illustrate multi-soliton solutions of the *Korteweg-de Vries (KdV) equation* (Korteweg and Vries 1895; Lax 1968). The corresponding TISE is

$$u'' + \frac{2m}{\hbar^2}(E + V_0 \, \text{sech}^2 \alpha x)u = 0. \tag{2.61}$$

In order to find a solution of this equation, it is convenient to make the substitutions

$$u \equiv \frac{w}{\cosh^s \alpha x}, \qquad s \equiv \frac{1}{2}\left(-1 + \sqrt{1 + \frac{8mV_0}{\alpha^2 \hbar^2}}\right).$$

In terms of these new quantities, TISE then becomes

$$w'' - 2s\alpha(\tanh \alpha x)w' + \left(\alpha^2 s^2 + \frac{2mE}{\hbar^2}\right)w = 0.$$

On using the further new variables

$$y \equiv \sinh^2 \alpha x, \qquad k \equiv \frac{1}{\alpha}\sqrt{-\frac{2mE}{\hbar^2}},$$

the equation becomes the *hypergeometric equation*

$$y(1+y)\frac{d^2w}{dy^2} + \left((1-s)y + \frac{1}{2}\right)\frac{dw}{dy} + \frac{1}{4}(s^2 - k^2)w = 0.$$

Since $V(x) = V(-x)$, then the solutions will have even and odd parity: they are

$$w_1 = F\left(-\frac{1}{2}s + \frac{1}{2}k, -\frac{1}{2}s - \frac{1}{2}k, \frac{1}{2}, -y\right), \tag{2.62}$$

$$w_2 = \sqrt{y}F\left(-\frac{1}{2}s + \frac{1}{2}k + \frac{1}{2}, -\frac{1}{2}s - \frac{1}{2}k + \frac{1}{2}, \frac{3}{2}, -y\right), \tag{2.63}$$

which are given in terms of the hypergeometric function $F$. We must be careful with the sign of $x$ because as $x$ changes sign, $y$ retains the same sign but $\sqrt{y}$ changes sign.

The cases $E < 0$ and $E > 0$, and the KdV equation, will be considered separately:

*Case $E < 0$*

Look for a discrete spectrum. Now $u = (1+y)^{-s/2}w_1 \to 0$ as $y \to \infty$. This implies that $\frac{1}{2}k - \frac{1}{2}s$ is a negative integer or zero. The function $F$ is then a polynomial of degree $\frac{1}{2}s - \frac{1}{2}k$. Similarly, $u = (1+y)^{-s/2}w_2 \to 0$ as $y \to \infty$, and this implies that $\frac{1}{2}k - \frac{1}{2}s + \frac{1}{2}$ is a negative integer. Hence $s - k = n$ for $n = 0, 1, 2, \ldots, N$ for some integer $N$. Since $s$ is fixed and $k > 0$, it follows that $s - n > 0$, and so $n < s$. Hence

$$2n < \sqrt{1 + \frac{8mV_0}{\alpha^2\hbar^2}} - 1,$$

and this determines the value of the integer $N$. Since

$$n = s - k = \frac{1}{2}\left(-1 + \sqrt{1 + \frac{8mV_0}{\alpha^2\hbar^2}}\right) - \frac{1}{\alpha}\sqrt{-\frac{2mE}{\hbar^2}},$$

then the energy levels are given by

$$E \equiv E_n = -\frac{\hbar^2\alpha^2}{8m}\left(-(1+2n) + \sqrt{1 + \frac{8mV_0}{\alpha^2\hbar^2}}\right)^2. \tag{2.64}$$

*Case $E > 0$*

Define $\kappa \equiv \sqrt{2mE}/\hbar^2$. Then $k = i\kappa/\alpha$, and the general solution becomes

$$w = \cosh^s \alpha x \left(C_1 F\left(-\frac{1}{2}s + \frac{1}{2}i\frac{\kappa}{\alpha}, -\frac{1}{2}s - \frac{1}{2}i\frac{\kappa}{\alpha}, \frac{1}{2}, -\sinh^2 \alpha x\right)\right.$$

$$+ C_2(\sinh \alpha x) F\left(-\frac{1}{2}s + \frac{1}{2}i\frac{\kappa}{\alpha} + \frac{1}{2}, -\frac{1}{2}s - \frac{1}{2}i\frac{\kappa}{\alpha} + \frac{1}{2}, \frac{3}{2}, -\sinh^2 \alpha x\right)\right).$$

The ratio of $C_1$ to $C_2$ is chosen such that $w$ has only the asymptotic form $e^{i\kappa x}$ as $x \to \infty$ (Gol'dman and Krivchenkov 1961). The reflection coefficient may then be calculated as

$$R = \frac{\cos^2(\frac{1}{2}\pi\sqrt{1 + \frac{8mV_0}{\alpha^2\hbar^2}})}{\sinh^2 \frac{\pi\kappa}{\alpha} + \cos^2(\frac{1}{2}\pi\sqrt{1 + \frac{8mV_0}{\alpha^2\hbar^2}})}. \tag{2.65}$$

### The KdV equation

The KdV equation can be obtained from the TISE by the replacement $2m/\hbar^2 \to 1$. Then in this case,

$$E_n = -\frac{\alpha^2}{4}\left(-(1 + 2n) + \sqrt{1 + \frac{4V_0}{\alpha^2}}\right)^2$$

with

$$2n < \sqrt{1 + \frac{4V_0}{\alpha^2}} - 1$$

and

$$R = \frac{\cos^2(\frac{1}{2}\pi\sqrt{1 + \frac{4V_0}{\alpha^2}})}{\sinh^2 \frac{\pi\kappa}{\alpha} + \cos^2(\frac{1}{2}\pi\sqrt{1 + \frac{4V_0}{\alpha^2}})}.$$

Two special cases are commonly associated with these results:

1.  $V(x) = -6 \operatorname{sech}^2 x$, giving $\alpha = 1$ and $V_0 = 6$. Then $2n < \sqrt{25} - 1 = 4$, or $n = 0, 1$. The energy levels are

$$E_0 = -\frac{1}{4}(-1 + 5)^2 = -4, \qquad E_1 = -\frac{1}{4}(-(1 + 2) + 5)^2 = -1,$$

and

$$R = \cos^2\left(\frac{1}{2}\pi\sqrt{1 + 24}\right) = 0.$$

2.  $V(x) = -2 \operatorname{sech}^2 x$, giving $\alpha = 1$ and $V_0 = 2$. Then $2n < \sqrt{9} - 1 = 2$, or $n = 0$. Hence there is only one energy value

$$E_0 = -\frac{1}{4}(-1 + \sqrt{9})^2 = -1,$$

and

$$R = \cos^2\left(\frac{1}{2}\pi\sqrt{1 + 8}\right) = 0.$$

## 2.6 Operators and observables

Equations (2.6) and (2.7), namely

$$\left(i\hbar\frac{\partial}{\partial t}\right)\Psi = E\Psi \quad \text{and} \tag{2.66}$$

$$\left(-i\hbar\frac{\partial}{\partial x}\right)\Psi = p\Psi, \tag{2.67}$$

suggest that to each observable there corresponds an *operator*, such that for each observable, the operator acting on a function produces the corresponding measure multiplied by that function. If the suffix *op* denotes the corresponding operator, the observables met so far suggest the following associations:

$$\text{momentum:} \quad \mathbf{p}_{op} = -i\hbar\nabla$$

$$\text{energy:} \quad E_{op} = i\hbar\frac{\partial}{\partial t}$$

$$\text{kinetic energy:} \quad (K)_{op} = \frac{1}{2m}\mathbf{p}_{op}^2 = -\frac{\hbar^2}{2m}\nabla^2 \quad \text{(constant mass)}$$

$$\text{potential energy:} \quad V_{op} = V \quad \text{(multiplication operator)}$$

$$\text{position:} \quad \mathbf{r}_{op} = \mathbf{r} \quad \text{(multiplication operator)}$$

$$\text{time:} \quad t_{op} = t \quad \text{(multiplication operator)}$$

From now on, it will be assumed that the observables are operators unless otherwise stated, and the suffix *op* will be omitted.

Let $A$, $B$, etc. denote operators. Only *linear* operators $A$, $B$, etc. will be considered. These have the properties that, for any complex numbers $c$ and $d$ and any functions $\phi$ and $\psi$,

$$(cA + dB)\psi \equiv cA\psi + dB\psi$$
$$A(\psi + \phi) = A\psi + A\phi \tag{2.68}$$
$$A(c\psi) = cA\psi.$$

The power $A^q$ of an operator is defined inductively as $(A^q)\psi = A^{q-1}(A\psi)$. In this way it is possible to define a function $f(A)$ of an operator $A$ as a series expansion $(\sum_{i=0}^{N} g_i A^i)$ of $A$.

Equations (2.66) and (2.67) are in the form of *eigenvalue equations*, in which an operator acting on a function (the *eigenfunction*) produces a number (the *eigenvalue*) multiplied by that function. Since appropriate boundary conditions on the eigenfunctions mean that the equation is valid for only a discrete set of eigenvalues, this is an appropriate way of building the discreteness of certain observable quantities into our theory.

Consider the eigenvalue equation

$$A\phi = a\phi \qquad (2.69)$$

where $A$ is an operator, $\phi$ is a function, and $a$ is a number. In general, there will be a set of eigenfunctions and eigenvalues corresponding to $A$ when appropriate boundary conditions are applied to the solution:

$$A\phi_n = a_n\phi_n, \quad n = 1, 2, \ldots \qquad (2.70)$$

Hence, for a linear operator,

$$A\phi_n = a_n\phi_n \quad \Rightarrow \quad A^q\phi_n = a_n{}^q\phi_n, \qquad (2.71)$$

and

$$f(A)\phi_n = \left(\sum_{i=0}^{N} g_i A^i\right)\phi_n = \sum_{i=0}^{N} g_i a_n{}^i\phi_n = f(a_n)\phi_n.$$

Hence

$$A\phi_n = a_n\phi_n \quad \Rightarrow \quad f(A)\phi_n = f(a_n)\phi_n. \qquad (2.72)$$

**Definition 2.1.** The *inner product* $(\psi, \phi)$ of any two functions $\phi$ and $\psi$ is

$$(\psi, \phi) \equiv \int \overline{\psi}\phi d^3\mathbf{r}. \qquad (2.73)$$

This inner product has the property that, for all functions $\phi$ and $\psi$,

$$\overline{(\psi, \phi)} = (\phi, \psi).$$

In particular, this result shows that the quantity $(\psi, \psi)$ is real.

**Definition 2.2.** Functions $\phi$ and $\psi$ are *orthogonal* if $\phi \neq 0$, $\psi \neq 0$ and $(\phi, \psi) = 0$.

**Definition 2.3.** An operator $A$ is *hermitian* if, for any functions $\phi$ and $\psi$, we have

$$(\psi, A\phi) = (A\psi, \phi). \qquad (2.74)$$

From the definition (2.73), it is easily verified that for any complex constants $c$ and $d$ and any functions $\psi$, $\psi_1$, $\psi_2$, $\phi$, $\phi_1$ and $\phi_2$, then

$$(\psi, c\phi) = c(\psi, \phi) \qquad (2.75)$$
$$(d\psi, \phi) = \overline{d}(\psi, \phi) \qquad (2.76)$$
$$(\psi_1 + \psi_2, \phi_1 + \psi_2) = (\psi_1, \phi_1) + (\psi_1, \phi_2) + (\psi_2, \phi_1) + (\psi_2, \phi_2) \qquad (2.77)$$
$$\overline{(\psi, \phi)} = (\phi, \psi) \qquad (2.78)$$
$$(\psi, \psi) = 0 \quad \text{if and only if} \quad \psi = 0. \qquad (2.79)$$

Now suppose that the operator $A$ is hermitian. Then $\overline{(\phi, A\phi)} = (A\phi, \phi) = (\phi, A\phi)$. Hence

$$A \text{ hermitian} \quad \Rightarrow \quad (\phi, A\phi) \text{ is real.} \qquad (2.80)$$

Further, if $A$ is hermitian and

$$A\phi_n = a_n\phi_n \quad \text{with} \quad (\phi_n, \phi_n) \neq 0, \quad n = 1, 2, \ldots,$$

then it follows that

$$a_n(\phi_n, \phi_n) = (\phi_n, a_n\phi_n) = (\phi_n, A\phi_n) = (A\phi_n, \phi_n) = (a_n\phi_n, \phi_n) = \overline{a_n}(\phi_n, \phi_n).$$

Hence $\overline{a_n} = a_n$, and so the eigenvalues of an hermitian operator are real. Further,

$$0 = (\phi_m, A\phi_n) - (A\phi_m, \phi_n) = (\phi_m, a_n\phi_n) - (a_m\phi_m, \phi_n) = (a_n - a_m)(\phi_m, \phi_n).$$

Hence if $a_n \neq a_m$, then $(\phi_m, \phi_n) = 0$. It follows that the eigenfunctions corresponding to distinct eigenvalues are orthogonal.

It is easily shown that the operators $\mathbf{r}$, $\mathbf{p}$ and the kinetic energy are hermitian. For example, since $x$ is real, then

$$(\psi, x\phi) = \int \overline{\psi}x\phi d^3\mathbf{r} = \int \overline{x\psi}\phi d^3\mathbf{r} = (x\psi, \phi),$$

and

$$(\psi, p_x\phi) = \int \overline{\psi}(-i\hbar\partial_x)\phi d^3\mathbf{r} = -i\hbar[\overline{\psi}\phi]_B + i\hbar \int (\partial_x\overline{\psi})\phi d^3\mathbf{r}$$

$$= \int (\overline{-i\hbar\partial_x\psi})\phi d^3\mathbf{r} = (p_x\psi, \phi)$$

where we have used the fact that, for any real system, the functions $\psi$ and $\phi$ vanish at the boundaries $B$. Further, since $K = \mathbf{p}^2/(2m)$ for a constant mass $m$, then

$$(\psi, K\phi) = \left(\psi, \frac{1}{2m}\mathbf{pp}\phi\right) = \frac{1}{2m}(\psi, \mathbf{pp}\phi) = \frac{1}{2m}(\mathbf{pp}\psi, \phi) = (K\psi, \phi)$$

since $m$ is real and $\mathbf{p}$ is hermitian. Hence the kinetic energy operator $K$ is hermitian.

Hermitian operators are very important in quantum mechanics. If we associate the eigenvalues and expectation values of operators with observed values, then we want these eigenvalues to be real. Hermitian operators have real eigenvalues, and we will see later that we postulate that an observable may be represented by a linear hermitian operator. However, an operator does not need to be hermitian to have real eigenvalues: for example, the operator $H \equiv p^2 + ix^3$ is not hermitian (prove this) but it does have real eigenvalues. Later, however, when we seek an operator which represent an observable (in particular, the operator which represents kinetic energy for a particle with a position-dependent mass), we will always look for a hermitian operator. In that way, we can be sure that the operator will give real eigenvalues.

If the eigenfunctions $\phi_n$ are normalised, that is, $\int |\phi_n|^2 d^3\mathbf{r} = 1$, then $(\phi_n, \phi_n) = 1$. Hence if $\phi_n$ and $\phi_m$ are normalised and belong to different eigenvalues, we have

$$(\phi_n, \phi_m) = \int \overline{\phi_n} \phi_m d^3\mathbf{r} = \delta_{mn}. \tag{2.81}$$

In this case, we say that the $\{\phi_n : n = 1, 2, \ldots\}$ form an *orthonormal set.*

**Definition 2.4.** An eigenvalue is *non-degenerate* when there corresponds to it only one eigenfunction.

**Definition 2.5.** An eigenvalue is *degenerate* when there corresponds to it more than one eigenfunction. If, to the eigenvalue $a_n$ there correspond the distinct eigenfunctions $\phi_{n1}, \phi_{n2}, \ldots, \phi_{np}$, that is, $A\phi_{ni} = a_n\phi_{ni}, (i = 1, \ldots, p)$, then we say that $a_n$ is *p-fold degenerate.*

Any linear combination of the members of the set $\{\phi_{ni}\}$ is also an eigenfunction corresponding to the eigenvalue $a_n$. The members of this set are not necessarily orthogonal among themselves, but a set of orthogonal eigenfunctions can be constructed from them using the *Schmidt Orthogonalisation Procedure.* For example, suppose that $p = 2$, and that $(\phi_{n1}, \phi_{n2}) \neq 0$. We require that $\phi_{n1}$ and the linear combination $\phi_{n1} + \alpha\phi_{n2}$ are orthogonal. Then

$$0 = (\phi_{n1}, \phi_{n1} + \alpha\phi_{n2}) = (\phi_{n1}, \phi_{n1}) + \alpha(\phi_{n1}, \phi_{n2})$$

giving $\alpha = -(\phi_{n1}, \phi_{n1})/(\phi_{n1}, \phi_{n2})$. Hence the eigenfunctions

$$\phi_{n1} \quad \text{and} \quad \phi_{n1} - \frac{(\phi_{n1}, \phi_{n1})}{(\phi_{n1}, \phi_{n2})}\phi_{n2}$$

are orthogonal.

**Definition 2.6.** The set of functions $\phi_1, \phi_2, \ldots, \phi_N$ is a *complete set* of orthonormal functions if

$$\sum_{i=1}^{N} d_i\phi_i = 0 \quad \Rightarrow \quad d_i = 0 \quad (i = 1, \ldots, N)$$

and any arbitrary function can be expanded in terms of them:

$$\psi = \sum_{i=1}^{N} c_i\phi_i. \tag{2.82}$$

The coefficients $c_i$ can be calculated, since

$$(\phi_j, \psi) = \left(\phi_j, \sum_i c_i\phi_i\right) = \sum_i c_i(\phi_j, \phi_i) = \sum_i c_i\delta_{ij} = c_j.$$

Hence Equation (2.82) becomes

$$\psi = \sum_{i=1}^{N} (\phi_i, \psi)\phi_i. \tag{2.83}$$

Suppose that the system is in state $\Psi$, and write $\Psi = \sum_{i=1}^{N} c_i \phi_i$. If $\Psi$ is normalised, then

$$1 = (\Psi, \Psi) = \left( \sum_i c_i \phi_i, \sum_j c_j \phi_j \right) = \sum_i \sum_j \overline{c_i} c_j (\phi_i, \phi_j) = \sum_i |c_i|^2.$$

**Definition 2.7.** The *expectation value* $\langle A \rangle$ of the operator $A$ is defined as

$$\langle A \rangle \equiv (\Psi, A\Psi) = \left( \sum_i c_i \phi_i, A \sum_j c_j \phi_j \right) = \sum_i |c_i|^2 a_i. \qquad (2.84)$$

This expectation value is real if $A$ is hermitian since all the $a_i$ will then be real. Note that the expectation value of any operator $A$ is a linear combination of its eigenvalues; the significance of this linear combination will be discussed in Sect. 2.8.

**Definition 2.8.** For any operator $A$ which is not necessarily hermitian, define the *adjoint* operator $A^\times$ by

$$(\phi, A\psi) = (A^\times \phi, \psi) \quad \text{for all functions } \phi \text{ and } \psi. \qquad (2.85)$$

Then for any operators $A$ and $B$ and any constant $c$, it is easily proved that

$$(A^\times)^\times = A, \qquad (cA)^\times = \overline{c} A^\times, \qquad (AB)^\times = B^\times A^\times,$$
$$(A + A^\times)^\times = (A + A^\times). \qquad (2.86)$$

Further, if $f(x)$ is any complex function of $x$, then

$$[f(A)]^\times = \overline{f}(A^\times). \qquad (2.87)$$

In particular, if $A$ and $B$ are both hermitian, then $A^\times = A$, $B^\times = B$ and $(AB)^\times = BA$.

**Definition 2.9.** An operator $U$ is *unitary* if $UU^\times = U^\times U = I$.

Hence in this case, $U^\times$ acts as the inverse operator. Also, for any functions $\phi$ and $\psi$,

$$(U\phi, U\psi) = (U^\times U\phi, \psi) = (\phi, \psi) \quad (U \text{ unitary}). \qquad (2.88)$$

**Definition 2.10.** For any two operators $A$ and $B$, their *commutator* is defined as

$$[A, B] \equiv AB - BA. \qquad (2.89)$$

It is easily proved that if $[A, B] = 0$ then $[f(A), g(B)] = 0$ for functions $f$ and $g$. For example, taking $f(A) \equiv A^2$ and $f(B) \equiv B$, then the relation $[A, B] = 0$ implies that

$$[A^2, B] = A^2 B - BA^2 = AAB - BAA = ABA - BAA = BAA - BAA = 0.$$

An important example of such a commutator occurs between position and momentum operators. For any function $\psi$,

$$[x, p_x]\psi = x(-i\hbar\partial_x)\psi - (-i\hbar\partial_x)(x\psi) = -i\hbar x\partial_x\psi + i\hbar x\partial_x\psi + i\hbar\psi = i\hbar\psi.$$

A similar result holds between the energy and time operators. Some of the more important results are:

$$[x, p_x] = [y, p_y] = [z, p_z] = i\hbar$$
$$[x, p_y] = 0 \quad \text{etc.}$$
$$[x, y] = [y, z] = [z, x] = 0$$
$$[p_x, p_y] = [p_y, p_z] = [p_z, p_x] = 0$$
$$[x, t] = [y, t] = [z, t] = 0$$
$$[E, t] = i\hbar.$$

The following result will have an important consequence when we come to discuss the process of measurement in quantum mechanics in the next Section.

**Theorem 2.1.** *A necessary and sufficient condition for two operators to commute is that they possess a complete set of simultaneous orthogonal eigenfunctions.*

*Proof.* (i) Suppose that two linear operators $A$ and $B$ commute. Then $AB = BA$, and $A$ commutes with any function $f(B)$ of $B$: $Af(B) = f(B)A$. Further, let the eigenfunction $\phi_n$ of $A$ correspond to the eigenvalue $a_n$, and let the eigenfunction $\xi_n$ of $B$ correspond to the eigenvalue $b_n$. Since the set $\{\xi_n\}$ is a complete set, then $\phi_n$ can be expanded in terms of this set:

$$\phi_n = \sum_i c_{ni}\xi_i.$$

Let $f(B)$ be an arbitrary function of $B$. Then

$$0 = Af(B)\phi_n - f(B)A\phi_n = Af(B)\phi_n - f(B)a_n\phi_n$$
$$= (A - a_n)f(B)\sum_i c_{ni}\xi_i = (A - a_n)\sum_i c_{ni}f(b_i)\xi_i$$
$$= \sum_i c_{ni}f(b_i)(A - a_n)\xi_i.$$

Now, the function $f$ is arbitrary. In particular, choose $f$ such that $f(x) = \delta_{x,b_k}$ for some eigenvalue $b_k$ of $B$. Then

$$0 = \sum_{i(k)} c_{ni}(A - a_n)\xi_i$$

where $\sum_{i(k)}$ denotes the summation over values of $i$ for which $b_i = b_k$. Hence $A\xi_i = a_n\xi_i$ for all those eigenfunctions $\xi_i$ of $B$ for which $b_i = b_k$.

(ii) Suppose that $A$ and $B$ possess a complete set of simultaneous eigenfunctions $\phi_n$:

$$A\phi_n = a_n\phi_n, \qquad B\phi_n = b_n\phi_n.$$

Let $\psi = \sum_i c_i\phi_i$ be an arbitrary function. Then

$$[A, B]\psi = AB\psi - BA\psi = AB\sum_i c_i\phi_i - BA\sum_i c_i\phi_i$$

$$= A\sum_i c_ib_i\phi_i - B\sum_i c_ia_i\phi_i = \sum_i c_ib_ia_i\phi_i - \sum_i c_ia_ib_i\phi_i$$

$$= 0.$$

Since the function $\psi$ is arbitrary, then $[A, B] = 0$. This completes the proof. $\square$

## 2.7 The Uncertainty Principle

Let $A$ and $B$ be two operators representing observables. Then $A$ and $B$ are hermitian. Let

$$[A, B] = iC \tag{2.90}$$

where $C$ is an operator. Then for any functions $\theta$ and $\phi$,

$$(\theta, C\phi) = \left(\theta, \frac{1}{i}(AB - BA)\phi\right) = \frac{1}{i}((BA - AB)\theta, \phi) = \left(-\frac{1}{i}(BA - AB)\theta, \phi\right)$$

$$= (C\theta, \phi).$$

Hence the operator $C$ is hermitian. If $\Psi$ is the state function of the system, then

$$\langle C\rangle = (\Psi, C\Psi) = (C\Psi, \Psi) = \overline{(\Psi, C\Psi)} = \overline{\langle C\rangle}$$

so that the expectation value $\langle C\rangle$ of $C$ is real. Now define the operators

$$\alpha \equiv A - \langle A\rangle \quad \text{and} \quad \beta \equiv B - \langle B\rangle$$

—these will both be hermitian, and

$$\alpha\beta - \beta\alpha = iC.$$

Now define $\Delta a \equiv \sqrt{\langle\alpha^2\rangle}$ and $\Delta b \equiv \sqrt{\langle\beta^2\rangle}$ to be the uncertainties in the measurements of the observables to which $A$ and $B$ correspond. Since $\overline{(\beta\Psi, \alpha\Psi)} = (\alpha\psi, \beta\Psi)$, it follows that

$$2i\,\mathrm{Imag}(\beta\Psi, \alpha\Psi) = (\beta\Psi, \alpha\Psi) - (\alpha\Psi, \beta\Psi)$$

$$= (\Psi, \beta\alpha\Psi) - (\Psi, \alpha\beta\Psi)$$

$$= (\Psi, [\beta, \alpha]\Psi) = -\langle[\alpha, \beta]\rangle$$

$$= -i\langle C\rangle$$

Hence

$$|\langle C \rangle| = 2|\text{Imag}(\beta\Psi, \alpha\Psi)| \leq 2|(\beta\Psi, \alpha\Psi)|. \tag{2.91}$$

Now for any functions $f$ and $g$ we have the *Schwartz inequality* (written in one dimension only)

$$\left(\int |f|^2 dx\right)\left(\int |g|^2 dx\right) \geq \left|\int \overline{f}g\,dx\right|^2,$$

or $(f, f)(g, g) \geq |(f, g)|^2$. Using this inequality with $f \equiv \beta\Psi$ and $g \equiv \alpha\Psi$, Equation (2.91) becomes

$$\frac{1}{4}|\langle C \rangle|^2 \leq |(\beta\Psi, \alpha\Psi)|^2$$

$$\leq (\beta\Psi, \beta\Psi)(\alpha\Psi, \alpha\Psi) = (\Psi, \beta^2\Psi)(\Psi, \alpha^2\Psi) = (\Delta a)^2(\Delta b)^2.$$

Hence

$$\Delta a \cdot \Delta b \geq \frac{1}{2}|\langle C \rangle|. \tag{2.92}$$

This result shows that if $\langle C \rangle \neq 0$, then $\Delta a$ and $\Delta b$ cannot simultaneously both be zero. In terms of measurement, this is interpreted as meaning that the observables to which the operators $A$ and $B$ correspond cannot be measured simultaneously and precisely.

It has been shown in Sect. 2.6 that a necessary and sufficient condition for two linear operators to commute is that they possess a complete set of simultaneous eigenfunctions. Hence it follows from Equation (2.92) that two observables whose operators possess a complete set of simultaneous eigenfunctions can be simultaneously observable.

For example, we know that $[x, p_x] = i\hbar$, so that $C = \hbar$ in this case. Result (2.92) then gives

$$\Delta x \cdot \Delta p_x \geq \frac{1}{2}\hbar,$$

so that $x$ and $p_x$ cannot be measured simultaneously and precisely. A similar result holds for the pairs $\{y, p_y\}$ and $\{z, p_z\}$. However, since $[x, p_y] = 0$ then we have $\Delta x \cdot \Delta p_y \geq 0$ and so it is possible to measure the pair $\{x, p_y\}$ simultaneously and precisely, with similar results for the remaining pairs. Further, since $[E, t] = i\hbar$, then $\Delta E \cdot \Delta t \geq 0$, and so the energy of a system and the time at which that energy is measured cannot be measured simultaneously and precisely.

Two simple examples will illustrate the application of the Uncertainty Principle.

1. Consider the creation of virtual $\pi$-mesons when a proton decays: $p \rightarrow p + \pi^0$. This process violates the classical principle of conservation of energy, since mass is seemingly created. However, the virtual particle lives for such a short time that the uncertainty principle does not allow us to see that the conservation principle is being violated. Writing $\Delta t \cdot \Delta E \sim \hbar$ and taking $\Delta E$ to be the pion mass

(= 135 MeV), we find that $\Delta t \sim \hbar/(\Delta E) \sim 5.2 \times 10^{-24}$ seconds: this is taken as a measure of the mean lifetime of the virtual pion.

2. When atoms radiate, electrons jump from an orbit with energy $E_2$ to an orbit with lower energy $E_1$. The frequency of the emitted radiation is $\nu$ where $h\nu = E_2 - E_1$. An excited state of an atom lasts approximately for a time $\Delta t \sim 10^{-8}$ seconds. Hence by the uncertainty principle, $\Delta E \neq 0$, or equivalently $\Delta \nu \neq 0$. This uncertainty in the energy measurement manifests itself in the production of spectral lines of nonzero width.

## 2.8 The postulates of quantum mechanics

Having built up a somewhat loose picture of how quantum mechanics operates, and having found a mathematical framework for carrying these ideas, we may now formalise everything into a set of postulates which encapsulate these ideas.

### POSTULATE 1

A system is completely specified by a normalised state function $\Psi$ which contains an arbitrary factor of modulus unity. This state function allows us to determine everything that can be known about the system.

The word "everything" in this postulates does not mean that we can know as much as we have come to expect from non-quantum classical mechanics. The determinism of classical mechanics is now replaced by a much lower expectation of knowledge based on the Uncertainty Principle. Now, the most we can predict are probabilities rather than certainties. The state function is normalised, which means that

$$|\Psi|^2 = (\Psi, \Psi) = 1, \tag{2.93}$$

and an arbitrary factor of modulus unity means that the state function remains normalised:

$$|e^{i\alpha}\Psi|^2 = |\Psi|^2. \tag{2.94}$$

### POSTULATE 2

To every observable there corresponds a linear hermitian operator for which a complete set of orthonormal eigenfunctions may be derived.

If the eigenvalues of the operator are non-degenerate, then the eigenfunctions are already orthogonal: all that is required is that this set be normalised. On the other hand, if the eigenvalues are degenerate, then a complete set of orthogonal eigenfunctions may be derived using the Schmidt Orthogonalisation procedure: this derived set may then be normalised.

## POSTULATE 3

Let $A$ be an operator representing an observable, and let $a_n$ and $\phi_n$ ($n = 1, 2, \ldots, N$) be the set of eigenvalues and eigenfunctions:

$$A\phi_n = a_n\phi_n, \quad (n = 1, 2, \ldots, N). \tag{2.95}$$

Then the only possible outcome of a precise measurement of the observable is one of the eigenvalues $a_n$.

We need the $a_n$ to be real if they are the outcomes of physical measurement: we have ensured this by taking the operator $A$ to be hermitian. Further, this postulate does not say which of the eigenvalues will be observed in a precise measurement— only that one of them will be.

## POSTULATE 4

An hermitian operator $H$, called the *Hamiltonian operator*, exists for each system. The time development of the system is governed by the time-dependent Schrödinger equation

$$i\hbar\frac{\partial\Psi}{\partial t} = H\Psi, \tag{2.96}$$

provided that the system is not disturbed.

In practice, this turns out to be a partial differential equation for which a solution is impossible to find except in the simplest and most artificial of cases. We very often end up having to make approximations, both mathematical and numerical, in order to get some sort of solution. The words "provided that the system is not disturbed" look harmless, until it is remembered that the process of measurement disturbs a system. Hence Equation (2.96) is valid for a system provided that we are not looking at it. We need some other way of describing the development of the system when we are looking at it.

## POSTULATE 5

Let operator $A$ correspond to an observable, and let $a_n$ and $\phi_n$ ($n = 1, 2, \ldots, N$) be the complete set of eigenvalues and orthonormal eigenfunctions of $A$. Order the eigenvalues so that $a_1 \leq a_2 \leq a_3 \leq \cdots \leq a_N$. Let the state function be $\Psi = \sum_{n=1}^{N} c_n\phi_n$. Then the probability of finding a measured value of the observable in the given range $a' \leq a \leq a''$ is

$$P(a', a'') \equiv \sum_i |c_i|^2 \tag{2.97}$$

where the summation is taken only over those values of $i$ for which $a_i$ lies in the given range.

One consequence of this postulate is that, if the eigenvalues are non-degenerate, then the probability that the eigenvalue $a_i$ itself is measured is $P(a_i, a_i) = |c_i|^2$.

Further, with all summations running from 1 to $N$, we have

$$1 = (\Psi, \Psi) = \left( \sum_n c_n \phi_n, \sum_i c_i \phi_i \right) = \sum_n \sum_i \overline{c_n} c_i (\phi_n, \phi_i)$$

$$= \sum_n \sum_i \overline{c_n} c_i \delta_{ni} = \sum_i |c_i|^2.$$

Hence $\sum_{i=1}^N |c_i|^2 = 1$, and so the normalisation of the state function ensures that all the probabilities add to unity. We may evaluate the inner product

$$(\Psi, A\Psi) = \left( \sum_i \phi_i, \sum_j c_j A\phi_j \right) = \sum_i \sum_j \overline{c_i} c_j a_j (\phi_i, \phi_j)$$

$$= \sum_{i=1}^N |c_i|^2 a_i = \sum_{i=1}^N P(a_i, a_i) a_i. \tag{2.98}$$

This quantity corresponds in some sense to the average value of the operator $A$, and is the expectation value defined in Equation (2.84). Finally, suppose that we know that the system is definitely in state $\Psi = \phi_n$. Then

$$\phi_n = \sum_{i=1}^N c_i \phi_i \quad \Rightarrow \quad c_n = 1 \text{ and } c_{(i \neq n)} = 0.$$

Hence the probability of measuring the value $a_n$ is $|c_n|^2 = 1$: that is, we are certain to measure the value $a_n$.

## POSTULATE 6

Suppose a measurement is made of the observable to which the operator $A$ corresponds, and the result is seen to lie in the range $a' \leq a \leq a''$. Then immediately after the measurement, the system is in the state

$$\psi' = \frac{\sum_i c_j \phi_j}{\sqrt{\sum_j |c_j|^2}} \tag{2.99}$$

where the summations are over those values of $j$ for which $a' \leq a_j \leq a''$.

Equation (2.96) describes how the system develops when we are not looking at it. Postulate 6 describes how the system develops when we are looking at it: the effect of this measurement on the state function is called the *collapse of the wave packet*. In practise, the differential equation Equation (2.96) governs the smooth development of the system when initial conditions are given. If a measurement is then made on the system at a later time, the wave packet collapses as described. The system subsequently develops according to Equation (2.96) using the new initial condition given by Equation (2.99). Note that the denominator on the right hand side of Equation (2.99) is merely the inverse of the normalisation factor. Finally, if $a' = a'' = a_i$, then we have measured $a_i$ exactly, and then

$$\Psi' = \frac{\sum_j c_j \phi_j}{\sqrt{\sum_j |c_j|^2}}$$

(where the summations are over those values of $j$ for which $\phi_j$) is an eigenfunction corresponding to the eigenvalue $a_i$. In particular, for non-degenerate $a_i$, the summation reduces to one term $\Psi' = (c_i/|c_i|)\phi_i$.

## 2.9 The harmonic oscillator

The study of the quantised harmonic oscillator is important in such fields as atomic vibrations and quantised wave fields. Classically, the one dimensional harmonic oscillator consists of a particle of mass $m$ moving along the $x$-axis under the action of a force $\mu|x|$ directed towards the origin $O$, where $\mu$ is a positive constant. Its equation of motion is $m\ddot{x} = -\mu x$, or $\ddot{x} = -\omega_c^2 x$ where $\omega_c \equiv \sqrt{\mu/m}$ is the classical circular frequency. Its potential is $V(x)$ where $dV/dx = \mu x$, or

$$V(x) = \frac{1}{2}\mu x^2 \tag{2.100}$$

which is determined up to an additive constant.

The equivalent quantum problem gives rise to the TISE

$$-\frac{\hbar^2}{2m}\frac{d^2 u(x)}{dx^2} + \frac{1}{2}\mu x^2 u(x) = E u(x). \tag{2.101}$$

It is required to find the energy eigenvalues $E$ and eigenfunctions $u(x)$ which satisfy the conditions that $u(x)$ is finite everywhere, and $u(x) \to 0$ as $|x| \to \infty$.

Although this equation looks simple, getting the solution is not a straightforward matter. Two methods will be described in order to generate the solution. The first method will be to solve the differential equation in terms of Hermite polynomials. This method will be used to generate properties of the solution which are useful when we come to evaluate integrals based on the solution. The second method will use the elegant *ladder operator* approach.

### 2.9.1 Solution of the differential equation

In this section, the purely algebraic manipulations will be left to the reader to complete. Define positive constants $\alpha$ and $\lambda$, and a new independent variable $\xi$, by

$$\alpha^4 \equiv \frac{m\mu}{\hbar^2}, \qquad \lambda^2 \equiv \frac{4E^2m}{\hbar^2\mu}, \qquad \xi \equiv \alpha x. \tag{2.102}$$

Then writing $w(\xi) \equiv u(x)$, Equation (2.101) becomes

$$w'' + (\lambda - \xi^2)w = 0 \tag{2.103}$$

where primes denote differentiation with respect to $\xi$. Now make the substitution

$$w(\xi) \equiv H(\xi)e^{-\frac{1}{2}\xi^2} \tag{2.104}$$

into Equation (2.103) and obtain an equation for $H(\xi)$. Note that the function $H(\xi)$ is not to be confused with the hamiltonian operator: no conflict of notation should arise in this section. The resulting equation is

$$H'' - 2\xi H' + (\lambda - 1)H = 0. \tag{2.105}$$

We now proceed by looking for a series solution of the form

$$H(\xi) = \xi^s(a_0 + a_1\xi + a_2\xi^2 + \ldots) \tag{2.106}$$

where $a_0 \neq 0$ and $s \geq 0$. This last condition ensures that $H$ does not diverge at $\xi = 0$. The series is substituted into Equation (2.105), and coefficients of the various powers of $\xi$ are set to zero. The first two coefficients give

$$s(s - 1)a_0 = 0, \qquad (s + 1)sa_1 = 0,$$

while the evaluation of the general coefficient gives

$$(s + r + 2)(s + r + 1)a_{r+2} - (2s + 2r + 1 - \lambda)a_r = 0. \tag{2.107}$$

The first of these equations shows that $s = 0$ or $s = 1$. The second equation shows that $a_1 = 0$ if $s = 1$, or that both $s$ and $a_1$ can be zero. Note that Equation (2.107) will give all the even-indexed coefficients $a_2$, $a_4$, $a_6$ etc. proportional to $a_0$, and will give all of the odd-indexed coefficients proportional to $a_1$. Hence the series for $H(\xi)$ splits into two alternating series—one proportional to $a_0$ and the other proportional to $a_1$. Now

$$\frac{a_{r+2}}{a_r} = \frac{2s + 2r + 1 - \lambda}{(s + r + 2)(s + r + 1)}$$

$$\to \frac{2}{r} \quad \text{as } r \to \infty,$$

which is exactly the behaviour of the corresponding series for the function $\xi^n e^{\xi^2}$. Hence this comparison test shows that the series for $H(\xi)$ would diverge if it were allowed to continue as an infinite series. In order to avoid this, the values of $\lambda$ must be such that the series terminates. Since $a_0$ is not zero, we must terminate this part of the series to a polynomial, and we must not allow the other alternating series to start, otherwise it would leapfrog the $a_0$ series and continue indefinitely. Hence we must take $a_1 = 0$, and if $a_R$ is the last coefficient, we must have $a_{R+2} = 0$. It then follows from Equation (2.107) that

$$2s + 2R + 1 - \lambda = 0,$$

or $\lambda = 2s + 2R + 1$. Hence taking both of the cases $s = 0$ or $s = 1$ into account, $\lambda$ must be of the form

$$\lambda = 2n + 1, \quad (n = 0, 1, 2, \ldots). \tag{2.108}$$

Recalling that $\lambda = 2E/(\hbar\omega_c)$, it follows that the quantised energy levels are

$$E \equiv E_n = \left(n + \frac{1}{2}\right)\hbar\omega_c, \quad (n = 0, 1, 2, \ldots). \tag{2.109}$$

The lowest, or *groundstate*, energy is

$$E_0 = \frac{1}{2}\hbar\omega_c. \tag{2.110}$$

From Equation (2.105), it can be seen that the function $H_n(\xi)$ associated with the energy eigenvalue $E_n$ will satisfy the equation

$$H_n'' - 2\xi H_n' + 2n H_n = 0, \tag{2.111}$$

and the eigenfunction corresponding to the energy level will be

$$u_n(x) = N_n H_n(\alpha x)e^{-\frac{1}{2}\alpha^2 x^2} \tag{2.112}$$

where $N_n$ is the normalisation factor which will be found later.

The function $H_n$ is called the $n$th *Hermite Polynomial*. Because of the leapfrog properties of the coefficients $a_r$ in the expansion of $H_n$, the expansion splits into a part which is even in $\xi$ and a part which is odd. One of these parts must terminate, while the other must be set to zero. Hence $H_n(\xi)$ is either an even or odd function of $\xi$. Since the term $e^{\frac{1}{2}\xi^2}$ is an even function of $\xi$, then the it can be seen from Equation (2.104) that the eigenfunction $u_n(x)$ is either even or odd.

Properties of the Hermite polynomials will now be derived which will be useful in evaluating the integrals involved in the determination of expectation values of operators. Consider the *generating function* $G$ which generates a set of functions $A_n(\xi)$:

$$G(\xi, \sigma) \equiv e^{\xi^2 - (\sigma - \xi)^2} = e^{-\sigma^2 + 2\sigma\xi}$$

$$\equiv \sum_{n=0}^{\infty} \frac{1}{n!} A_n(\xi)\sigma^n, \tag{2.113}$$

where $0! = 1$ is taken by convention. By evaluating $\partial G/\partial \xi$, it may be shown that

$$\sum_{n=0}^{\infty} \frac{2}{n!} A_n(\xi)\sigma^{n+1} = \sum_{n=0}^{\infty} \frac{1}{n!} A_n'(\xi)\sigma^n,$$

from which it follows that

$$A_n' = 2n A_{n-1}. \tag{2.114}$$

Similarly, by evaluating $\partial G/\partial \sigma$, it may be shown that

$$\sum_{n=0}^{\infty} \frac{1}{n!} (-2\sigma + 2\xi) A_n(\xi)\sigma^n = \sum_{n=0}^{\infty} \frac{1}{(n-1)!} A_n(\xi)\sigma^{n-1},$$

from which it follows that

$$A_{n+1} = 2\xi A_n - 2n A_{n-1} = 2\xi A_n - A_n'.$$

On differentiating this result, and using Equation (2.114), it follows that

$$A_n'' - 2\xi A_n' + 2n A_n = 0$$

—this is just the Equation (2.111) which was found for the Hermite polynomials. Hence $G(\xi, \sigma)$ is the generating function for these polynomials. Now

$$\frac{\partial^n G}{\partial \sigma^n} = e^{\xi^2} \frac{\partial^n}{\partial \sigma^n} e^{-(\sigma-\xi)^2} = (-1)^n e^{\xi^2} \frac{\partial^n}{\partial \xi^n} e^{-(\sigma-\xi)^2},$$

and so it follows from Equation (2.113) that

$$H_n(\xi) = \left( \frac{\partial^n G}{\partial \sigma^n} \right)_{\sigma=0} = (-1)^n e^{\xi^2} \frac{\partial^n}{\partial \xi^n} e^{-\xi^2}. \tag{2.115}$$

The first four polynomials are easily calculated as

$$H_0(\xi) = 1$$
$$H_1(\xi) = 2\xi$$
$$H_2(\xi) = 4\xi^2 - 2$$
$$H_3(\xi) = 8\xi^3 - 12\xi.$$

Note how each polynomial is either an even or an odd function of $\xi$.

Finally, the orthonormal properties of the energy eigenfunctions can be derived. By evaluating

$$\int_{-\infty}^{\infty} G(\xi, \sigma) G(\xi, \tau) e^{-\xi^2} d\xi$$

using Equation (2.113), we obtain

$$\sum_{n=0}^{\infty}\sum_{m=0}^{\infty}\frac{\sigma^n\tau^m}{n!m!}\int_{-\infty}^{\infty}H_n(\xi)H_m(\xi)e^{-\xi^2}d\xi$$

$$=\int_{-\infty}^{\infty}e^{-\sigma^2+2\sigma\xi}\cdot e^{-\tau^2+2\tau\xi}\cdot e^{-\xi^2}$$

$$=\int_{-\infty}^{\infty}e^{-(\xi-\sigma-\tau)^2+2\sigma\tau}d\xi=\sqrt{\pi}e^{2\sigma\tau}$$

$$=\sqrt{\pi}\sum_{r=0}^{\infty}\frac{1}{r!}(2\sigma\tau)^r.$$

Then, by equating coefficients of various powers of $\sigma$ and $\tau$, we get

$$\int_{-\infty}^{\infty}H_n(\xi)H_m(\xi)e^{-\xi^2}d\xi=\sqrt{\pi}2^n n!\delta_{mn}. \tag{2.116}$$

The normalisation factor $N_n$ can now be found, remembering that $0!=1$. We require

$$1=\int_{-\infty}^{\infty}|u_n(x)|^2dx=\frac{N_n{}^2}{\alpha}\int_{-\infty}^{\infty}H_n(\xi)^2e^{-\xi^2}d\xi=\frac{N_n{}^2}{\alpha}\sqrt{\pi}2^n n!$$

and hence

$$N_n=\sqrt{\frac{\alpha}{\sqrt{\pi}2^n n!}}. \tag{2.117}$$

Therefore the complete normalised solution of the TISE for the harmonic oscillator is

$$u_n(x)=\sqrt{\frac{\alpha}{\sqrt{\pi}2^n n!}}H_n(\alpha x)e^{-\frac{1}{2}\alpha^2 x^2}, \tag{2.118}$$

$$E_n=\left(n+\frac{1}{2}\right)\hbar\omega_c,\quad\text{where} \tag{2.119}$$

$$\alpha^4=\frac{\mu m}{\hbar^2}. \tag{2.120}$$

Note that $n$ is uniquely determined for a fixed value of $E_n$, and $H_n(\xi)$ is uniquely determined by the value of $n$. Hence the eigenvalues $E_n$ are non degenerate in this one dimensional case, but we shall see how degeneracy may occur for an oscillator in more than one dimension.

## 2.9.2 *The ladder operator method*

The ladder operator method is an extremely elegant method of obtaining the energy eigenvalues of Equation (2.101). The method will be encountered again when dealing with spin angular momentum.

Operators $r_+$ and $r_-$ will be used to climb up and down the ladder whose rungs are the energy eigenvalues. Equation (2.101) can be written in the form

$$Hu_n = E_n u_n$$

where the Hamiltonian operator $H$ (now not to be confused with the Hermite polynomial) is given by

$$H \equiv \frac{\hbar^2}{2m}\frac{d^2}{dx^2} + \frac{1}{2}\mu x^2 = \frac{1}{2m}p^2 + \frac{1}{2}\mu x^2$$

and $p = -i\hbar\partial/\partial x$ is the momentum operator in the $x$ direction. Now define the ladder operators

$$r_+ \equiv \frac{1}{\sqrt{2m}}p + i\sqrt{\frac{\mu}{2}}x \qquad (2.121)$$

$$r_- \equiv \frac{1}{\sqrt{2m}}p - i\sqrt{\frac{\mu}{2}}x. \qquad (2.122)$$

Recalling that

$$[x, p] \equiv xp - px = i\hbar, \qquad (2.123)$$

then it can be easily shown that

$$(r_+)^\times = r_-, \qquad (r_-)^\times = r_+, \qquad (2.124)$$

$$H = r_+ r_- + \frac{1}{2}\hbar\omega_c = r_- r_+ - \frac{1}{2}\hbar\omega_c, \qquad (2.125)$$

$$[r_+, r_-] = -\hbar\omega_c, \qquad (2.126)$$

$$[H, r_\pm] = \pm\hbar\omega_c r_\pm. \qquad (2.127)$$

It can now be shown that the effect of the ladder operator $r_+$ on $u_n$, which corresponds to the energy eigenvalue $E_n$, is to produce the eigenvalue $E_n + \hbar\omega_c$, while the effect of the operator $r_-$ produces the eigenvalue $E_n - \hbar\omega_c$. Since $Hu_n = E_n u_n$, and since $r_+$ is a linear operator, then

$$\begin{aligned}
H(r_+ u_n) &= (Hr_+)u_n \\
&= (r_+ H + \hbar\omega_c r_+)u_n \\
&= r_+ E_n u_n + \hbar\omega_c r_+ u_n \\
&= (E_n + \hbar\omega_c)r_+ u_n. \qquad (2.128)
\end{aligned}$$

Similarly, it can be shown that

$$H(r_-u_n) = (E_n - \hbar\omega_c)r_-u_n. \tag{2.129}$$

This demonstrates that the difference in consecutive energy eigenvalues is $\hbar\omega_c$, giving the result

$$E_n = E_0 + n\hbar\omega_c. \tag{2.130}$$

It now remains to determine the groundstate energy $E_0$, given by $Hu_0 = E_0 u_0$. Applying the result of Equation (2.129), we have

$$H(r_-u_0) = (E_0 - \hbar\omega_c)r_-u_0.$$

However, since $E_0$ is the lowest eigenvalue, this equation has seemingly produced an even lower one. This contradiction is resolved if we take $r_-u_0 = 0$. Then $r_+(r_-u_0) = 0$, and it follows from Equation (2.125) that

$$Hu_0 - \frac{1}{2}\hbar\omega_c u_0 = 0, \quad \text{or} \quad \left(E_0 - \frac{1}{2}\hbar\omega_c\right)u_0 = 0.$$

Since $u_0$ is not identically zero, this result can only hold if $E_0 = \frac{1}{2}\hbar\omega_c$. The energy levels are then given from Equation (2.130) as

$$E_n = \frac{1}{2}\hbar\omega_c + n\hbar\omega_c = \left(n + \frac{1}{2}\right)\hbar\omega_c$$

as before.

Since the energy levels are non-degenerate, then Equation (2.128) shows that $u_{n+1}$ is some multiple of $r_+u_n$:

$$u_{n+1} = c_{n+1}r_+u_n. \tag{2.131}$$

This multiple $c_{n+1}$ can be found by normalisation:

$$1 = \int_{-\infty}^{\infty} \overline{u_n}u_n dx,$$

$$1 = \int_{-\infty}^{\infty} \overline{u_{n+1}}u_{n+1}dx = \int_{-\infty}^{\infty} \overline{c_{n+1}r_+u_n}c_{n+1}r_+u_n dx$$

$$= |c_{n+1}|^2 \int_{-\infty}^{\infty} \overline{u_n}(r_+)^\times r_+u_n dx = |c_{n+1}|^2 \int_{-\infty}^{\infty} \overline{u_n}(r_-r_+)u_n dx$$

$$= |c_{n+1}|^2 \int_{-\infty}^{\infty} \overline{u_n}\left(H + \frac{1}{2}\hbar\omega_c\right)u_n dx = |c_{n+1}|^2 \int_{-\infty}^{\infty} \overline{u_n}\left(E_n + \frac{1}{2}\hbar\omega_c\right)u_n dx$$

$$= |c_{n+1}|^2(n + 1)\hbar\omega_c.$$

Hence

$$c_{n+1} = \frac{1}{\sqrt{\hbar \omega_c (n+1)}},$$

giving

$$u_{n+1} = c_{n+1} r_+ u_n = \frac{1}{\sqrt{\hbar \omega_c (n+1)}} r_+ u_n,$$

from which it follows that

$$u_n = \frac{1}{\sqrt{(\hbar \omega_c)^n n!}} (r_+)^n u_0. \tag{2.132}$$

### 2.9.3  Oscillations in more than one dimension

For oscillations in three dimensions, the generalisation of the potential energy given by Equation (2.100) is

$$V(x, y, z) = \frac{1}{2} \mu_1 x^2 + \frac{1}{2} \mu_2 y^2 + \frac{1}{2} \mu_3 z^2$$

where the positive constants $\mu_1$, $\mu_2$ and $\mu_3$ are not necessarily equal. The TISE becomes

$$-\frac{\hbar^2}{2m} \left( \frac{\partial^2 u}{\partial x^2} + \frac{\partial^2 u}{\partial y^2} + \frac{\partial^2 u}{\partial z^2} \right) + \frac{1}{2} (\mu_1 x^2 + \mu_2 y^2 + \mu_3 z^2) u = E u, \tag{2.133}$$

where $u \equiv u(x, y, z)$. We look for a separable solution

$$u(x, y, z) = X(x) Y(y) Z(z)$$

and substitute into Equation (2.133):

$$-\frac{\hbar^2}{2m} \left( \frac{X''}{X} + \frac{Y''}{Y} + \frac{Z''}{Z} \right) + \frac{1}{2} (\mu_1 x^2 + \mu_2 y^2 + \mu_3 z^2) = E,$$

or

$$-\frac{\hbar^2}{2m} \frac{X''}{X} + \frac{1}{2} \mu_1 x^2 = E + \frac{\hbar^2}{2m} \left( \frac{Y''}{Y} + \frac{Z''}{Z} \right) - \frac{1}{2} (\mu_2 y^2 + \mu_3 z^2).$$

Since the left hand side is a function of $x$ only, and the right hand side is a function of $y$ and $z$ only, then each side must be a constant $E_1$, giving

$$-\frac{\hbar^2}{2m} X'' + \frac{1}{2} \mu_1 x^2 X = E_1 X.$$

The function $X(x)$ must have the same boundary conditions as the solution of the one dimensional oscillator, and hence, as before, the allowed energy levels associated with motion along the $x$-direction are

$$E_{1n_1} = \left(n_1 + \frac{1}{2}\right)\hbar\sqrt{\frac{\mu_1}{m}}, \quad (n_1 = 0, 1, 2, \ldots)$$

with associated energy eigenfunctions

$$X_{n_1}(x) = \sqrt{\frac{\alpha_1}{\sqrt{\pi}\, 2^{n_1} n_1!}} H_{n_1}(\alpha_1 x) e^{-\frac{1}{2}\alpha_1^2 x^2}$$

where

$$\alpha_1^4 = \frac{\mu_1 m}{\hbar^2}.$$

Similar solutions are associated with the $y$- and $z$-directions, with $E_{2n_2}$ and $E_{3n_3}$ being the energy eigenvalues associated with the motions in these directions. The total energy is therefore

$$E_{n_1 n_2 n_3} = E_{1n_1} + E_{2n_2} + E_{3n_3}$$

$$= \left(n_1 + \frac{1}{2}\right)\hbar\sqrt{\frac{\mu_1}{m}} + \left(n_2 + \frac{1}{2}\right)\hbar\sqrt{\frac{\mu_2}{m}} + \left(n_3 + \frac{1}{2}\right)\hbar\sqrt{\frac{\mu_3}{m}}$$

where $n_1$, $n_2$ and $n_3$ are all non-negative integers.

Note that degeneracy occurs when $\mu_1 = \mu_2 = \mu_3 \equiv \mu$, for then

$$E_{n_1 n_2 n_3} = \left(n_1 + n_2 + n_3 + \frac{3}{2}\right)\hbar\sqrt{\frac{\mu}{m}},$$

giving

$$N \equiv n_1 + n_2 + n_3 = \frac{1}{\hbar}\sqrt{\frac{m}{\mu}} E_{n_1 n_2 n_3} - \frac{3}{2}.$$

Then for a given energy $E_{n_1 n_2 n_3}$, all eigensolutions with integers $n_1$, $n_2$ and $n_3$ which total to $N$ will correspond to this given value. By counting these permutations, it is easy to show that the degeneracy of the energy level $E$ is

$$\frac{1}{2}(N+1)(N+2),$$

where

$$N \equiv \frac{1}{\hbar}\sqrt{\frac{m}{\mu}} E - \frac{3}{2}.$$

Note that the symmetry imposes degeneracy on the energy levels. We shall see this process happening later when evaluating energy levels in the quantum wells found at the material interfaces in semiconductor devices.

### 2.9.4 The displaced harmonic oscillator

It is now shown that the potential energy of the harmonic oscillator may have linear terms added to it, resulting in the overall nature of the solution being unchanged. Returning to the one dimensional case, the potential energy can be generalised to the form

$$V(x) = \frac{1}{2}\mu x^2 + \varepsilon \mu x + b\mu \tag{2.134}$$

where $\varepsilon$ and $b$ are constants. The TISE can then be written

$$-\frac{\hbar^2}{2m}\frac{d^2u}{dx^2} + \left(\frac{1}{2}\mu(x+\varepsilon)^2 + b\mu - \varepsilon^2\right)u = Eu,$$

or

$$-\frac{\hbar^2}{2m}\frac{d^2u}{dx^2} + \frac{1}{2}\mu(x+\varepsilon)^2 u = (E+\varepsilon^2 - b\mu)u.$$

On changing the independent variable from $x$ to $x_1 \equiv x + \varepsilon$, with $v(x_1) \equiv u(x)$, this equation becomes

$$-\frac{\hbar^2}{2m}\frac{d^2v}{dx_1^2} + \frac{1}{2}\mu x_1^2 v = (E+\varepsilon^2 - b\mu)v.$$

This has the same form as Equation (2.101), and hence a similar method of solution will give the condition

$$E_n + \varepsilon^2 - b\mu = \left(n + \frac{1}{2}\right)\hbar\sqrt{\frac{\mu}{m}}.$$

The energy levels are then

$$E_n = \left(n + \frac{1}{2}\right)\hbar\sqrt{\frac{\mu}{m}} - \varepsilon^2 + b\mu, \quad (n = 0, 1, 2, \ldots).$$

Hence the energy levels are all displaced by the same amount $b\mu - \varepsilon^2$, and the centre of the oscillation is displaced from $x = 0$ to $x = -\varepsilon$.

## 2.10 Spherically symmetric potentials

An important class of situations arises for those problems in which the potential depends only on the distance from a fixed point, and not on any angular distribution about that fixed point. In particular, the discussion of the hydrogen atom falls into this category.

**Fig. 2.10** The spherical polar
coordinates $r$, $\theta$, and $\phi$.

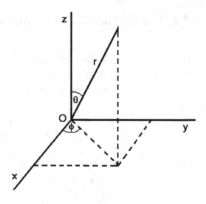

### 2.10.1 The Schrödinger equation in spherical polar coordinates

In problems involving a potential $V(r)$ which depends only on the distance $r$ from
the centre $O$, it is most natural to use the spherical polar coordinates $r$, $\theta$, and $\phi$
shown in Fig. 2.10. The ranges of these coordinates are

$$0 \le r, \qquad 0 \le \theta \le \pi, \qquad 0 \le \phi \le 2\pi.$$

The transformation between cartesian and spherical polar coordinates is given by

$$x = r \sin\theta \cos\phi, \qquad y = r \sin\theta \sin\phi, \qquad z = r \cos\theta, \qquad (2.135)$$

and the expression for $\nabla^2$ in spherical polar coordinates is given by

$$\nabla^2 = \frac{1}{r^2} \frac{\partial}{\partial r} \left( r^2 \frac{\partial}{\partial r} \right) + \frac{1}{r^2 \sin\theta} \frac{\partial}{\partial \theta} \left( \sin\theta \frac{\partial}{\partial \theta} \right) + \frac{1}{r^2 \sin^2\theta} \frac{\partial^2}{\partial \phi^2}. \qquad (2.136)$$

In quantum mechanical problems involving a spherically symmetric potential, the
quantity $m$ normally denotes an integer. Consequently, particle mass will be denoted by $m_e$: the suffix $e$ can denote the terms "electron" or, more generally, "effective".

Using this expression in Equation (2.136), the TISE is

$$-\frac{\hbar^2}{2m_e} \left\{ \frac{1}{r^2} \frac{\partial}{\partial r} \left( r^2 \frac{\partial}{\partial r} \right) + \frac{1}{r^2 \sin\theta} \frac{\partial}{\partial \theta} \left( \sin\theta \frac{\partial}{\partial \theta} \right) + \frac{1}{r^2 \sin^2\theta} \frac{\partial^2}{\partial \phi^2} \right\} u(r, \theta, \phi)$$
$$+ V(r)u(r, \theta, \phi)$$
$$= Eu(r, \theta, \phi). \qquad (2.137)$$

It is required that the solution $u(r, \theta, \phi)$ be finite and continuous everywhere. We
look for a solution of the form

$$u(r, \theta, \phi) = R(r)\Theta(\theta)\Phi(\phi)$$

and use separation of variables to obtain the three equations

$$\frac{d^2\Phi}{d\phi^2} + f\Phi = 0, \qquad (2.138)$$

$$\frac{1}{\sin\theta}\frac{d}{d\theta}\left(\sin\theta\frac{d\Theta}{d\theta}\right) + \left(g - \frac{f}{\sin^2\theta}\right)\Theta = 0, \qquad (2.139)$$

$$\frac{1}{r^2}\frac{d}{dr}\left(r^2\frac{dR}{dr}\right) + \left\{\frac{2m_e}{\hbar^2}(E - V(r)) - \frac{g}{r^2}\right\}R = 0 \qquad (2.140)$$

where $f$ and $g$ are constants of separation. Since the element of volume in spherical polar coordinates is $r^2\sin\theta\,dr\,d\theta\,d\phi$, the normalisation condition is

$$\int_0^\infty dr \int_0^\pi d\theta \int_0^{2\pi} d\phi\, r^2\sin\theta |u(r,\theta,\phi)|^2 = 1,$$

or

$$\int_0^\infty r^2|R(r)|^2 dr \int_0^\pi |\Theta(\theta)|^2 \sin\theta d\theta \int_0^{2\pi} d\phi|\Phi(\phi)|^2 = 1.$$

It can be seen from this result that each coordinate solution may be normalised separately.

Note that the potential function $V(r)$ is contained only in the radial Equation (2.140). The angular solutions $\Theta(\theta)$ and $\Phi(\phi)$ are given in Equations (2.138) and (2.139). Consequently, the total angular part of the solution can be found without having to specify the form of $V(r)$.

## 2.10.2 Solution of the angular components

The solution of Equation (2.138) is easily obtained. For continuity, we require that $\Phi(\phi + 2\pi) = \Phi(\phi)$ for all values of $\phi$. If $f < 0$, the solution must be written in terms of real exponentials $\exp(\pm\sqrt{-f}\phi)$, which will not satisfy this continuity requirement. Hence, $f$ must be non-negative: $f = m^2$ where $m$ is real. The solution is then

$$\Phi(\phi) = Ae^{im\phi} + Be^{-im\phi}$$

where $A$ and $B$ are constants, and the above continuity condition will require that $m = 0, \pm 1, \pm 2, \ldots$. Separate normalisation of this part of the solution means that

$$1 = \int_0^{2\pi} |\Phi(\phi)|^2 d\phi = 2\pi(|A|^2 + |B|^2).$$

Since there is no other condition which will provide the values of $A$ and $B$ separately, then without loss of generality we may take $B = 0$ and $A$ to be real, giving

$$\Phi(\phi) \equiv \Phi_n(\phi) = \frac{1}{\sqrt{2\pi}} e^{im\phi}, \quad (m = 0, \pm 1, \pm 2, \ldots). \tag{2.141}$$

The solution of the Equation (2.139) for $\Theta(\theta)$ is much more complicated. It involves a method using a series solution, much like the method which generated the Hermite polynomials for the linear harmonic oscillator. Using the fact that $f = m^2$ where $m$ is an integer, then Equation (2.139) is

$$\frac{1}{\sin\theta}\frac{d}{d\theta}\left(\sin\theta\frac{d\Theta}{d\theta}\right) + \left(g - \frac{m^2}{\sin^2\theta}\right)\Theta = 0,$$

and using the substitution $\omega \equiv \cos\theta$, $P(\omega) \equiv \Theta(\theta)$, this equation becomes

$$\frac{d}{d\omega}\left((1 - \omega^2)\frac{dP}{d\omega}\right) + \left(g - \frac{m^2}{1 - \omega^2}\right)P = 0. \tag{2.142}$$

Since the value $\omega^2 = 1$ relates to points on the $z$-axis, which is an arbitrary direction in space, we must ensure that the solution of this equation remains finite at $\omega = \pm 1$.

Of course, the function $P(\omega)$ will depend on the value of $m$ and so it will carry this index, but it will be omitted for the time being. The calculation leading to the solution for a general value of $m$ is quite complicated, although the calculation for the case $m = 0$ is relatively easy. Further, it turns out that the solution for a general value of $m$ can be written in terms of the solution for which $m = 0$, and this simpler case will be considered first. For $m = 0$, Equation (2.142) becomes

$$\frac{d}{d\omega}\left((1 - \omega^2)\frac{dP}{d\omega}\right) + gP = 0, \tag{2.143}$$

where $g$ is a constant. Now look for a series solution

$$P(\omega) = \sum_{r=0}^{\infty} a_r \omega^r, \tag{2.144}$$

where the coefficients $a_r$ are to be determined. Substitution into Equation (2.143) gives

$$\sum_{r=0}^{\infty} \{a_{r+2}(r + 1)(r + 2) - a_r[r(r + 1) - g]\}\omega^r = 0.$$

Since this must be valid for all values of $\omega$, each coefficient must be zero, giving

$$\frac{a_{r+2}}{a_r} = \frac{r(r + 1) - g}{(r + 1)(r + 2)}. \tag{2.145}$$

Note that this relation shows that the even coefficients $a_2, a_4, a_6, \ldots$ will be given in terms of $a_0$, and the odd coefficients $a_3, a_5, a_7, \ldots$ will be given in terms of $a_1$. Hence the series solution for $P$ splits up into two separate series—one proportional to $a_0$ and the other proportional to $a_1$. It can be seen from Equation (2.145) that

$a_{r+2}/a_r \to 1$ as $r \to \infty$, and so the series would diverge if this limit were allowed to happen. As in the case of the series solution for the Hermite polynomials, we must stop this happening by taking either $a_0$ or $a_1$ to be zero, and terminating the remaining series to a polynomial. If $a_{r'}$ is the last term in the series, then it can be seen from Equation (2.145) that $a_{r'+2} = 0$ giving $g = r'(r' + 1)$. Hence $g$ has the form

$$g = l(l + 1) \qquad (2.146)$$

where $l \geq 0$ is an integer. We must take $a_0 = 0$ if $l$ is odd, and $a_1 = 0$ if $l$ is even, in order to avoid the leapfrogging of one sub-series to infinity. Hence the solution is either an even or an odd function of $\omega$. Equation (2.143) then becomes

$$\frac{d}{d\omega}\left((1 - \omega^2)\frac{dP_l(\omega)}{d\omega}\right) + l(l + 1)P_l(\omega) = 0 \qquad (2.147)$$

where we have now attached the suffix $l$ to the solution. The polynomials $P_l(\omega)$ are called *Legendre polynomials*, and it can be verified (Sneddon 1961) that their generating function is

$$\frac{1}{\sqrt{1 - 2\omega x + x^2}} = \sum_{l=0}^{\infty} P_l(\omega)x^l, \quad x < 1. \qquad (2.148)$$

The first four polynomials can be calculated from Equations (2.145) and (2.146) as

$$P_0(\omega) = 1$$
$$P_1(\omega) = \omega$$
$$P_2(\omega) = \frac{1}{2}(3\omega^2 - 1)$$
$$P_3(\omega) = \frac{1}{2}(5\omega^3 - 3\omega).$$

Note how each polynomial is either an even or odd function of $\omega$.

We now return to the full Equation (2.142), remembering that its solution $P(\omega)$ now depends on the integers $l$ and $m$: $P(\omega) \equiv P_l^m(\omega)$. This equation is

$$\frac{d}{d\omega}\left((1 - \omega^2)\frac{dP_l^m}{d\omega}\right) + \left(l(l + 1) - \frac{m^2}{1 - \omega^2}\right)P_l^m = 0,$$

$$m = 0, \pm 1, \pm 2, \ldots; \ l = 0, 1, 2, \ldots. \qquad (2.149)$$

We again seek a series solution. A calculation, which is much more complex than for the case $m = 0$, gives the result (Sneddon 1961)

$$P_l^m(\omega) = (1 - \omega^2)^{\frac{1}{2}|m|}\frac{d^{|m|}}{d\omega^{|m|}}P_l(\omega), \qquad (2.150)$$

and is called the *Associated Legendre polynomial*. Now $P_l(\omega)$ is a polynomial of degree $l$. Hence the $|m|$th derivative of $P_l(\omega)$ would be zero if $|m| > l$, since the derivative would vanish. This would give the physically unrealistic situation that the wave function, given in terms of $P_l(\omega)$, would vanish. Hence we must have $|m| \leq l$. This polynomial has the property

$$\int_0^\pi \overline{\Theta_l^m} \Theta_l^m \sin\theta d\theta = \int_{-1}^1 \overline{P_l^m} P_l^m d\omega = \frac{2}{2l+1} \frac{(l+|m|)!}{(l-|m|)!},$$

which provides the normalisation factor

$$\left[ \frac{2l+1}{2} \frac{(l-|m|)!}{(l+|m|)!} \right]^{\frac{1}{2}}.$$

Hence the total angular part of the solution, denoted by $Y_{lm}(\theta, \phi)$, to the spherically symmetric problem is

$$Y_{lm}(\theta, \phi) = \left[ \frac{2l+1}{4\pi} \frac{(l-|m|)!}{(l+|m|)!} \right]^{\frac{1}{2}} P_l^m(\cos\theta) e^{im\phi}, \quad l \geq |m|. \tag{2.151}$$

These functions are called *spherical harmonics*. It is easily verified that the first few terms of the $Y_{lm}$ are given by

$$Y_{00} = \sqrt{\frac{1}{4\pi}}$$

$$Y_{10} = \sqrt{\frac{3}{4\pi}} \cos\theta$$

$$Y_{11} = \sqrt{\frac{3}{8\pi}} \sin\theta e^{i\phi}$$

$$Y_{1-1} = \sqrt{\frac{3}{8\pi}} \sin\theta e^{-i\phi}.$$

See Problem 2.4.

### 2.10.3 Angular momentum

Classically, a particle with position vector $\mathbf{r}$ and momentum $\mathbf{p}$ has angular momentum $\mathbf{M} = \mathbf{r} \times \mathbf{p}$. Therefore the quantum mechanical operator corresponding to this classical angular momentum will be given by

$$\mathbf{M} = \mathbf{r} \times \mathbf{p}$$
$$= (x, y, z) \times \left( -i\hbar \frac{\partial}{\partial x}, -i\hbar \frac{\partial}{\partial y}, -i\hbar \frac{\partial}{\partial z} \right)$$

$$= -i\hbar \left( y\frac{\partial}{\partial z} - z\frac{\partial}{\partial y}, z\frac{\partial}{\partial x} - x\frac{\partial}{\partial z}, x\frac{\partial}{\partial y} - y\frac{\partial}{\partial x} \right)$$

$$\equiv (M_x, M_y, M_z). \tag{2.152}$$

Using the coordinate transformation equations (2.135), these expressions can be written in terms of derivatives with respect to $r$, $\theta$ and $\phi$. For example,

$$\frac{\partial}{\partial \phi} = \frac{\partial x}{\partial \phi}\frac{\partial}{\partial x} + \frac{\partial y}{\partial \phi}\frac{\partial}{\partial y} + \frac{\partial z}{\partial \phi}\frac{\partial}{\partial z}$$

$$= -r\sin\theta\sin\phi\frac{\partial}{\partial x} + r\sin\theta\cos\phi\frac{\partial}{\partial y} + 0$$

$$= -y\frac{\partial}{\partial x} + x\frac{\partial}{\partial y}$$

$$= \frac{M_x}{-i\hbar}.$$

Hence $M_z = -i\hbar\partial/\partial\phi$. Similar expressions can be found for $M_x$ and $M_y$. The complete list is

$$M_x = i\hbar \left( \sin\phi\frac{\partial}{\partial\theta} + \cot\theta\cos\phi\frac{\partial}{\partial\phi} \right) \tag{2.153}$$

$$M_y = i\hbar \left( -\cos\phi\frac{\partial}{\partial\theta} + \cot\theta\sin\phi\frac{\partial}{\partial\phi} \right) \tag{2.154}$$

$$M_z = -i\hbar\frac{\partial}{\partial\phi} \tag{2.155}$$

$$M^2 = M_x{}^2 + M_y{}^2 + M_z{}^2$$

$$= -\hbar^2 \left[ \frac{1}{\sin\theta}\frac{\partial}{\partial\theta}\left( \sin\theta\frac{\partial}{\partial\theta} \right) + \frac{1}{\sin^2\theta}\frac{\partial^2}{\partial\phi^2} \right]. \tag{2.156}$$

Further, it follows that

$$[M_x, M_y] = -\hbar^2 \left( y\frac{\partial}{\partial z} - z\frac{\partial}{\partial y} \right)\left( z\frac{\partial}{\partial x} - x\frac{\partial}{\partial z} \right)$$

$$+ \hbar^2 \left( z\frac{\partial}{\partial x} - x\frac{\partial}{\partial z} \right)\left( y\frac{\partial}{\partial z} - z\frac{\partial}{\partial y} \right)$$

$$= -\hbar^2 \left( y\frac{\partial}{\partial x} - x\frac{\partial}{\partial y} \right)$$

$$= i\hbar M_z.$$

Hence the angular momentum operators satisfy the commutation relations

$$[M_x, M_y] = i\hbar M_z$$
$$[M_y, M_z] = i\hbar M_x$$
$$[M_z, M_x] = i\hbar M_y \tag{2.157}$$
$$[M_x, M^2] = [M_y, M^2] = [M_z, M^2] = 0.$$

We are now in a position to find the eigenvalues and eigenfunctions of the angular momentum operators. Using the expression for the spherical harmonic in Equation (2.151), we have

$$M_z Y_{lm}(\theta, \phi) = -i\hbar \frac{\partial}{\partial \phi} Y_{lm}(\theta, \phi) = m\hbar Y_{lm}(\theta, \phi). \tag{2.158}$$

Hence the $Y_{lm}(\theta, \phi)$ are eigenfunctions of the operator $M_z$ corresponding to the eigenvalue $m\hbar$. Further, using Equations (2.139) and (2.156), it can be shown that

$$M^2 Y_{lm}(\theta, \phi) = l(l+1)\hbar^2 Y_{lm}(\theta, \phi). \tag{2.159}$$

Hence the $Y_{lm}(\theta, \phi)$ are also eigenfunctions of the operator $M^2$ corresponding to the eigenvalue $l(l+1)\hbar^2$. Since the value of $m$ must be such that $|m| \leq l$, it follows that there are $2l + 1$ values of $m$ corresponding to the eigenvalue $l(l+1)\hbar^2$ of $M^2$. Therefore the eigenvalue $l(l+1)\hbar^2$ of $M^2$ is $(2l+1)$-fold degenerate. Since the operators $M^2$ and $M_z$ possess a set of simultaneous eigenfunctions, then they are simultaneously measurable. This fact is also evident from the relation $[M^2, M_z] = 0$ of Equation (2.157). By symmetry, the operators $M^2$ and $M_x$ will also possess a set of simultaneous eigenfunctions, and so will the operators $M^2$ and $M_y$. However, these will be *different* sets of simultaneous eigenfunctions, and the observables to which the pair $M_x$ and $M_y$ correspond (and $M_y$ and $M_z$) will not be simultaneously observable because each of these pairs does not possess a set of simultaneous eigenfunctions.

### 2.10.4 The hydrogen atom

We now consider the hydrogen atom as the quantum equivalent of the classical central orbit problem. Let $q$ denote the magnitude of the electron charge. The atom consists of an electron with mass $m_e$ and negative charge $-q$ interacting with a nucleus with mass $m_{nuc}$ and a positive charge $Zq$, where $Z$ is a positive constant. For the hydrogen atom, we will take $Z = 1$, but we take $Z = 2$ for the helium atom. Let the nucleus have coordinates $(x_{nuc}, y_{nuc}, z_{nuc})$, and let the electron have coordinates $(x_e, y_e, z_e)$. The potential will be

$$V(\mathbf{r}) = \frac{(Zq)(-q)}{r} = \frac{-Zq^2}{\sqrt{(x_e - x_{nuc})^2 + (y_e - y_{nuc})^2 + (z_e - z_{nuc})^2}}. \tag{2.160}$$

The hamiltonian operator for the system is then

$$H = -\frac{\hbar^2}{2m_{nuc}}\nabla^2_{nuc} - \frac{\hbar^2}{2m_e}\nabla^2_e + V(x_e - x_{nuc}, y_e - y_{nuc}, z_e - z_{nuc}) \quad (2.161)$$

where $\nabla^2_{nuc}$ and $\nabla^2_e$ operate with respect to the coordinates of the nucleus and electron respectively.

The standard procedure in any two-body problem is to introduce the coordinates $(X, Y, Z)$ of the centre of mass, and the position coordinates $(x, y, z)$ relative to the centre of mass. The total mass is $M = m_{nuc} + m_e$. These coordinates are defined by

$$(X, Y, Z) \equiv \frac{m_e(x_e, y_e, z_e) + m_{nuc}(x_{nuc}, y_{nuc}, z_{nuc})}{M} \quad (2.162)$$

$$(x, y, z) \equiv (x_e, y_e, z_e) - (x_{nuc}, y_{nuc}, z_{nuc}). \quad (2.163)$$

Using the standard rules of partial differentiation, it can then be verified that the expression for $H$ given by Equation (2.161) becomes

$$H = -\frac{\hbar^2}{2M}\nabla^2_{X,Y,Z} - \frac{\hbar^2}{2\mu}\nabla^2_{x,y,z} + V(x, y, z)$$

where

$$\mu \equiv \frac{m_e m_{nuc}}{m_e + m_{nuc}} \quad (2.164)$$

is called the *reduced mass* (so-called because it is less than both $m_e$ and $m_{nuc}$). We now look for a solution of the TISE in terms of separation of variables

$$u(x, y, z)U(X, Y, Z).$$

This produces the two separated equations

$$-\frac{\hbar^2}{2M}\nabla^2_{X,Y,Z}U = E'U, \quad (2.165)$$

$$-\frac{\hbar^2}{2\mu}\nabla^2_{x,y,z}u + V(r)u = Eu \quad (2.166)$$

where the total energy is given by $E + E'$.

Equation (2.165) is the equation of motion of the centre of mass, and shows that it moves as a free particle in a zero potential. Equation (2.166) is the equation of motion about the centre of mass, and it has the same form as Equation (2.137). This equation can then be separated into Equations (2.138)–(2.140) as before, with the total angular part of the solution being the spherical harmonics $Y_{lm}(\theta, \phi)$. The radial equation (2.140) becomes

$$\frac{1}{r^2}\frac{d}{dr}\left(r^2\frac{dR}{dr}\right) + \left\{\frac{2\mu}{\hbar^2}\left(E + \frac{Zq^2}{r}\right) - \frac{l(l+1)}{r^2}\right\}R = 0. \quad (2.167)$$

As we would expect for a bound state, the energy $E$ is negative (because positive energy would have to be supplied in order to drag the two particles an infinite distance apart). It follows that we can define real positive constants $\alpha$ and $\gamma$ by

$$\alpha^2 \equiv -\frac{8\mu E}{\hbar^2}, \quad \alpha > 0,$$

$$\gamma \equiv \frac{Zq^2}{\hbar}\left(-\frac{\mu}{2E}\right)^{\frac{1}{2}}. \tag{2.168}$$

In order to proceed with the solution of Equation (2.167), define a new independent variable $\rho$ and a new dependent solution $L(\rho)$ such that

$$\rho \equiv \alpha r,$$

$$R(r) \equiv e^{-\frac{1}{2}\rho}\rho^s L(\rho), \quad s \geq 0 \tag{2.169}$$

for some constant $s$ which must be non-negative to avoid divergence of the solution at $\rho = 0$. Then substitution of these quantities into Equation (2.167) can be shown to give the equation

$$\rho^2 L'' + \rho[2(s+1) - \rho]L' + [\rho(\gamma - s - 1) + s(s+1) - l(l+1)]L = 0. \tag{2.170}$$

Now look for a series solution of the form

$$L(\rho) = \sum_{i=0}^{\infty} a_i \rho^i, \quad a_0 \neq 0 \tag{2.171}$$

—note that no generality has been lost in taking $a_0 \neq 0$ since the value of $s$ is to be determined using this condition. This series is substituted into Equation (2.170) and the coefficients of $\rho^k$ are equated to zero for all values of $k$. For $k = 0$, this process gives

$$s(s+1) = l(l+1)$$

which has solution $s = l$ and $s = -(l+1)$. The second solution must be discarded in view of the condition $s \geq 0$. Hence on dividing by $\rho$, Equation (2.170) becomes

$$\rho L'' + [2(l+1) - \rho]L' + (\gamma - l - 1)L = 0. \tag{2.172}$$

The series solution Equation (2.171) is substituted into this equation and an analysis is made of the coefficients, just as we made in obtaining the Hermite polynomials for the linear harmonic oscillator, and in obtaining the Legendre polynomials for the $\theta$-dependence. It is found that $L(\rho)$ has to be terminated to a polynomial to avoid the solution diverging as $\rho \to \infty$. This leads to the condition that $(\gamma - l - 1)$ has to be a non-negative integer. Hence $\gamma$ must be a positive integer $n$:

$$\gamma = \frac{Zq^2}{\hbar}\left(-\frac{\mu}{2E}\right)^{\frac{1}{2}} = n, \quad n = 1, 2, 3, \ldots,$$

giving

$$E \equiv E_n = -\frac{\mu Z^2 q^4}{2\hbar^2 (n+1)^2}, \quad n = 0, 1, 2, \ldots. \tag{2.173}$$

The groundstate energy $E_0$ is then given by

$$E_0 = -\frac{\mu Z^2 q^4}{2\hbar^2}, \tag{2.174}$$

so that

$$E_n = \frac{E_0}{(n+1)^2}, \quad n = 0, 1, 2, \ldots. \tag{2.175}$$

When the electron jumps from an orbit with energy $E_{n_1}$ to an orbit with a lower energy $E_{n_2}$, a photon with frequency $\lambda$ is emitted where

$$\frac{hc}{\lambda} = E_{n_1} - E_{n_2},$$

where $c$ is the speed of light. This gives

$$\frac{1}{\lambda} = \frac{1}{ch} \cdot \frac{\mu Z^2 q^4}{2\hbar^2} \left[ \frac{1}{(n_2+1)^2} - \frac{1}{(n_1+1)^2} \right]$$

$$= R_y \left[ \frac{1}{(n_2+1)^2} - \frac{1}{(n_1+1)^2} \right]$$

where $R_y \equiv \mu Z^2 q^4 / (4\pi \hbar^3 c)$ is called the *Rydberg constant*, and has the value $R_y = 1.0974 \times 10^7 \text{ m}^{-1}$ for the hydrogen atom.

This formula generates a number of series transitions:

(i) The *Lyman series* is generated for transitions into the groundstate $n_2 = 0$:

$$\frac{1}{\lambda} = R_y \left[ 1 - \frac{1}{(n_1+1)^2} \right], \quad n_1 = 1, 2, 3, \ldots.$$

(ii) The *Balmer series* is generated for transitions into the first excited state $n_2 = 1$:

$$\frac{1}{\lambda} = R_y \left[ \frac{1}{4} - \frac{1}{(n_1+1)^2} \right], \quad n_1 = 2, 3, 4, \ldots.$$

(iii) The *Paschen series* is generated for transitions into the second excited state $n_2 = 2$:

$$\frac{1}{\lambda} = R_y \left[ \frac{1}{9} - \frac{1}{(n_1+1)^2} \right], \quad n_1 = 3, 4, 5, \ldots.$$

The quantity $\lambda^{-1}$ is directly proportional to the reduced mass $\mu$ which appears in the expression for $R_y$. For a nucleus consisting of a proton, the ratio $m_{nuc}/m_e \approx 1836$, and so it follows from Equation (2.164) that

$$\mu = \frac{m_e}{1 + \frac{m_e}{m_{nuc}}} \approx 0.99946 m_e.$$

Although this value is very close to $m_e$, the effect of using $\mu$ rather than $m_e$ in the expression for $R_y$ is observed in a shift of the spectral lines.

Since $\gamma$ has to be a positive integer, Equation (2.172) becomes

$$\rho L'' + [2(l+1) - \rho]L' + (n - l - 1)L = 0 \tag{2.176}$$

with $(n - l - 1)$ being a non-negative integer. Hence $n \geq l + 1$. The solution of this equation is the *Associated Laguerre polynomial* given by

$$L_{n+l}^{2l+1}(\rho) = \sum_{k=0}^{n-l-1} \frac{(-1)^{k+1}[(n+l)!]^2}{(n-l-1-k)!(2l+1+k)!k!} \rho^k, \tag{2.177}$$

for which the generating function is

$$\frac{(-x)^p \exp(-\rho \frac{x}{1-x})}{(1-x)^{p+1}} = \sum_{q=p}^{\infty} \frac{1}{q!} L_q^p(\rho) x^q. \tag{2.178}$$

The functions

$$e^{-\frac{1}{2}\rho} \rho^s L_{n+1}^{2l+1}(\rho)$$

which appear in Equation (2.169) are called the *Laguerre functions*. Their properties are well known, and normalisation gives (Sneddon 1961)

$$R_{nl}(r) = \left[ \left( \frac{2Z\mu q^2}{n\hbar^2} \right)^3 \frac{(n-l-1)!}{2n[(n+l)!]^3} \right]^{\frac{1}{2}} e^{-\frac{1}{2}\alpha_n r} (\alpha_n r)^l L_{n+l}^{2l+1}(\alpha_n r) \tag{2.179}$$

where

$$\alpha_n^2 \equiv -\frac{8\mu E_{n-1}}{\hbar^2}, \quad n = 1, 2, 3, \ldots. \tag{2.180}$$

Finally, the complete solution of Equation (2.166) is then

$$u_{nlm}(r, \theta, \phi) = R_{nl} Y_{lm}(\theta, \phi). \tag{2.181}$$

The degeneracy of the energy eigenvalues is easily found. We have $H u_{nlm} = E_{n-1} \times u_{nlm}$ (remember, $n$ starts from 1, but we count energy levels starting from the groundstate $E_0$ by convention). The value of $l$ must lie in the range $0 \leq l \leq n - 1$, and there are $(2l + 1)$ values of $m$ satisfying the condition $|m| \leq l$. Hence the total degeneracy of the energy $E_{n-1}$ is

$$\sum_{l=0}^{n-1} (2l + 1) = n^2. \tag{2.182}$$

## 2.11 Angular momentum and spin

The eigenvalues of the orbital angular momentum have been found in Sect. 2.10.3. However, observations show that these eigenvalues are split into many more levels than can be accounted for by orbital angular momentum alone. This leads to a more general definition of angular momentum, and to the introduction of *spin* angular momentum.

### 2.11.1 The necessity for extra energy levels

Suppose that an electron moves in a spherically symmetric potential, and a magnetic field $B_z$ is applied in the $z$ direction. Classically the hamiltonian becomes

$$H' = H^{(0)} + \frac{q}{2\mu c} BL_z \tag{2.183}$$

where $\mu$ is the reduced mass of the electron, $L_z$ is the $z$ component of the angular momentum, the hamiltonian $H^{(0)}$ is given by

$$H^{(0)} = -\frac{\hbar^2}{2\mu} \nabla^2 + V(r), \tag{2.184}$$

and

$$H^{(0)} u_{nlm}(r, \theta, \phi) = E_n u_{nlm}(r, \theta, \phi) \tag{2.185}$$

where $n$, $l$ and $m$ take integer values. It is known that

$$L_z u_{nlm}(r, \theta, \phi) = -i\hbar \frac{\partial}{\partial \phi} Y_{lm}(\theta, \phi) R_{nl}(r) = n\hbar u_{nlm}(r, \theta, \phi), \tag{2.186}$$

and therefore

$$H' u_{nlm}(r, \theta, \phi) = \left( E_n + \frac{q\hbar m}{2\mu c} B_z \right) u_{nlm}(r, \theta, \phi) \tag{2.187}$$

for $m = -l, -l + 1, \ldots, l - 1, l$. Hence on this calculation, the energy level $E_n$ is split by adding any of $(2l + 1)$ values to it—this is the Normal Zeeman effect. However, it is observed that there is much more splitting than this, pointing to the fact that the electron possesses another degree of freedom. To introduce this extra freedom, we will generalise our notion of angular momentum.

The orbital angular momentum is defined as $\mathbf{L} = \mathbf{r} \times \mathbf{p}$, and using the commutation relations $[x, p_x] = i\hbar$ and $[x, p_y] = 0$ etc., it is easily shown that

$$[L_x, L_y] = i\hbar L_z, \qquad [L_y, L_z] = i\hbar L_x, \qquad [L_z, L_x] = i\hbar L_y,$$
$$[L^2, L_x] = [L^2, L_y] = [L^2, L_z] = 0,$$

and that $L^2$ has eigenvalues $l(l + 1)\hbar^2$ for $l = 0, 1, 2, \ldots$.

## 2.11.2 Generalised angular momentum

We now use these commutation relations to *define* a more general angular momentum $\mathbf{M}$ which is hermitian (that is, $\mathbf{M}^{\times} = \mathbf{M}$) such that

$$[M_x, M_y] = i\hbar M_z, \qquad [M_y, M_z] = i\hbar M_x, \qquad [M_z, M_x] = i\hbar M_y, \qquad (2.188)$$

and

$$[M^2, M_x] = [M^2, M_y] = [M^2, M_z] = 0. \qquad (2.189)$$

Now let $W$ represent an operator, or a set of operators, such that $W$, $M^2$ and $M_z$ form a set of commuting operators, with $W$ commuting with $M_x$ and $M_y$ also: this last condition represents spherical symmetry. Let $w$, $\mu^2$ and $\mu_z$ represent the eigenvalues of $W$, $M^2$ and $M_z$ respectively (with $\mu \geq 0$), and let $\psi(w, \mu^2, \mu_z)$ be a simultaneous eigenfunction:

$$W\psi(w, \mu^2, \mu_z) = q\psi(w, \mu^2, \mu_z),$$
$$M^2\psi(w, \mu^2, \mu_z) = \mu^2\psi(w, \mu^2, \mu_z),$$
$$M_z\psi(w, \mu^2, \mu_z) = \mu_z\psi(w, \mu^2, \mu_z).$$

It follows that

$$(M_x{}^2 + M_y{}^2)\psi(w, \mu^2, \mu_z) = (M^2 - M_z{}^2)\psi(w, \mu^2, \mu_z)$$
$$= (\mu^2 - \mu_z{}^2)\psi(w, \mu^2, \mu_z),$$

and so $\psi(w, \mu^2, \mu_z)$ is also an eigenfunction of the operator $(M_x{}^2 + M_y{}^2)$ corresponding to the eigenvalue $(\mu^2 - \mu_z{}^2)$. This operator will have non-negative eigenvalues, and so $\mu^2 \geq \mu_z{}^2$, or $|\mu_z| \leq \mu$. Hence for a given value of $\mu$, the eigenvalues of $M_z$ are bounded above and below. Now define the two operators

$$A_+ \equiv M_x + iM_y \qquad (2.190)$$
$$A_- \equiv M_x - iM_y \qquad (2.191)$$

Then $A_+{}^{\times} = A_-$. Using Equations (2.188) and (2.189), it is easily verified that

$$[A_\pm, M_z] = \mp\hbar A_\pm \qquad (2.192)$$
$$[A_\pm, M^2] = 0 \qquad (2.193)$$
$$A_+A_- = M^2 - M_z{}^2 + \hbar M_z \qquad (2.194)$$
$$A_-A_+ = M^2 - M_z{}^2 - \hbar M_z. \qquad (2.195)$$

It follows that

$$WA_\pm\psi(w, \mu^2, \mu_z) = A_\pm W\psi(w, \mu^2, \mu_z)$$
$$= wA_\pm\psi(w, \mu^2, \mu_z), \qquad (2.196)$$

$$M^2 A_\pm \psi(w, \mu^2, \mu_z) = A_\pm M^2 \psi(w, \mu^2, \mu_z)$$
$$= \mu^2 A_\pm \psi(w, \mu^2, \mu_z), \qquad (2.197)$$
$$M_z A_\pm \psi(w, \mu^2, \mu_z) = (A_\pm M_z \pm \hbar A_\pm) \psi(w, \mu^2, \mu_z)$$
$$= (\mu_z \pm \hbar) A_\pm \psi(w, \mu^2, \mu_z). \qquad (2.198)$$

These results show that the functions $A_\pm \psi(q, \mu^2, \mu_z)$ are themselves simultaneous eigenfunctions of the operator $W$ with eigenvalue $w$, of the operator $M^2$ with eigenvalue $\mu^2$ as before, and of the operator $M_z$ with displaced eigenvalues $(\mu_z \pm \hbar)$.

Hence by repeatedly operating on the function $\psi(w, \mu^2, \mu_z)$ with the operator $A_+$ we generate a whole set of eigenvalues $\mu^2$ of $M^2$, and eigenvalues $\mu_z, \mu_z + \hbar, \mu_z + 2\hbar, \mu_z + 3\hbar, \ldots$ of $M_z$. Similarly, by operating on the function $\psi(w, \mu^2, \mu_z)$ with the operator $A_-$ we generate a whole set of eigenvalues $\mu^2$ of $M^2$, and eigenvalues $\mu_z, \mu_z - \hbar, \mu_z - 2\hbar, \mu_z - 3\hbar, \ldots$ of $M_z$. The operators $A_\pm$ are ladder operators of a type encountered in the case of the harmonic oscillator.

Since the eigenvalues of $M_z$ are bounded above and below, the eigenvalues of $M_z$ cannot be generated indefinitely for a fixed value of $\mu$. Let the maximum eigenvalue of $M_z$ be $l\hbar$ for some value $l$:

$$M_z \psi(w, \mu^2, l\hbar) = l\hbar \psi(w, \mu^2, l\hbar).$$

Operating with $A_+$ would then give

$$M_z A_+ \psi(w, \mu^2, l\hbar) = (l\hbar + \hbar) A_+ \psi(w, \mu^2, l\hbar). \qquad (2.199)$$

However, this has generated an even larger eigenvalue of $M_z$, which is not allowed. Hence Equation (2.199) can hold only if

$$A_+ \psi(w, \mu^2, l\hbar) = 0,$$

or

$$A_- A_+ \psi(w, \mu^2, l\hbar) = 0.$$

Using Equation (2.195) this becomes

$$(M^2 - M_z{}^2 - \hbar M_z) \psi(w, \mu^2, l\hbar) = 0,$$

or

$$(\mu^2 - l^2 \hbar^2 - l\hbar^2) \psi(w, \mu^2, l\hbar) = 0.$$

Hence

$$\mu^2 = l(l + 1)\hbar^2. \qquad (2.200)$$

Similarly, there must be a least eigenvalue $l'\hbar$ of $M_z$ with

$$A_- \psi(w, \mu^2, l'\hbar) = 0,$$

or

$$A_+ A_- \psi(w, \mu^2, l'\hbar) = 0.$$

Using Equation (2.194), this becomes

$$(M^2 - M_z{}^2 + \hbar M_z)\psi(w, \mu^2, l'\hbar) = 0,$$

or

$$(\mu^2 - l'^2\hbar^2 + l'\hbar^2)\psi(w, \mu^2, l'\hbar) = 0.$$

Hence it follows that

$$l'(l' - 1)\hbar^2 = \mu^2 = l(l + 1)\hbar^2$$

which has solutions $l' = l + 1$ (which is not physically acceptable, otherwise the lowest eigenvalue would be larger than the highest), or $l' = -l$. Hence for a given value of $l$ the eigenvalues of $M^2$ are $\mu^2 = l(l + 1)\hbar^2$, and the eigenvalues of $M_z$ are $n\hbar$ ($n = -l, -l + 1, \ldots, l - 1, l$). Since there must be an integral number of such values, then the only possibility is that

$$l = \frac{1}{2} \times \text{integer}$$

$$= 0, \frac{1}{2}, 1, \frac{3}{2}, 2, \ldots. \qquad (2.201)$$

Therefore the quantity $l$ can now take half-integer values as well as integer values. The half-integer values do not describe the orbital angular momentum **L** (remember, these are described by integer values). This means that we have found an additional degree of freedom **S**, called the *spin* angular momentum, when $l = \frac{1}{2} \times$ integer. The total angular momentum is

$$\mathbf{M} = \mathbf{L} + \mathbf{S} \qquad (2.202)$$

where both **L** and **S** obey the commutation rules of Equations (2.188) and (2.189).

The spin represents a new internal degree of freedom in the system. It is found that the behaviour of an electron in a magnetic field requires that the electron has spin $l = \frac{1}{2}$, so that $S_z$ can have one of the two eigenvalues $\pm\frac{1}{2}\hbar$ and $S^2$ has eigenvalue $\frac{3}{4}\hbar^2$. Both $S_x$ and $S_y$ can also have values $\pm\frac{1}{2}\hbar$ since there is nothing special about the direction of the $z$ axis. However, we can specify $S^2$ and only one of the quantities $S_x$, $S_y$ and $S_z$ simultaneously, since these last three operators do not commute.

### 2.11.3 Particles with spin $\frac{1}{2}$

It is found that electrons, protons and neutrons have spin $l = \frac{1}{2}$, with $n = (-\frac{1}{2}, \frac{1}{2})$. Let the eigenfunctions of the operator $S_z$ corresponding to the eigenvalues $-\frac{1}{2}\hbar$ and $\frac{1}{2}\hbar$ be $a_-$ and $a_+$ respectively. Then

$$S_z a_- = -\frac{1}{2}\hbar a_-, \qquad S_z a_+ = \frac{1}{2}\hbar a_+.$$

The spin state of a particle can be described by a vector in the two dimensional spin space $S$. Particles without spin are represented by wave functions in the *Hilbert space* $\mathcal{H}$ spanned by a complete orthonormal set of functions $\xi(\mathbf{r})$. Hence in general, electron wave functions can be thought of as vectors in the product space $\mathcal{H} \times S$, and can be written as

$$\psi = a_-\psi_-(\mathbf{r}) + a_+\psi_+(\mathbf{r})$$

where

$$\psi_\pm(\mathbf{r}) \equiv \sum_k c^\pm{}_k \xi_k(\mathbf{r}).$$

The probability of finding an electron in volume $V$ with $z$-component of spin either $+\frac{1}{2}\hbar$ or $-\frac{1}{2}\hbar$ is

$$\int_V |\psi_\pm(\mathbf{r})|^2 d^3\mathbf{r}.$$

The probability of finding the electron in volume $V$ regardless of spin is

$$\int_V \left[ |\psi_+(\mathbf{r})|^2 + |\psi_-(\mathbf{r})|^2 \right] d^3\mathbf{r},$$

with normalisation given by

$$\int \left[ |\psi_+(\mathbf{r})|^2 + |\psi_-(\mathbf{r})|^2 \right] d^3\mathbf{r} = 1,$$

where this integral is taken over all space. The electron wave function may also be written in the form

$$\psi(\mathbf{r}) = \begin{pmatrix} \psi_+(\mathbf{r}) \\ \psi_-(\mathbf{r}) \end{pmatrix}$$

$$= \psi_+(\mathbf{r}) \begin{pmatrix} 1 \\ 0 \end{pmatrix} + \psi_-(\mathbf{r}) \begin{pmatrix} 0 \\ 1 \end{pmatrix}.$$

Hence the eigenfunctions $a_-$ and $a_+$ can be represented by $\begin{pmatrix} 0 \\ 1 \end{pmatrix}$ and $\begin{pmatrix} 1 \\ 0 \end{pmatrix}$ respectively. The inner product in this representation of any two vectors $\phi = \begin{pmatrix} \phi_+ \\ \phi_- \end{pmatrix}$ and $\chi = \begin{pmatrix} \chi_+ \\ \chi_- \end{pmatrix}$ in $\mathcal{H} \times S$ is defined as

$$(\phi, \chi) = \int \left\{ (\overline{\phi_+}, \overline{\phi_-}) \begin{pmatrix} \chi_+ \\ \chi_- \end{pmatrix} \right\} d^3\mathbf{r}$$

$$= \int \left\{ \overline{\phi_+}\chi_+ + \overline{\phi_-}\chi_- \right\} d^3\mathbf{r}$$

where the integrals are again taken over all space. In $S$ only, if we write $\phi \equiv \begin{pmatrix} p \\ q \end{pmatrix}$, then $(\phi, \phi) = p^2 + q^2$, and the corresponding normalised vector is $\frac{1}{\sqrt{p^2+q^2}} \begin{pmatrix} p \\ q \end{pmatrix}$.

Operators on these two-component vectors will be $2 \times 2$ matrices with complex elements. Let the matrix $\begin{pmatrix} a & b \\ c & d \end{pmatrix}$ correspond to the operator $S_z$. Then

$$\begin{pmatrix} a & b \\ c & d \end{pmatrix} \begin{pmatrix} 1 \\ 0 \end{pmatrix} = \frac{1}{2}\hbar \begin{pmatrix} 1 \\ 0 \end{pmatrix} \quad \text{and} \quad \begin{pmatrix} a & b \\ c & d \end{pmatrix} \begin{pmatrix} 0 \\ 1 \end{pmatrix} = -\frac{1}{2}\hbar \begin{pmatrix} 0 \\ 1 \end{pmatrix}.$$

These equations have solution $b = c = 0$, $a = \hbar/2$ and $d = -\hbar/2$. Hence

$$S_z = \frac{1}{2}\hbar \begin{pmatrix} 1 & 0 \\ 0 & -1 \end{pmatrix}.$$

Now $S^2$, $S_x$, $S_y$ and $S_z$ must obey the commutation relations in Equations (2.188) and (2.189). It is easily verified that these relations are satisfied if we take

$$S_x = \frac{1}{2}\hbar \begin{pmatrix} 0 & 1 \\ 1 & 0 \end{pmatrix} \tag{2.203}$$

$$S_y = \frac{1}{2}\hbar \begin{pmatrix} 0 & -i \\ i & 0 \end{pmatrix} \tag{2.204}$$

$$S_z = \frac{1}{2}\hbar \begin{pmatrix} 1 & 0 \\ 0 & -1 \end{pmatrix}. \tag{2.205}$$

These results may be written in the form

$$\mathbf{S} = \frac{1}{2}\hbar \boldsymbol{\sigma} \tag{2.206}$$

where

$$\sigma_x = \begin{pmatrix} 0 & 1 \\ 1 & 0 \end{pmatrix} \tag{2.207}$$

$$\sigma_y = \begin{pmatrix} 0 & -i \\ i & 0 \end{pmatrix} \tag{2.208}$$

$$\sigma_z = \begin{pmatrix} 1 & 0 \\ 0 & -1 \end{pmatrix} \tag{2.209}$$

are called the *Pauli spin matrices*. Note that the matrices corresponding to these operators are hermitian. Further, it follows that

$$S^2 = S_x{}^2 + S_y{}^2 + S_z{}^2 = \frac{3}{4}\hbar^2 \begin{pmatrix} 1 & 0 \\ 0 & 1 \end{pmatrix}. \tag{2.210}$$

### 2.11.4 Energy splitting using spin

We can now demonstrate how the spin splits the energy levels even further. Equation (2.187) can be written as

$$H'u_{nlm}(r,\theta,\phi) = E'_{nm}u_{nlm}(r,\theta,\phi). \tag{2.211}$$

Now consider the modified hamiltonian

$$H = H' + \frac{qB}{\mu c}S_z = \begin{pmatrix} 1 & 0 \\ 0 & 1 \end{pmatrix}H' + \frac{q\hbar B}{2\mu c}\begin{pmatrix} 1 & 0 \\ 0 & -1 \end{pmatrix}. \tag{2.212}$$

If this spin coupling modifies the energies to $E'_{nm}+\Delta E$, suppose that the new energy eigenfunctions are $\begin{pmatrix} \alpha u_{nlm}(r) \\ \beta u_{nlm}(r) \end{pmatrix}$, with

$$\left\{\begin{pmatrix} 1 & 0 \\ 0 & 1 \end{pmatrix}H' + \frac{q\hbar B}{2\mu c}\begin{pmatrix} 1 & 0 \\ 0 & -1 \end{pmatrix}\right\}\begin{pmatrix} \alpha u_{nlm}(r) \\ \beta u_{nlm}(r) \end{pmatrix} = (E'_{nm}+\Delta E)\begin{pmatrix} \alpha u_{nlm}(r) \\ \beta u_{nlm}(r) \end{pmatrix}.$$

Hence

$$\frac{q\hbar B}{2\mu c}\begin{pmatrix} 1 & 0 \\ 0 & -1 \end{pmatrix}\begin{pmatrix} \alpha \\ \beta \end{pmatrix}u_{nlm}(r) = \Delta E\begin{pmatrix} \alpha \\ \beta \end{pmatrix}u_{nlm}(r).$$

Since the matrices have no effect on the $u_{nlm}(r)$, these can be cancelled to give

$$\left(\frac{q\hbar B}{2\mu c} - \Delta E\right)\alpha = 0 \quad \text{and} \quad \left(-\frac{q\hbar B}{2\mu c} - \Delta E\right)\beta = 0$$

from which it follows that

$$\begin{vmatrix} q\hbar B/(2\mu c) - \Delta E & 0 \\ 0 & -q\hbar B/(2\mu c) - \Delta E \end{vmatrix} = 0$$

which has the solution

$$\Delta E = \pm\frac{q\hbar B}{2\mu c}.$$

Considering the positive and negative solutions separately, we have

$$\Delta E = +\frac{q\hbar B}{2\mu c} \quad \Rightarrow \quad \begin{pmatrix} \alpha \\ \beta \end{pmatrix} = \begin{pmatrix} 1 \\ 0 \end{pmatrix}$$

$$\Delta E = -\frac{q\hbar B}{2\mu c} \quad \Rightarrow \quad \begin{pmatrix} \alpha \\ \beta \end{pmatrix} = \begin{pmatrix} 0 \\ 1 \end{pmatrix}.$$

Hence the (already split) energy eigenvalues are further split into

$$E_{nm} + \frac{q\hbar B}{2\mu c} \quad \text{and} \quad E_{nm} - \frac{q\hbar B}{2\mu c}$$

with corresponding eigenfunctions

$$\begin{pmatrix} u_{nlm}(r) \\ 0 \end{pmatrix} \quad \text{and} \quad \begin{pmatrix} 0 \\ u_{nlm}(r) \end{pmatrix}.$$

**Fig. 2.11** The helium atom showing the nucleus with charge $2q$, and two identical electrons $e_1$ and $e_2$ with charges $-q$.

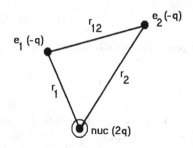

## 2.12 Systems of identical particles: BE and FD statistics

We now consider a system of $N$ identical particles, in which we can specify the positions of the particles, but we cannot identify which particle is at which point. Let $H(1, 2, \ldots, N)$ be the Hamiltonian operator of the system, and let $P_{ij}$ be the permutation operator which interchanges the two particles $i$ and $j$. This permutation operator has the properties that

$$(P_{ij})^2 = 1, \qquad P_{ij}(fg) = (P_{ij}f)(P_{ij}g) \tag{2.213}$$

where $f$ and $g$ are any two functions of the particles $1, 2, \ldots, N$. The fact that the particles are identical means that a permutation of the particles leaves the Hamiltonian unchanged:

$$P_{ij}H(1, 2, \ldots, N) = H(1, 2, \ldots, N). \tag{2.214}$$

For example, the helium atom shown in Fig. 2.11 has two identical electrons orbiting a nucleus. The Hamiltonian operator for this system is

$$H = -\frac{\hbar^2}{2m_e}(\nabla_1{}^2 + \nabla_2{}^2) - 2q^2 \left( \frac{1}{r_1} + \frac{1}{r_2} \right) + \frac{q^2}{|\mathbf{r_1} - \mathbf{r_2}|} \tag{2.215}$$

and is clearly symmetric under the interchange of labels 1 and 2.

### 2.12.1 Symmetric and antisymmetric wave functions

In what follows, we will write $H$ and $\psi$ as shorthand to denote $H(1, 2, \ldots, N)$ and the solution $\psi(1, 2, \ldots, N)$ of the TISE respectively. Now

$$H(1, 2, \ldots, N)\psi(1, 2, \ldots, N) = E\psi(1, 2, \ldots, N). \tag{2.216}$$

Hence applying the operator $P_{ij}$ to both sides, we get

$$P_{ij}(H\psi) = P_{ij}(E\psi) = E(P_{ij}\psi)$$

giving

$$E(P_{ij}\psi) = (P_{ij}H)P_{ij}\psi = H(P_{ij}\psi).$$

Hence

$$H(P_{ij}\psi) = E(P_{ij}\psi),$$

which shows that $P_{ij}\psi$ is also an eigenfunction of $H$ corresponding to the eigenvalue $E$. One possible special case which follows from this result is that

$$P_{ij}\psi = \lambda_{ij}\psi$$

for some scalar quantity $\lambda_{ij}$. Hence

$$\psi = (P_{ij})^2\psi = P_{ij}(\lambda_{ij}\psi) = \lambda_{ij}P_{ij}\psi = \lambda_{ij}^2\psi$$

giving

$$\lambda_{ij}^2 = 1, \quad \text{or} \quad \lambda_{ij} = \pm 1.$$

Hence

$$P_{ij}\psi = \pm\psi. \tag{2.217}$$

This equation shows that the wave function for a system of identical particles can be either an even or an odd function of the particle coordinates under a permutation of two particles. Particles whose wave function is symmetric satisfy *Bose-Einstein* statistics, and are called *bosons*—for example, $\pi$-mesons and photons. Particles whose wave function is antisymmetric satisfy *Fermi-Dirac* statistics, and are called *fermions*—for example, electrons, protons and neutrons.

Equation (2.217) also shows that $\psi$ is an eigenfunction of the permutation operator $P_{ij}$ with eigenvalues $\pm 1$. Further, for any function $f \equiv f(1, 2, \ldots, N)$, we have

$$[P_{ij}, H]f = P_{ij}(Hf) - HP_{ij}f = (P_{ij}H)(P_{ij}f) - HP_{ij}f$$
$$= HP_{ij}f - HP_{ij}f = 0,$$

showing that $[P_{ij}, H] = 0$. Hence

$$\frac{d}{dt}\langle P_{ij} \rangle = \frac{1}{i\hbar}\langle [P_{ij}, H] \rangle = 0, \tag{2.218}$$

indicating that the even or odd characteristic of the wave function $\psi$ is preserved in time—fermions cannot flip over to become bosons, and vice versa.

## 2.12.2  The Pauli exclusion principle

Suppose that the particles are fermions, and therefore have antisymmetric wave functions under the action of the operator $P_{ij}$. Further, suppose that the particles

$i$ and $j$ are in the same state. Then

$$\psi = P_{ij}\psi = -\psi$$

Hence

$$\psi = 0, \quad \text{or} \quad |\psi|^2 = 0,$$

showing that there is a zero probability of finding two fermions in the same state. The two $z$-components of spin angular momentum for the electron are $\pm\hbar/2$. Recall from Sect. 2.10.4 in the case of the hydrogen atom that the TISE

$$Hu_{nlm} = E_n u_{nlm}$$

provided an energy level $E_n$ which was $n^2$-fold degenerate. That is, there were $n^2$ distinct quantum states corresponding to the energy $E_n$. More generally for a system with $N$ identical electrons, the state will be specified by the function $u_{nlm...pq...\pm\frac{1}{2}}$ corresponding to the energy level $E_n$. Such a collection of states forms an *energy shell*, and the quantum states $nlm \ldots pq \ldots$ (that is, excluding the $\pm 1/2$) form an *orbit* within that shell. The exclusion principle then asserts that if two electrons are in the same orbit, then they must have opposite spins. Once a shell is full, any other electron must go into a different shell, and this rule gives rise to the periodic table of elements.

Later, in Chapter 3 on the treatment of quantum states in statistical mechanics, we will need to sum over the particle numbers of various states. For fermions, the sum will therefore only involve the particle numbers 0 and 1. However, there is no such restriction in the case of bosons, and particle numbers in a state will range from 0 to infinity.

### 2.12.3 Non-interacting identical particles

When the identical particles are non-interacting, the Hamiltonian $H(1, 2, \ldots, N)$ splits up into a sum of $N$ separate terms, each one corresponding to a separate particle:

$$H(1, 2, \ldots, N) = H(1) + H(2) + \ldots + H(N). \tag{2.219}$$

For example, if the interaction term in Equation (2.215) for the Hamiltonian of the helium atom is ignored, then it can be written

$$H = \left(-\frac{\hbar^2}{2m_e}\nabla_1{}^2 - \frac{2q^2}{r_1}\right) + \left(-\frac{\hbar^2}{2m_e}\nabla_2{}^2 - \frac{2q^2}{r_2}\right).$$

The TISE corresponding to the $k$th term in the Hamiltonian is then

$$H(k)\psi_{\alpha_k}(k) = E_{\alpha_k}(k)\psi_{\alpha_k}(k), \quad (k = 1, \ldots, N; \alpha_k = 0, 1, 2, \ldots), \tag{2.220}$$

and the total energy will be

$$E = \sum_{k=1}^{N} E_{\alpha_k}(k),$$

with the eigenfunction corresponding to this energy being

$$\psi_{\alpha_1 \alpha_2 \dots \alpha_N}(1, 2, \dots, N) = \prod_{k=1}^{N} \psi_{\alpha_k}(k). \tag{2.221}$$

In fact, since the particles are identical, any permutation of the particles will give an eigenfunction corresponding to this energy, and so will any linear combination of the $N!$ possible permutations. This leads us to consider the function

$$\psi_{\alpha_1 \beta_2 \dots \gamma_N}(1, 2, \dots, N) \equiv \frac{1}{\sqrt{N!}} \begin{vmatrix} \psi_{\alpha_1}(1) & \psi_{\alpha_1}(2) & \dots & \psi_{\alpha_N}(N) \\ \psi_{\beta_1}(1) & \psi_{\beta_1}(2) & \dots & \psi_{\beta_N}(N) \\ \dots & \dots & \dots & \dots \\ \psi_{\gamma_1}(1) & \psi_{\gamma_1}(2) & \dots & \psi_{\gamma_N}(N) \end{vmatrix}. \tag{2.222}$$

An interchange of any two coordinates in the determinant will interchange two columns, and hence change the sign of the function. Further, if any two coordinate numbers are identical then two columns will be equal, and so the function will be zero. Again, the action of the Hamiltonian in Equation (2.219) will produce the equation

$$H\psi_{\alpha_1 \beta_2 \dots \gamma_N} = (E_{\alpha_1} + E_{\beta_2} + \dots + E_{\gamma_N})\psi_{\alpha_1 \beta_2 \dots \gamma_N}.$$

Hence the function $\psi_{\alpha_1 \beta_2 \dots \gamma_N}$ given by the determinant in Equation (2.222) is the wave function of $N$ non-interacting fermions. If the individual functions $\psi_{\alpha_k}(k)$ are normalised, then the factor $1/\sqrt{N!}$ provides the correct normalisation factor.

## 2.13 The Schrödinger equation in device modelling

A number of separate modelling equations are used to model semiconductor devices. These will be described fully in Chapter 6. One of these equations is the *Poisson equation* which must be solved for the electrostatic potential $\psi$. The minimum of the conduction band is $E_c$ where

$$E_c = E_h - q\psi. \tag{2.223}$$

Here $q$ is the magnitude of the electron charge, and $E_h$ (measured in eV) is the abrupt conduction band discontinuity at the interface between different material layers. With a change of notation, Equation (2.16) is written

$$-\frac{\hbar^2}{2} \nabla \cdot \left( \frac{1}{m} \nabla \xi_k \right) + (V_{xc} + E_c)\xi_k = \lambda_k \xi_k \tag{2.224}$$

in which the quantities $\xi_k$ are the energy eigenfunctions and the $\lambda_k$, measured in eV, are the energy eigenvalues. The quantity $V_{xc}$ is the *exchange correlation energy*, measured in eV, which arises from the Coulomb repulsion between electrons, and will be described more fully in Sect. 6.8. The calculation of the normalisation factors in the solutions $\xi_k$ will depend on the numbers of spatial degrees of freedom allowed to the electrons. In particular,

- For the one dimensional solution, $\int dy |\xi_k(y)|^2 = 1$. In this case, $\xi_k(y)$ has dimension $m^{-\frac{1}{2}}$.
- For the two dimensional solution, $\int dx \int dy |\xi_k(x, y)|^2 = 1$. In this case, $\xi_k(x, y)$ has dimension $m^{-1}$.

## Problems

**Problem 2.1.** Normalise the solutions

$$u(x) = \begin{cases} A \sin kx + B \cos kx & \text{for } |x| \leq a \\ Ce^{-k_1 x} + De^{k_1 x} & \text{for } |x| \geq a \end{cases}$$

given in Equation (2.44) for the finite square well potential.

**Problem 2.2.** Investigate the case of a particle approaching the finite square well potential given in Sect 2.5.6 from the left hand side with energy $E > V_0$. Obtain the reflection coefficient.

**Problem 2.3.** Prove in the case $E > 0$ for the $\delta$-function potential given by

$$V(x) = V_0 \delta(x)$$

that the reflection coefficient is

$$R = \frac{|E|}{|E| + |E_0|}.$$

**Problem 2.4.** Calculate the spherical harmonic function $Y_{13}(\theta, \phi)$.

**Problem 2.5.** If the hamiltonian $H$ is independent of time $t$, show that

$$\mathbf{p} = \frac{im}{\hbar}[H, \mathbf{r}].$$

**Problem 2.6.** Use the result in Problem 2.5 to show that

$$\frac{d}{dt}\langle x^2 \rangle = \frac{1}{m}(\langle xp_x \rangle + \langle p_x x \rangle).$$

**Problem 2.7.** A linear harmonic oscillator $(m, \mu)$ is in the $n$th energy state. Obtain an expression for the probability that the particle will be found outside its classical limits along the $x$-axis in terms of the quantity

$$I_n(a) \equiv \frac{1}{2^n \pi^{\frac{1}{2}} n!} \int_{-a}^{a} H_n(\xi)^2 e^{-\xi^2} \, d\xi.$$

Calculate this value for the groundstate, given that $I_0(1) = 0.84$.

**Problem 2.8.** For the linear harmonic oscillator in the $n$th energy state, calculate $\langle x \rangle$, $\langle x^2 - \langle x \rangle^2 \rangle$, $\langle p_x \rangle$ and $\langle p_x^2 - \langle p_x \rangle^2 \rangle$.

# Chapter 3
# Equilibrium thermodynamics and statistical mechanics

The subject of thermodynamics deals with general systems and the relationships between the different macroscopic variables which describe each system. The first three Laws of thermodynamics will be introduced. In particular, the Second Law of thermodynamics leads to the related ideas of absolute temperature and entropy. Using the fact that the entropy of a system cannot decrease as it approaches equilibrium, the statistical entropy is introduced with properties which are related to those of the thermodynamic entropy. This statistical entropy is based on the probabilities of finding the system in its various microscopic states, and is shown to have the same form as Shannon's information content. The maximisation of this entropy, subject to certain constraints, leads to Bose-Einstein and Fermi-Dirac statistics. The applications of thermodynamics and statistical mechanics are rich and varied, but in Chapters 4 and 5 we will concentrate only on those aspects which are directly relevant to device modelling.

## 3.1 The scope and laws of thermodynamics

The set of variables $\mathbf{x}$ describing a system must be

- *macroscopic*, and will include measurable quantities such as pressure, temperature, volume;
- a *complete set*, which is found experimentally by different observers.

Systems will be in *equilibrium*. That is, the averages of the macroscopic variables will not change with time if these averages are taken over a sufficiently long time.

**Definition 3.1.** The *phase space $E$* is the set of points $\mathbf{x}$ which represent equilibrium states. This space is embedded in a larger space, whose extra axes represent spatial gradients and time derivatives of the variables. The *state* of the equilibrium system will be represented by a point in $E$.

E.A.B. Cole, *Mathematical and Numerical Modelling of*
*Heterostructure Semiconductor Devices: From Theory to Programming*,
DOI 10.1007/978-1-84882-937-4_3, © Springer-Verlag London Limited 2009

Thermodynamics deals with the bulk properties of a system, and essentially ignores surface effects. Statistical mechanics deals with the underlying microscopic processes, and tries to predict these bulk properties by studying the microscopic effects under different types of microscopic process, which can be either classical or quantum.

**Definition 3.2.** An *adiabatic enclosure* allows only the transmission of mechanical energy, electric current, and externally applied long range forces such as electromagnetic and gravitational forces. It does not allow the transmission of radiation, mass, or thermal energy.

**Definition 3.3.** An *adiabatic process* is a process which is experienced by a system totally enclosed in an adiabatic enclosure.

Adiabatic processes are not necessarily reversible; if an adiabatic process exists which takes state $i \in E$ into state $j \in E$, there is not necessarily an adiabatic process which takes state $j$ into state $i$. Further, a path linking the internal states $i$ and $j$ of $E$ of this adiabatic process may not lie wholly inside $E$: some of the intermediate states on the path may be nonequilibrium states which will lie outside $E$.

The first three Laws of thermodynamics will now be introduced. The Zeroth Law will imply the existence of empirical temperature, the First Law will imply the existence of internal energy, and the Second Law will imply the existence of entropy and absolute temperature.

### 3.1.1 The Zeroth Law of thermodynamics

The Zeroth Law of thermodynamics consists of a two-part statement:

1. If two different systems are in equilibrium with a third, then they are in equilibrium with each other;
2. If states $i, j, k \in E$ are such that the pairs $i$ and $j$, and $j$ and $k$ can be linked by an adiabatic process, then the pairs $i$ and $k$ can be linked by an adiabatic process.

**Theorem 3.1.** *The Zeroth Law of thermodynamics implies the existence of empirical temperature.*

*Proof.* Let systems 1, 2 and 3 be such that the pairs 1 and 2, 2 and 3 are in equilibrium. Let $\mathbf{x}_1$, $\mathbf{x}_2$ and $\mathbf{x}_3$ be the variables representing the system in these states. Then there exist functions $f_{12}$ and $f_{23}$ such that

$$f_{12}(\mathbf{x}_1, \mathbf{x}_2) = 0 \quad \text{and} \quad f_{23}(\mathbf{x}_2, \mathbf{x}_3) = 0. \tag{3.1}$$

Part 1 of the statement then implies that there exists a function $f_{13}$ such that

$$f_{13}(\mathbf{x}_1, \mathbf{x}_3) = 0. \tag{3.2}$$

Hence Equation (3.1) must imply Equation (3.2). Let $x_2 = (w_2, y_2)$, where the variable $w_2$ appears explicitly in both of the Equations (3.1). These equations may be solved explicitly for $w_2$ in the form

$$w_2 = \phi_1(x_1, y_2) = \phi_2(y_2, x_3) \tag{3.3}$$

for some functions $\phi_1$ and $\phi_2$. Then taking any function $\Psi$ of $w_2$ and $y_2$,

$$\Psi(w_2, y_2) = \Psi(\phi_1(x_1, y_2), y_2) = \Psi(\phi_2(y_2, x_3), y_2). \tag{3.4}$$

One consequence of this last equation must be Equation (3.2) in which $y_2$ does not appear. Hence a function $\Psi$ exists such that $y_2$ drops out of Equation (3.4). This can be done only if there exist functions $t_1, t_3, \psi_1$ and $\psi_2$ such that

$$\Psi(\phi_1(x_1, y_2), y_2) \equiv t_1(x_1)\psi_1(y_2) + \psi_2(y_2) \quad \text{and}$$
$$\Psi(\phi_2(y_2, x_3), y_2) \equiv t_3(x_3)\psi_1(y_2) + \psi_2(y_2).$$

Note that $\psi_1 \neq 0$, otherwise Equation (3.4) would give $0 = 0$ which, although not incorrect, does not imply Equation (3.2). Hence from Equation (3.4) we get

$$t_1(x_1) = t_3(x_3). \tag{3.5}$$

Writing

$$t_2(x_2) \equiv \frac{\Psi(w_2, y_2) - \psi_2(y_2)}{\psi_1(y_2)},$$

then Equation (3.4) implies that

$$t_2(x_2) = \frac{t_1(x_1)\psi_1(y_2) + \psi_2(y_2) - \psi_2(y_2)}{\psi_1(y_2)} = t_1(x_1).$$

Hence there exist functions $t_1(x_1)$, $t_2(x_2)$ and $t_3(x_3)$ such that

$$t_1(x_1) = t_2(x_2) = t_3(x_3) \tag{3.6}$$

when the systems are in equilibrium. These functions are called the *empirical temperatures*. This completes the proof. $\square$

There is no unique empirical temperature for a system, but if a temperature function is defined for one system, for example, the height of a column of mercury, then it can be used to define the temperatures of many other systems through Equation (3.6).

Part 2 of the Zeroth Law may be illustrated using the simple example of heating a simple liquid in a container by stirring only (an adiabatic process). States $A$, $B$ and $C$ are represented by temperatures $t_A < t_B < t_C$. Then the processes $A \to B$, $B \to C$ and $A \to C$ are adiabatic, but the reverse processes are not (the liquid cannot be "unstirred"). Further, since $A \to B$ and $A \to C$ are adiabatic, part (b) of the law says that at least one of the processes $B \to C$ and $C \to B$ must be adiabatic; clearly the first of these is adiabatic.

**Definition 3.4.** As a consequence of Part 2 of the Zeroth Law, a subset of equilibrium points $\beta \subset E$ can be defined such that any two points in $\beta$ can be linked by an adiabatic process, and no point that can be in the set is excluded.

Hence if $\beta_1$ and $\beta_2$ are two such sets, either $\beta_1 = \beta_2$, or $\beta_1$ and $\beta_2$ do not intersect. See Problem 3.1.

## 3.1.2 The First Law of thermodynamics

This states that the energy $W_a(i, j)$ supplied to a closed system in an adiabatic process $i \rightarrow k$ is such that

$$W_a(i, k) = W_a(i, j) + W_a(j, k) \tag{3.7}$$

whenever $i \rightarrow j$ and $j \rightarrow k$ are adiabatic, where $j$ is an intermediate state.

Then Equation (3.7) implies that for each state $i$ there exists a function $U(i)$ such that

$$W_a(i, j) = U(j) - U(i). \tag{3.8}$$

The quantity $U(i)$ is called the *internal energy* of the system in state $i$. Clearly, it is defined to within an arbitrary additive constant.

Using this Law, we may now introduce the concept of *heat*. If there exists an adiabatic process $i \rightarrow j$, but the process $i \rightarrow j$ actually performed is not necessarily adiabatic, with mechanical work $W(i, j)$ done *on* the system, define

$$Q(i, j) \equiv W_a(i, j) - W(i, j). \tag{3.9}$$

Then

$$Q(i, j) = U(j) - U(i) - W(i, j). \tag{3.10}$$

Hence

1. $Q = 0$ for an adiabatic process;
2. $Q \neq 0$ implies that the process is not adiabatic;
3. $Q$ is defined in mechanical terms: it makes up the deficit in mechanical work;
4. $Q$ is called the *heat* added to the system in the process.

It follows that

$$U(j) - U(i) = Q(i, j) + W(i, j) \tag{3.11}$$

or, in terms of infinitesimals,

$$dU = d^*Q + d^*W \tag{3.12}$$

where the notation $d^* f$ means that the quantity is not a complete differential. This arises because both $Q$ and $W$ are not functions of state, but can be made up arbitrarily for a transition between two states. Note that $d^*Q$ and $d^*W$ are the heat and work applied *to* the system. These quantities are additive: suppose that two systems, denoted by 1 and 2, are combined. Then the relevant quantities applied to the whole system are

$$d^*Q = d^*Q^{(1)} + d^*Q^{(2)}$$
$$d^*W = d^*W^{(1)} + d^*W^{(2)}.$$

In general, a typical increment of mechanical work done *by* a system can be written

$$-d^*W = \sum_q f_q da_q$$

where the $f_q$ are a set of *generalised forces* and the $a_q$ are suitable deformation coordinates. Then

$$d^*Q = dU + \sum_q f_q da_q. \tag{3.13}$$

Three simple examples will illustrate these ideas.

- for a gas under pressure $p$ with volume $V$ in a cylinder fitted with a piston, a force applied to the piston does an amount $d^*W = -pdV$ of work on the gas. Hence $d^*Q = dU + pdV$.
- A wire of length $l$ is held under tension by a force $f$. Then $d^*W = fdl$.
- A membrane of area $A$ is held in tension $\gamma$. Then $d^*W = \gamma dA$.

*Examples*

The notation will be such that the quantity $(\frac{\partial f}{\partial x})_y$ will denote the partial derivative of the function $f(x, y)$ with respect to the variable $x$, while keeping $y$ constant.

1. *General.* For a gas specified by pressure $p$, volume $V$, temperature $t$ and internal energy $U$, there are usually two equations of state linking these four quantities, so there will be only two independent variables. For example, as we shall see later, the equations of state for an ideal classical gas are $pV = At$ and $(\frac{\partial U}{\partial V})_t = 0$, where $A$ is a constant. Hence we may write

$$d^*Q = C_V dt + l_V dV \tag{3.14}$$
$$= C_p dt + l_p dp \tag{3.15}$$
$$= dU + pdV \tag{3.16}$$

where

$$C_V \equiv \lim_{\delta t \to 0} \left( \frac{\delta Q}{\delta t} \right)_V = \text{specific heat at constant volume}$$

$$C_p \equiv \lim_{\delta t \to 0} \left( \frac{\delta Q}{\delta t} \right)_p = \text{specific heat at constant pressure}$$

$$l_V \equiv \lim_{\delta V \to 0} \left( \frac{\delta Q}{\delta V} \right)_t = \text{latent heat of volume increase}$$

$$l_p \equiv \lim_{\delta p \to 0} \left( \frac{\delta Q}{\delta p} \right)_t = \text{latent heat of pressure increase.}$$

From Equations (3.14) and (3.15), we have $l_p dp = l_V dV + (C_V - C_p)dt$. Hence it follows that

$$l_p = l_V \left( \frac{\partial V}{\partial p} \right)_t . \tag{3.17}$$

For an adiabatic change $(d^*Q = 0)$, with the suffix $a$ on certain quantities denoting their adiabatic form, we have

$$\gamma \equiv \frac{C_p}{C_V} = \frac{l_p}{l_V} \left( \frac{\partial p}{\partial V} \right)_a . \tag{3.18}$$

Now $U \equiv U(p,t)$ or $U \equiv U(V,t)$. Hence

$$dU = \left( \frac{\partial U}{\partial p} \right)_t dp + \left( \frac{\partial U}{\partial t} \right)_p dt = \left( \frac{\partial U}{\partial V} \right)_t dV + \left( \frac{\partial U}{\partial t} \right)_V dt.$$

Hence at constant pressure $(dp = 0)$, it follows that

$$\left( \frac{\partial U}{\partial t} \right)_p = \left( \frac{\partial U}{\partial V} \right)_t \left( \frac{\partial V}{\partial t} \right)_p + \left( \frac{\partial U}{\partial t} \right)_V . \tag{3.19}$$

Therefore, the specific heats are linked by the equations

$$C_p \equiv \lim_{\delta t \to 0} \left( \frac{\delta Q}{\delta t} \right)_p = \left( \frac{\partial U}{\partial t} \right)_p + p \left( \frac{\partial V}{\partial t} \right)_p$$

$$= \left( \frac{\partial U}{\partial V} \right)_t \left( \frac{\partial V}{\partial t} \right)_p + \left( \frac{\partial U}{\partial t} \right)_V + p \left( \frac{\partial V}{\partial t} \right)_p , \tag{3.20}$$

$$C_V = \left( \frac{\partial U}{\partial t} \right)_V , \tag{3.21}$$

$$C_p - C_V = \left[ \left( \frac{\partial U}{\partial V} \right)_t + p \right] \left( \frac{\partial V}{\partial t} \right)_p . \tag{3.22}$$

2. *The Boyle's Law fluid*. This has equation of state

$$pV = f(t) \tag{3.23}$$

where $f(t)$ is some function of temperature. Therefore, the latent heats are linked by the equations

$$\frac{l_p}{l_V} = \left( \frac{\partial V}{\partial p} \right)_t = -\frac{f(t)}{p^2} = -\frac{V}{p} .$$

Hence for an adiabatic change, it follows that

$$\gamma = \frac{l_p}{l_V}\left(\frac{\partial p}{\partial V}\right)_a = -\frac{V}{p}\left(\frac{\partial p}{\partial V}\right)_a.$$

In particular, if $\gamma$ is a constant, then

$$\gamma p + V\left(\frac{\partial p}{\partial V}\right)_a = 0, \quad \text{or} \quad \left(\frac{\partial}{\partial V}(pV^\gamma)\right)_a = 0,$$

giving

$$pV^\gamma = \text{constant}. \tag{3.24}$$

3. *The Ideal Classical gas.* This applies to dilute gases. It is defined by the two relations

$$pV = At \quad (A = \text{constant}) \tag{3.25}$$

$$\left(\frac{\partial U}{\partial V}\right)_t = 0. \tag{3.26}$$

It follows from Equation (3.25) that $p(\frac{\partial V}{\partial t})_p = A$, and then Equation (3.22) gives

$$C_p - C_V = A. \tag{3.27}$$

In particular, for a weak gas consisting of $N$ molecules, $A = Nk_B$ where $k_B = 1.38 \times 10^{-23}\,\mathrm{J\,K^{-1}}$ is Boltzmann's constant.

4. *Fallacy!* Writing $U \equiv U(V, t)$, then

$$d^*Q = dU + pdV = \left(\frac{\partial U}{\partial t}\right)_V dt + \left(\frac{\partial U}{\partial V}\right)_t dV + pdV$$

$$= \left(\frac{\partial U}{\partial t}\right)_V dt + \left[p + \left(\frac{\partial U}{\partial V}\right)_t\right]dV.$$

On the other hand,

$$d^*Q = \left(\frac{\partial Q}{\partial t}\right)_V dt + \left(\frac{\partial Q}{\partial V}\right)_t dV. \tag{3.28}$$

Then equating coefficients of $dt$ and $dV$ we have

$$\left(\frac{\partial Q}{\partial t}\right)_V = \left(\frac{\partial U}{\partial t}\right)_V, \quad \left(\frac{\partial Q}{\partial V}\right)_t = p + \left(\frac{\partial U}{\partial V}\right)_t.$$

Further differentiation then gives

$$\frac{\partial^2 U}{\partial V \partial t} = \frac{\partial^2 Q}{\partial V \partial t} = \frac{\partial^2 Q}{\partial t \partial V} = \left(\frac{\partial p}{\partial t}\right)_V + \frac{\partial^2 U}{\partial t \partial V},$$

giving $(\frac{\partial p}{\partial t})_V = 0$. This result is clearly physically unrealistic—for example, this result predicts that if a gas in a rigid container is heated then its pressure does not rise. This fallacy arises from the mistaken assumption in Equation (3.28) that $d^*Q$ is a perfect differential.

### 3.1.3 The Second Law of thermodynamics

In order to describe the Second Law of thermodynamics, it is necessary to introduce the idea of an idealised process called a *quasistatic process*.

**Definition 3.5.** A quasistatic process linking two states $i, j \in E$ is a *reversible* process such that all the states through which the system passes are equilibrium states.

Therefore, a quasistatic process is represented by a curve lying completely in $E$. A non-quasistatic process may link the same two states $i$ and $j$, but some of the intermediate points will not be equilibrium points, and will therefore lie outside $E$. Quasistatic processes are never realised in practice, but can be approximated to any degree of accuracy. For example, the temperature of a system may be raised by placing it in contact with a succession of large heat reservoirs, such that the temperature differences between consecutive reservoirs are made suitably small.

The set $\beta$ has been defined as the set of points in $E$ such that any two points in $\beta$ can be joined by an adiabatic process. Let $\Gamma$ be the set of interior points of $\beta$, that is, those points of $\beta$ which do not lie on its boundary. Then it follows that all points of $\Gamma$ have a $\beta$-neighbourhood, and any two points of $\Gamma$ can be joined by an adiabatic process.

**Definition 3.6.** A point $y$ of $\Gamma$ is an *i-point* if and only if every $\Gamma$-neighbourhood of $y$ contains a point which is adiabatically inaccessible from $y$.

**Definition 3.7.** A point $y$ of $\Gamma$ is an *a-point* if and only if it is not an i-point.

Hence each point of $\Gamma$ is either an i-point or an a-point, and

1. if, using quasistatic processes only, a point of $\Gamma$ is found to be an a-point, then that point is an a-point;
2. if, using quasistatic processes only, a point of $\Gamma$ is found to be an i-point, then that point may be an i-point or an a-point.

**Theorem 3.2.** *If*

$$d^*Q = \sum_{j=1}^{n} X_j da_j, \quad (n \geq 2) \tag{3.29}$$

*where the $X_j(a_1, \ldots, a_n)$ are analytic functions in the set $\Gamma$, then as judged by quasistatic processes only,*

(i) *The points of $\Gamma$ are either all a-points or all i-points;*
(ii) *A point of $\Gamma$ is an i-point if and only if $d^*Q$ has an integrating factor.*

*Proof.* The proof is given by Landsberg (1961). □

There are several different statement of the Second Law of thermodynamics, and we present three of them here. A study of the equivalence of the different statements under different assumptions has been made by Dunning Davies (1996).

1. *Kelvin's statement.* It is impossible to convert an amount of heat completely into work in a cyclic process without at the same time producing other changes.
2. *Clausius's statement.* It is impossible for heat to be transferred in a cyclic process from one body to a hotter one without , at the same time, producing other changes.
3. *Caratheodory's statement.* All points of $\Gamma$ are i-points.

Caratheodory's statement is certainly more opaque than either of the statements of Kelvin or Clausius. However, it follows from Caratheodory's statement and Theorem 3.2 that there exists an integrating factor $\lambda(a_1, \ldots, a_n)$ and a function $\phi(a_1, \ldots, a_n)$ such that

$$d^*Q = \lambda d\phi. \tag{3.30}$$

*Entropy and absolute temperature*

It can now be shown how Equation (3.30) can be written in such a way that the function $\lambda$ is the absolute temperature, and the corresponding function $\phi$ is the entropy, of the system. On taking $a_1 \equiv U$, so that $\phi \equiv \phi(U, a_2, \ldots, a_n)$, we have

$$d^*Q = dU - d^*W = dU + \sum_{q=2}^{n} f_q da_q = \lambda d\phi(U, a_2, \ldots, a_n)$$

$$= \lambda \frac{\partial \phi}{\partial U} dU + \lambda \sum_{q=2}^{n} \frac{\partial \phi}{\partial a_q} da_q.$$

Hence

$$\lambda \frac{\partial \phi}{\partial U} = 1, \quad \text{and} \quad \lambda \frac{\partial \phi}{\partial a_q} = f_q, \quad (q = 2, \ldots, n).$$

This means that $\partial \phi / \partial U \neq 0$, and so we may solve the equation $\phi \equiv \phi(U, a_2, \ldots, a_n)$ to give $U$ in terms of $\phi$. This also means that $\phi$ may be used as an independent variable in place of $U$. Further, if the set $\{a_q\}$ is ordered so that $\partial t / \partial a_2 \neq 0$, then the equation $t \equiv t(U, a_2, \ldots, a_n)$ may be solved to give $t$ in terms of $a_2$. This means that $t$ may be used as an independent variable in place of $a_2$. Hence the system may be specified by the $n$ independent variables $\phi, t, a_3, \ldots, a_n$.

Now consider two systems (1) and (2) which are specified by $m$ and $n$ variables $(t^{(1)}, \phi^{(1)}, a_3^{(1)}, \ldots, a_m^{(1)})$ and $(t^{(2)}, \phi^{(2)}, a_3^{(2)}, \ldots, a_n^{(2)})$ respectively. Suppose that the systems are in equilibrium, so that $t^{(1)} = t^{(2)} (\equiv t)$. Now the additive property of the heat increment means that

$$d^*Q = d^*Q^{(1)} + d^*Q^{(2)}$$

$$\lambda d\phi = \lambda^{(1)} d\phi^{(1)} + \lambda^{(2)} d\phi^{(2)}, \quad \text{or} \tag{3.31}$$

$$d\phi = \frac{\lambda^{(1)}}{\lambda} d\phi^{(1)} + \frac{\lambda^{(2)}}{\lambda} d\phi^{(2)}.$$

The quantity $\phi$ of the combined system is now a function of the $m + n - 1$ independent variables $(t, \phi^{(1)}, a_3^{(1)}, \ldots, a_m^{(1)}, \phi^{(2)}, a_3^{(2)}, \ldots, a_n^{(2)})$. Hence

$$d\phi = \frac{\partial \phi}{\partial t} dt + \frac{\partial \phi}{\partial \phi^{(1)}} d\phi^{(1)} + \sum_{q=3}^{m} \frac{\partial \phi}{\partial a_q^{(1)}} da_q^{(1)} + \frac{\partial \phi}{\partial \phi^{(2)}} d\phi^{(2)} + \sum_{q=3}^{n} \frac{\partial \phi}{\partial a_q^{(2)}} da_q^{(2)}. \tag{3.32}$$

Comparing Equations (3.31) and (3.32), we find that

$$\frac{\partial \phi}{\partial \phi^{(j)}} = \frac{\lambda^{(j)}}{\lambda}, \quad (j = 1, 2)$$

and

$$\frac{\partial \phi}{\partial t} = \frac{\partial \phi}{\partial a_q^{(1)}} = \frac{\partial \phi}{\partial a_r^{(2)}} = 0 \quad (q = 3, \ldots, m; r = 3, \ldots, n).$$

Hence

$$\frac{\partial}{\partial t} \left( \frac{\lambda^{(j)}}{\lambda} \right) = \frac{\partial}{\partial t} \left( \frac{\partial \phi}{\partial \phi^{(j)}} \right) = \frac{\partial}{\partial \phi^{(j)}} \left( \frac{\partial \phi}{\partial t} \right) = 0.$$

On expanding the first derivative, it is found that

$$\frac{1}{\lambda} \frac{\partial \lambda^{(j)}}{\partial t} - \frac{\lambda^{(j)}}{(\lambda)^2} \frac{\partial \lambda}{\partial t} = 0, \quad \text{or} \quad \frac{1}{\lambda^{(j)}} \frac{\partial \lambda^{(j)}}{\partial t} = \frac{1}{\lambda} \frac{\partial \lambda}{\partial t}, \quad (j = 1, 2).$$

Hence

$$\frac{1}{\lambda^{(1)}} \frac{\partial \lambda^{(1)}}{\partial t} = \frac{1}{\lambda^{(2)}} \frac{\partial \lambda^{(2)}}{\partial t} = \frac{1}{\lambda} \frac{\partial \lambda}{\partial t}$$

$$= \text{a function of only common variables}$$

$$\equiv g(t)$$

since the temperature $t$ is the only common variable of the systems 1 and 2. Hence it follows that

$$\lambda^{(j)}(t, \phi^{(j)}) = \Lambda^{(j)} \exp \left( \int_{t_0}^{t} g(t') dt' \right), \quad (j = 1, 2)$$

where $t_0$ is a standard empirical temperature. Therefore

$$d^* Q = \lambda d\phi = \Lambda(\phi) \exp \left( \int_{t_0}^{t} g(t') dt' \right) = C \exp \left( \int_{t_0}^{t} g(t') dt' \right) C^{-1} \Lambda(\phi) d\phi$$

$$= T dS \tag{3.33}$$

where $C$ can be taken as a constant for a whole class of systems, and the functions $T$ and $S$ are given by

$$T(t) \equiv C \exp\left(\int_{t_0}^{t} g(t')dt'\right)$$

$$S(\phi) \equiv C^{-1} \int_{\alpha}^{\phi} \Lambda(\phi')d\phi'.$$

The quantity $T$ is called the *absolute temperature*. It is the only integrating factor to depend *only* on the empirical temperature. When this particular integrating function is chosen, the function $S(\phi)$ is called the *entropy*, and depends on the lower limit $\alpha$ in the integration for which $S(\alpha) = 0$.

### 3.1.4 Properties of the thermodynamic entropy

1. If two systems (1) and (2) are in equilibrium, then $t^{(1)} = t^{(2)}$. Hence

$$T^{(1)} \equiv T(t^{(1)}) = T(t^{(2)}) \equiv T^{(2)} \equiv T.$$

It then follows from Equation (3.31) that $TdS = T^{(1)}dS^{(1)} + T^{(2)}dS^{(2)}$, giving $dS = dS^{(1)} + dS^{(2)}$. Hence $S = S^{(1)} + S^{(2)} + A$ where $A$ is a constant. The convention is to take $A = 0$, giving

$$S = S^{(1)} + S^{(2)}. \tag{3.34}$$

Hence the entropy is additive.

2. If the total number $N$ of particles in the system is allowed to change, define the *chemical potential* $\mu$ as

$$\mu \equiv -T\left(\frac{\partial S}{\partial N}\right)_{U,V}. \tag{3.35}$$

Then for $S \equiv S(U, V, N)$, it follows that

$$TdS = dU + pdV - \mu dN. \tag{3.36}$$

We now make use of Euler's theorem for homogeneous functions. If a function $f(x_1, \ldots, x_n)$ satisfies the condition $f(ax_1, \ldots, ax_n) = a^r f(x_1, \ldots, x_n)$ for any constant $a$, then we say that the function $f$ is homogeneous with degree $r$. Euler's theorem states that for such a function, then $\sum_i x_i \frac{\partial f}{\partial x_i} = rf$. Now the entropy $S$ is additive: $S(aU, aV, aN) = aS(U, V, N)$; that is, it is homogeneous with degree 1. Hence, applying Euler's theorem to the function $S(U, V, N)$, we have

$$U\left(\frac{\partial S}{\partial U}\right)_{V,N} + V\left(\frac{\partial S}{\partial V}\right)_{U,N} + N\left(\frac{\partial S}{\partial N}\right)_{U,V} = 1 \times S. \tag{3.37}$$

From Equation (3.36) we have

$$\left(\frac{\partial S}{\partial U}\right)_{V,N} = \frac{1}{T}, \qquad \left(\frac{\partial S}{\partial V}\right)_{U,N} = \frac{p}{T}, \qquad \left(\frac{\partial S}{\partial N}\right)_{U,V} = -\frac{\mu}{T}.$$

Applying these quantities to Equation (3.37) an re-arranging, we arrive at the result

$$TS = U + pV - \mu N. \tag{3.38}$$

3. If a system is displaced from equilibrium and then isolated, its entropy cannot decrease in its return to equilibrium (Dugdale 1966). When the system is in equilibrium, it is not being disturbed by measurement. In this sense, the less we know about the system, the greater is its entropy. Entropy gives us a measure of the information we have about the system.

## 3.2 The statistical entropy

Having obtained the thermodynamic entropy, we now wish to obtain a statistical entropy which is based on the microscopic states of the system. This statistical entropy must have some of the major the properties which have been found for the thermodynamic entropy.

Let $i = 1, 2, \ldots, n$ label the microscopic states of a system, and let $p_i$ be the probability of finding the system in the state $i$. Then $\sum_{i=1}^{n} p_i = 1$. Suppose that the average internal energy $U$ lies in a certain small range. The statistical entropy $S \equiv S(p_1, p_2, \ldots, p_n)$ of the non-equilibrium system is taken to be the only thermodynamic function which is a function of the probabilities only (as opposed to, for example, the quantity $U = \sum_i p_i E_i$ which depends also on the energies $E_i$ of the microscopic states). In the equilibrium situation, in which all the states are equally likely, the distribution will be, by the application of Laplace's Principle of Insufficient Reason,

$$p_i = \frac{1}{n}, \qquad (i = 1, 2, \ldots, n). \tag{3.39}$$

*Properties of the entropy*

In order to link the statistical entropy with the thermodynamic entropy, we would expect it to have the following properties:

1. The non-equilibrium entropy must be not greater than the entropy of the equilibrium situation:

$$S(p_1, p_2, \ldots, p_n) \leq S\left(\frac{1}{n}, \frac{1}{n}, \ldots, \frac{1}{n}\right) \equiv L(n). \tag{3.40}$$

2. The addition of an inaccessible state to the system does not alter the entropy:

$$S(p_1, p_2, \ldots, p_n) = S(p_1, p_2, \ldots, p_n, 0). \tag{3.41}$$

3. If systems $A$ and $B$ are combined to a system $AB$ and do not interact, then the entropy is additive:

$$S(AB) = S(A) + S(B). \tag{3.42}$$

The next step is to obtain an expression for $S(p_1, \ldots, p_n)$ in terms of the probabilities. Let a system $A$ have states $A_1, \ldots, A_n$ with probabilities $p_1, \ldots, p_n$. Suppose that the systems $A$ and $B$ do interact. If $A$ is in state $A_k$, suppose that the states accessible to $B$ are $B_{k1}, B_{k2}, \ldots$ with probabilities $q_{k1}, q_{k2}, \ldots$. Let $S_k(B) \equiv S(q_{k1}, q_{k2}, \ldots)$ be the entropy of $B$ conditional on $A$ being in state $A_k$. Then the mean conditional entropy of system $B$ is

$$S_A(B) \equiv \sum_{k=1}^{n} p_k S_k(B). \tag{3.43}$$

Then we generalise Equation (3.42) by writing

$$S(AB) = S(A) + S_A(B). \tag{3.44}$$

Note that if there is no interaction between the systems, then the $q_{kj}$ are independent of the index $k$, so that $S_k(B)$ is independent of the index $k$. Then $S_A(B) = \sum_k p_k S(B) = S(B)$, which gives $S(AB) = S(A) + S(B)$ once more.

**Theorem 3.3.**

$$L(n) = \lambda \log n, \quad \lambda \geq 0. \tag{3.45}$$

*Proof.* Now

$$L(n) \equiv S\left(\frac{1}{n}, \frac{1}{n}, \ldots, \frac{1}{n}\right) = S\left(\frac{1}{n}, \frac{1}{n}, \ldots, \frac{1}{n}, 0\right)$$

$$\leq S\left(\frac{1}{n+1}, \frac{1}{n_1}, \ldots, \frac{1}{n+1}\right) = L(n+1).$$

Hence $L(n) \leq L(n+1)$. Now consider a system $A$ with $r$ equally likely states, so that $S(A) = L(r)$. If $m$ such systems are lumped together, there are $r^m$ equally likely states of the compound system $mA$. Using property (3) above, it follows that $mL(r) = S(mA) = L(r^m)$. Hence

$$L(r^m) = mL(r) \quad \text{and} \quad L(r) \leq L(r+1),$$

and the result is proved. □

When logarithms are taken to the base $e$, we will show later that the equation of state for an ideal classical gas is $pV = N\lambda T = Nk_B T$. Hence we will identify $\lambda = k_B$ at this stage.

**Theorem 3.4.**

$$S(p_1, p_2, \ldots, p_n) = -k_B \sum_{i=1}^{n} p_i \ln p_i. \tag{3.46}$$

*Proof.* Let a system $A$ have states $A_1, \ldots, A_n$ with probabilities $p_i = g_i/g$, where the $g_i$ and $g$ are positive integers. Then $\sum_i g_i = g$ since $\sum_i p_i = 1$. Let $C$ denote a set of $n$ systems $C_i$ ($i = 1, \ldots, n$) which are dependent on $A$ in such a way that system $C_i$ has $g_i$ equally likely states $C_{i1}, C_{i2}, \ldots, C_{ig_i}$, and if $A$ is in state $A_i$ then system $C_i$ is chosen. Then $S_i(C) = L(g_i) = k_B \ln g_i$. Hence the mean conditional entropy of $C$ is $S_A(C) = k_B \sum_{i=1}^{n} p_i \ln g_i$. For a given state number $i$, the number of states of $C_i$ is $g_i$. Hence the probability of finding the combined system $AC$ with $A$ in state $A_i$ and $C$ in a specified state is $p_i \times 1/g_i = 1/g = $ constant. Hence the scheme $AC$ consists of $g$ equally likely states, and so $S(AC) = k_B \ln g$. It follows that

$$S(A) = S(AC) - S_A(C)$$

$$= k_B \ln g - k_B \sum_{i=1}^{n} p_i \ln g_i = -k_B \sum_{i=1}^{n} p_i (\ln g_i - \ln g)$$

$$= -k_B \sum_{i=1}^{n} p_i \ln p_i.$$

This completes the proof.  □

This result is valid for both equilibrium and non-equilibrium situations.

The result in Equation (3.46) is valid for systems which are large enough so that boundary effects are unimportant. However, these boundary effects are not negligible for nanosystems, and we will see in Chapter 14 how this result is generalised to take these effects into account through the use of *Tsallis statistics*.

## 3.3 Maximisation of entropy subject to constraints

We now seek the probability distribution which maximises the entropy in equilibrium. The entropy to be maximised is

$$S(p_1, p_2, \ldots, p_n) = -k_B \sum_{i=1}^{n} p_i \ln p_i, \tag{3.47}$$

and the distribution must satisfy the normalisation constraint

$$\sum_{i=1}^{n} p_i = 1. \tag{3.48}$$

There will be other constraints. For example, suppose that the volume $V$, the absolute temperature $T$ and the particle number $N$ are fixed, but the energy is allowed to fluctuate. If $E_i$ is the energy associated with the quantum state $i$, then we have the further constraint

$$\sum_{i=1}^{n} p_i E_i = U = \text{(constant)}.$$

On the other hand, if both energy and particle number are allowed to fluctuate while their averages remain constant, then we have the two constraints

$$\sum_{i=1}^{n} p_i E_i = U = \text{(constant)}_1 \quad \text{and} \quad \sum_{i=1}^{n} p_i n_i = N = \text{(constant)}_2$$

where $n_i$ is the number of particles in quantum state $i$.

We now investigate how a more general set of constraints may be applied. Suppose that $q$ variables are allowed to fluctuate, but their averages $A_i$ $(i = 1, \ldots, q)$ remain constant. Let these variables in the $i$th quantum state have the values $x_{ji}$ $(i = 1, \ldots, n; j = 1, \ldots, q)$, with

$$\sum_{i=1}^{n} p_i x_{ji} = A_j = \text{(constant)}_j, \quad (j = 1, \ldots, q). \tag{3.49}$$

For example, in the second example above, we can take $q = 2$, $A_1 \equiv U$, $x_{1i} \equiv E_i$, $A_2 \equiv N$, and $x_{2i} \equiv n_i$. Then the general problem is to maximise the entropy $S$ of Equation (3.47) subject to the set of constraints in Equations (3.48) and (3.49). In order to do this, define a function $S'$ by

$$S' \equiv S - k_B \alpha \left( \sum_{i=1}^{n} p_i - 1 \right) - k_B \sum_{j=1}^{q} \alpha_j \left( \sum_{i=1}^{n} p_i x_{ji} - A_j \right)$$

$$= -k_B \sum_{i=1}^{n} p_i \left( \ln p_i + \alpha + \sum_{j=1}^{q} \alpha_j x_{ji} \right) + C$$

where $C$ is a constant, and the quantities $\alpha$ and $\alpha_i$ $(i = 1, \ldots, q)$ are also constants called *Lagrange multipliers*. The maximisation condition $\partial S'/\partial p_i = 0$ $(i = 1, \ldots, n)$ gives

$$\ln p_i + \alpha + \sum_{j=1}^{q} \alpha_j x_{ji} + 1 = 0, \quad (i = 1, \ldots, n),$$

or

$$p_i = \exp \left\{ -\alpha - 1 - \sum_{j=1}^{q} \alpha_j x_{ji} \right\} = Q^{-1} \exp \left\{ -\sum_{j=1}^{q} \alpha_j x_{ji} \right\}$$

where $Q \equiv \exp(\alpha + 1)$ is itself a constant. Let the *degeneracy* $g(x_{1i}, x_{2i} \ldots, x_{qi})$ be the number of equally likely states which correspond to the set of variable values $x_{1i}, x_{2i}, \ldots, x_{qi}$. Then the probability of finding this set of variable values is

$$P(x_{1i}, x_{2i}, \ldots, x_{qi}) = g(x_{1i}, x_{2i}, \ldots, x_{qi}) p_i$$

$$= g(x_{1i}, x_{2i}, \ldots, x_{qi}) Q^{-1} \exp \left\{ -\sum_{j=1}^{q} \alpha_j x_{ji} \right\}. \quad (3.50)$$

On using the normalisation condition

$$\sum_{x_{1i}, x_{2i}, \ldots, x_{qi}} P(x_{1i}, x_{2i}, \ldots, x_{qi}) = 1,$$

the value of $Q$ can be found as

$$Q = \sum_{x_{1i}, x_{2i}, \ldots, x_{qi}} g(x_{1i}, x_{2i}, \ldots, x_{qi}) \exp \left\{ -\sum_{j=1}^{q} \alpha_j x_{ji} \right\}. \quad (3.51)$$

The quantity $Q$ is called the *partition function*. Hence the statistical entropy is given by

$$S = -k_B \sum_{i=1}^{n} p_i \ln p_i = -k_B \sum_{i=1}^{n} p_i \left( -\ln Q - \sum_{j=1}^{q} \alpha_j x_{ji} \right)$$

$$= k_B \ln Q + k_B \sum_{j=1}^{q} \alpha_j A_j. \quad (3.52)$$

In Equation (3.38) the thermodynamic entropy has been shown to be

$$S = \frac{1}{T}(pV + U - \mu N). \quad (3.53)$$

The connection between the microscopic and microscopic properties of a system comes in equating the two forms of the entropy given in Equations (3.52) and (3.53). This will enable the quantities $\alpha_j$ to be identified in terms of the macroscopic thermodynamic variables.

## 3.4 The distributions

Several distinct distributions arise when different sets of macroscopic quantities are kept constant. In each case, expressions for the macroscopic thermodynamic quantities can be found in terms of the partition function $Q$. As we shall see, each distribution has its own symbol for the partition function.

### 3.4.1 The Canonical distribution

The only fluctuating variable is the energy, and the partition function $Q$ is denoted by $Z$ in this case. The quantities $N$, $V$ and $T$ are fixed. Hence we take $q = 1$ with $x_{1i} \equiv E_i$ and $A_1 \equiv U$. Combining Equations (3.52) and (3.53) gives

$$S = k_B \ln Z + k_B \alpha_1 U = \frac{1}{T}(pV + U - \mu N),$$

and equating the coefficients gives

$$k_B T \ln Z = pV - \mu N = -(U - TS) \equiv -F \quad \text{and} \quad \alpha_1 = \frac{1}{k_B T}.$$

The quantity $F$ is called the *free energy*. Hence Equations (3.50) and (3.51) become

$$P(E) = g(E) Z^{-1} e^{-\frac{E}{k_B T}}, \quad \text{and} \tag{3.54}$$

$$Z = \sum_E g(E) e^{-\frac{E}{k_B T}}. \tag{3.55}$$

With angular brackets $\langle \ldots \rangle$ denoting average quantities, it follows that

$$U = \langle E \rangle = \sum_E E P(E) = Z^{-1} \sum_E g(E) E e^{-\frac{E}{k_B T}}$$

$$= \frac{\sum_E g(E) E e^{-\frac{E}{k_B T}}}{\sum_E g(E) e^{-\frac{E}{k_B T}}} = k_B T^2 \frac{d}{dT} \left[ \ln \sum_E g(E) E e^{-\frac{E}{k_B T}} \right]$$

$$= k_B T^2 \frac{d}{dT} \ln Z.$$

Hence

$$U = k_B T^2 \frac{d}{dT} \ln Z. \tag{3.56}$$

Further,

$$S = k_B \ln Z + \frac{U}{T}$$

$$= k_B \ln Z + k_B T \frac{d}{dT} \ln Z. \tag{3.57}$$

In this way, the thermodynamic quantities $U$ and $S$ are given in terms of the partition function $Z$.

## 3.4.2 The Grand Canonical distribution

We now have two fluctuating variables, which are the energy $E_i$ and the particle number $N_i$. The partition function $Q$ is denoted by $\Xi$ in this case. Hence we take $q = 2$ with $x_{1i} \equiv E_i$, $A_1 \equiv U$, $x_{2i} \equiv N_i$ and $A_2 \equiv N$. Then

$$S = k_B \ln \Xi + k_B \alpha_1 U + k_B \alpha_2 N = \frac{1}{T}(pV + U - \mu N),$$

and equating the coefficients gives

$$k_B T \ln \Xi = pV, \qquad \alpha_1 = \frac{1}{k_B T}, \qquad \alpha_2 = -\frac{\mu}{k_B T}.$$

Hence Equations (3.50) and (3.51) become

$$P(E, n) = g(E, n)\Xi^{-1} e^{\frac{1}{k_B T}(\mu n - E)}, \qquad \text{and} \tag{3.58}$$

$$\Xi = e^{\frac{pV}{k_B T}} = \sum_{E,n} g(E, n)e^{\frac{1}{k_B T}(\mu n - E)}. \tag{3.59}$$

Let $e_j$ be the known single-particle energy corresponding to the quantum state $j$, and let $n_j$ be the number of particles in this state. The total particle number and energy are

$$n = \sum_j n_j,$$

$$E = \sum_j e_j \quad \text{(for weak interaction)}.$$

With a change of notation, let $p(n_1, n_2, \ldots)$ be the probability of finding $n_1$ particles in state 2, $n_2$ particles in state 2, etc. The set of values $\{n_1, n_2, \ldots\}$ now specifies the microscopic state of the system, so that

$$P(E, n) = g(E, n)p(n_1, n_2, \ldots).$$

Then

$$p(n_1, n_2, \ldots) = \Xi^{-1} \exp\left\{\frac{1}{k_B T}(\mu n - E)\right\} = \Xi^{-1} \exp\left\{\frac{1}{k_B T}\sum_j (\mu - e_j)n_j\right\}$$

$$= \Xi^{-1} t_1^{n_1} t_2^{n_2} \ldots$$

where

$$t_j \equiv \exp\left\{\frac{1}{k_B T}(\mu - e_j)\right\}, \qquad (j = 1, \ldots).$$

The partition function is given by Equation (3.59):

$$\Xi = \sum_{n_1, n_2} t_1^{n_1} t_2^{n_2} \cdots = \left( \sum_{n_1} t_1^{n_1} \right) \left( \sum_{n_2} t_2^{n_2} \right) \cdots$$

Hence the probability of finding $n_j$ states in the state $j$ is

$$p_j(n_j) = \sum_{n_1, \ldots, n_{j-1}, n_{j+1}, \ldots} p(n_1, n_2, \ldots)$$

$$= \Xi^{-1} \left( \sum_{n_1} t_1^{n_1} \right) \left( \sum_{n_2} t_2^{n_2} \right) \cdots \left( \sum_{n_{j-1}} t_{j-1}^{n_{j-1}} \right) t_j^{n_j} \left( \sum_{n_{j+1}} t_{j+1}^{n_{j+1}} \right) \cdots$$

$$= \frac{t_j^{n_j}}{\sum_{n_j} t_j^{n_j}}. \tag{3.60}$$

Now

$$\frac{\partial \Xi}{\partial e_j} = \left( \sum_{n_1} t_1^{n_1} \right) \cdots \frac{\partial}{\partial e_j} \left( \sum_{n_j} t_j^{n_j} \right) \cdots$$

$$= \left( \sum_{n_1} t_1^{n_1} \right) \cdots \left( -\frac{1}{k_B T} \right) \left( \sum_{n_j} n_j t_j^{n_j} \right) \cdots .$$

It then follows that

$$\frac{\partial}{\partial e_j} \ln \Xi = \Xi^{-1} \frac{\partial \Xi}{\partial e_j} = -\frac{1}{k_B T} \Xi^{-1} \left( \sum_{n_1} t_1^{n_1} \right) \cdots \left( \sum_{n_j} n_j t_j^{n_j} \right) \cdots$$

$$= -\frac{1}{k_B T} \frac{\sum_{n_j} n_j t_j^{n_j}}{\sum_{n_k} t_k^{n_k}} = -\frac{1}{k_B T} \sum_{n_j} n_j p_j(n_j) = -\frac{1}{k_B T} \langle n_j \rangle$$

giving

$$\langle n_j \rangle = -k_B T \frac{\partial}{\partial e_j} \ln \Xi. \tag{3.61}$$

### 3.4.3 The Microcanonical distribution

In this case, there are no fluctuating variables, and formally we take all $\alpha_i \equiv 0$. The partition function $Q$ is denoted by $W$ in this case. If $\Omega$ is the number of microstates, so that $\sum_{i=1}^{\Omega}(1/\Omega) = 1$, then

$$p_i = W^{-1} = \text{constant} \tag{3.62}$$

$$S = -k_B \sum_{i=1}^{\Omega} \frac{1}{\Omega} \ln\left(\frac{1}{\Omega}\right) = k_B \ln \Omega. \tag{3.63}$$

## 3.5 Fermi-Dirac and Bose-Einstein statistics

So far, nothing has been said about the values which the particle numbers $n_i$ are allowed to take. The distribution functions for both Fermi-Dirac and Bose-Einstein statistics may now be found by allowing for different particle number totals in each case.

- For *fermions*, or those particles which obey Fermi-Dirac statistics, no single-particle quantum state can be occupied by more than one particle. Hence the values of $n_i$ are restricted to 0 and 1. It follows from Equation (3.60) that

$$p_j(n_j) = \frac{t_j^{n_j}}{\sum_{n_j} t_j^{n_j}} = \frac{t_j^{n_j}}{1 + t_j} = t_j^{n_j}(1 + t_j)^{-1}, \quad \text{and} \tag{3.64}$$

$$\Xi = \prod_j (1 + t_j). \tag{3.65}$$

- For *bosons*, or those particles obeying Bose-Einstein statistics, there is no restriction on the number of particles that can occupy a single particle quantum state. Hence the sum over $n_i$ extends over an infinite number of terms; it follows from Equation (3.60) that

$$p_j(n_j) = \frac{t_j^{n_j}}{\sum_{n_j=0}^{\infty} t_j^{n_j}} = t_j^{n_j}(1 - t_j), \tag{3.66}$$

$$\Xi = \prod_j (1 - t_j)^{-1}. \tag{3.67}$$

Note that in this case, we must have $t_j < 1$, or $e_j > \mu$, in order for the infinite sums to converge.

Both cases may conveniently be combined into the form

$$p_j(n_j) = t_j^{n_j}(1 + \alpha t_j)^{-\alpha}, \quad \text{and} \tag{3.68}$$

$$\Xi = \prod_j (1 + \alpha t_j)^{\alpha} \tag{3.69}$$

where $\alpha = +1$ for fermions and $\alpha = -1$ for bosons. It follows from Equation (3.61) that

$$\langle n_j \rangle = -k_B T \frac{\partial}{\partial e_j} \ln \Xi = -k_B T \alpha \frac{\partial}{\partial e_j} \sum_i \ln(1 + \alpha t_i)$$

$$= -k_B T \alpha \frac{\partial}{\partial e_j} \ln(1 + \alpha t_j) = -k_B T \frac{1}{1 + \alpha t_j} \frac{\partial t_j}{\partial e_j} \quad \text{(since } \alpha^2 = 1\text{)}$$

$$= -\frac{k_B T}{1 + \alpha t_j} \left( -\frac{t_j}{k_B T} \right) = \frac{t_j}{1 + \alpha t_j} = \frac{1}{t_j^{-1} + \alpha}$$

$$= \frac{1}{\exp\{\frac{1}{k_B T}(e_j - \mu)\} + \alpha}.$$

Hence the total mean number of particles is

$$N = \sum_j \langle n_j \rangle = \sum_j \frac{1}{\exp\{\frac{1}{k_B T}(e_j - \mu)\} + \alpha}, \qquad (\alpha = \pm 1). \qquad (3.70)$$

If the quantity $e_j$ is replaced by the continuous variable $E$, the mean occupation number at energy $E$ is

$$\langle n(E) \rangle = \frac{g(E)}{\exp\{\frac{1}{k_B T}(E - \mu)\} + \alpha}, \qquad (3.71)$$

where $g(E)$ is the degeneracy of energy $E$. In particular, if $E$ is large, then

$$\frac{\langle n(E) \rangle}{g(E)} = \frac{1}{\exp\{\frac{1}{k_B T}(E - \mu)\} + \alpha} \approx \exp\left\{ \frac{1}{k_B T}(\mu - E) \right\} = C e^{-\frac{E}{k_B T}} \qquad (3.72)$$

where $C$ is a constant. This is the classical Maxwell-Boltzmann distribution. This limit may also be obtained formally by taking the quantity $\mu/(k_B T)$ to be very large and negative in Equation (3.71).

## 3.6 The continuous approximation and the Ideal Quantum Gas

The total number $N$ of particles is expressed as a summation in Equation (3.70). However, the scientific community in general is much happier when evaluating integrals rather than summations, and it relies heavily on a comprehensive body of knowledge surrounding the integral and differential calculus. Accordingly, we will seek to write summations, which are involved in the calculation of $N$ and many other macroscopic quantities, in terms of integrals.

Assume that the number of single particle quantum states in the energy range $(E, E + dE)$ is $A_r E^r dE$, where $r$ is a constant and $A_r$ depends on the volume $V$ of the system but not on its temperature. For example, $r = 1/2$ for a particle moving in a zero potential, and $r = 2$ for electromagnetic waves. Then summations can be transformed into integrals through the replacement

$$\sum_i f(e_i) \rightarrow \int f(E) A_r E^r dE. \tag{3.73}$$

Hence for both fermions ($\alpha = +1$) and bosons ($\alpha = -1$), we have

$$\begin{aligned}
N &= \sum_{j=0}^{\infty} \langle n(e_j) \rangle = \sum_{j=0}^{\infty} \frac{1}{\exp\{\frac{1}{k_B T}(e_j - \mu)\} + \alpha} \\
&\rightarrow \int_0^{\infty} \frac{A_r E^r dE}{\exp\{\frac{1}{k_B T}(E - \mu)\} + \alpha} \\
&= A_r (k_B T)^{r+1} \int_0^{\infty} \frac{x^r dx}{\exp\{x - \frac{\mu}{k_B T}\} + \alpha} \\
&= A_r (k_B T)^{r+1} \Gamma(r+1) I\left(\frac{\mu}{k_B T}, r, \alpha\right)
\end{aligned} \tag{3.74}$$

where the integrals $\Gamma$ and $I$ are defined as

$$\Gamma(r+1) \equiv \int_0^{\infty} x^r e^{-x} dx, \tag{3.75}$$

$$I(\gamma, r, \alpha) \equiv \frac{1}{\Gamma(r+1)} \int_0^{\infty} \frac{x^r dx}{e^{x-\gamma} + \alpha}. \tag{3.76}$$

These integrals can be shown to have the following properties (see Problem 3.5):

$$\lim_{\gamma \to -\infty} I(\gamma, r, \alpha) = e^{\gamma} \quad \text{for all } r, \tag{3.77}$$

$$\frac{d}{d\gamma} I(\gamma, r, \alpha) = I(\gamma, r-1, \alpha) \quad \text{for all } r > 0, \tag{3.78}$$

$$\Gamma(r+1) = r\Gamma(r). \tag{3.79}$$

$$\left(\frac{\partial \gamma}{\partial T}\right)_{V,N} = -\frac{(r+1)}{T} \frac{I(\gamma, r, \alpha)}{I(\gamma, r-1, \alpha)}. \tag{3.80}$$

The result in Equation (3.77) is particularly useful when considering the classical limit. For, with $\gamma \equiv \mu/(k_B T)$, this classical limit is reached formally as $\mu \to -\infty$. Hence the $I$-integrals will all take the value $\exp(\mu/(k_B T))$ in the classical limit.

Applying the (summation) $\rightarrow$ (integration) prescription further, the free energy $F$ becomes

$$\begin{aligned}
F &\equiv U - TS = \mu N - pV = \mu N - k_B T \ln \Xi \\
&= \mu N - k_B T \ln \left( \prod_j \left[ 1 + \alpha e^{\frac{\mu - e_j}{k_B T}} \right]^{\alpha} \right) \\
&= \mu N - k_B T \alpha \sum_j \ln \left[ 1 + \alpha e^{\frac{\mu - e_j}{k_B T}} \right]
\end{aligned}$$

$$\rightarrow \mu N - k_B T \alpha \int_0^\infty A_r E^r \ln\left[1 + \alpha e^{\frac{\mu - E}{k_B T}}\right] dE$$

$$= \mu N - A_r (k_B T)^{r+2} \Gamma(r+1) I\left(\frac{\mu}{k_B T}, r+1, \alpha\right) \quad \text{(by parts)} \quad (3.81)$$

$$= A_r \Gamma(r+1)(k_B T)^{r+2}\left[\frac{\mu}{k_B T} I\left(\frac{\mu}{k_B T}, r, \alpha\right)\right.$$

$$\left. - I\left(\frac{\mu}{k_B T}, r+1, \alpha\right)\right]. \tag{3.82}$$

Now

$$dU = T dS - p dV - \mu dN,$$

and hence

$$\left(\frac{\partial F}{\partial T}\right)_{V,N} = \left(\frac{\partial U}{\partial T}\right)_{V,N} - S - T\left(\frac{\partial S}{\partial T}\right)_{V,N} = -S.$$

This relation can be used to calculate the entropy $S$, and it can be shown (see Problem 3.6) that

$$S = -\left(\frac{\partial F}{\partial T}\right)_{V,N} = k_B N\left((r+2)\frac{I(\frac{\mu}{k_B T}, r+1, \alpha)}{I(\frac{\mu}{k_B T}, r, \alpha)} - \frac{\mu}{k_B T}\right). \tag{3.83}$$

Further, the internal energy is given by

$$U = \sum_{j=0}^\infty e_j \langle n(e_j)\rangle \rightarrow \int_0^\infty E \frac{A_r E^r dE}{\exp\{\frac{1}{k_B T}(E - \mu)\} + \alpha}$$

$$= \int_0^\infty \frac{A_r E^{r+1} dE}{\exp\{\frac{1}{k_B T}(E - \mu)\} + \alpha} = A_r (k_B T)^{r+2} \int_0^\infty \frac{x^{r+1} dx}{\exp\{x - \frac{\mu}{k_B T}\} + \alpha}$$

$$= A_r (k_B T)^{r+2} \Gamma(r+2) I\left(\frac{\mu}{k_B T}, r+1, \alpha\right)$$

$$= A_r (r+1)(k_B T)^{r+2} \Gamma(r+1) I\left(\frac{\mu}{k_B T}, r+1, \alpha\right). \tag{3.84}$$

Hence the pressure is given by

$$p = \frac{\mu N - F}{V} = \frac{A_r (k_B T)^{r+2}}{V} \Gamma(r+1) I\left(\frac{\mu}{k_B T}, r+1, \alpha\right) = \frac{1}{r+1}\frac{U}{V}. \tag{3.85}$$

It follows that the equations of state for an *ideal quantum gas* are

$$pV = gU, \quad g \equiv \frac{1}{r+1} = \text{constant}. \tag{3.86}$$

It further follows from Equation (3.85) that

$$p = \frac{k_B T N}{V} \frac{I(\frac{\mu}{k_B T}, r+1, \alpha)}{I(\frac{\mu}{k_B T}, r, \alpha)} \tag{3.87}$$

which, in the classical limit in which $I(\frac{\mu}{k_B T}, r, \alpha) \to \exp(\frac{\mu}{k_B T})$, becomes

$$pV = k_B T N. \tag{3.88}$$

This is the justification for taking the constant $\lambda = k_B$ in Equation (3.45).

## Problems

**Problem 3.1.** Let $\beta \subset E$ be a subset of the set $E$ of equilibrium points such that any two points of $\beta$ can be joined by an adiabatic process. If $\beta_1$ and $\beta_2$ are two such subsets, prove that either $\beta_1 = \beta_2$, or $\beta_1$ and $\beta_2$ do not intersect.

**Problem 3.2.** Find the ratio of the latent heats for a van der Waals gas for which

$$\left(p + \frac{a}{V^2}\right)(V - b) = At$$

where $a$, $b$ and $A$ are constants.

**Problem 3.3.** Writing $d^*Q = m_V dV + m_p dp$, evaluate the coefficients $m_V$ and $m_p$ in terms of the specific and latent heats.

**Problem 3.4.** By using the result $TdS = dU + pdV$, obtain the Maxwell equation

$$\left(\frac{\partial T}{\partial V}\right)_S = -\left(\frac{\partial p}{\partial S}\right)_V$$

Further, by using the *free energy* $f \equiv U - TS$, the *enthalpy* $H \equiv U + pV$, and the *Gibbs function* $G \equiv U - TS + pV$, obtain the further Maxwell equations

$$\left(\frac{\partial S}{\partial V}\right)_T = \left(\frac{\partial p}{\partial T}\right)_V, \qquad \left(\frac{\partial T}{\partial p}\right)_S = \left(\frac{\partial V}{\partial S}\right)_p, \qquad \left(\frac{\partial V}{\partial T}\right)_p = -\left(\frac{\partial S}{\partial p}\right)_T.$$

**Problem 3.5.** Given that

$$I(c, r, \alpha) \equiv \frac{1}{\Gamma(r+1)} \int_0^\infty \frac{x^r}{e^{x-c} + \alpha} \quad \text{and} \quad \frac{d}{dc} I(c, r, \alpha) = I(c, r-1, \alpha),$$

show that

$$\left(\frac{\partial \gamma}{\partial T}\right)_{V,N} = -\frac{(r+1)}{T} \frac{I(\gamma, r, \alpha)}{I(\gamma, r-1, \alpha)}$$

where $\gamma = \frac{\mu}{k_B T}$.

*Solution.* From Equation (3.74) we have $N = A_r(k_B T)^{r+1}\Gamma(r+1)I(\gamma, r, \alpha)$. Differentiating partially with respect to $T$ keeping $V$ and $N$ constant, we have

$$
\begin{aligned}
0 &= A_r k_B^{r+1}(r+1)T^r\Gamma(r+1)I(\gamma, r, \alpha) \\
&\quad + A_r(k_B T)^{r+1}\Gamma(r+1)\frac{\partial I(\gamma, r, \alpha)}{\partial \gamma}\left(\frac{\partial \gamma}{\partial T}\right)_{V,N} \\
&= A_r(k_B T)^{r+1}\Gamma(r+1)\left[\frac{r+1}{T}I(\gamma, r, \alpha) + I(\gamma, r-1, \alpha)\left(\frac{\partial \gamma}{\partial T}\right)_{V,N}\right],
\end{aligned}
$$

and the result follows. $\square$

**Problem 3.6.** Verify that, for the ideal quantum gas, the entropy is given by

$$
S = k_B N\left((r+2)\frac{I(\frac{\mu}{k_B T}, r+1, \alpha)}{I(\frac{\mu}{k_B T}, r, \alpha)} - \frac{\mu}{k_B T}\right).
$$

*Solution.* From Equation (3.82) we have

$$
F = A_r\Gamma(r+1)(k_B T)^{r+2}[\gamma I(\gamma, r, \alpha) - I(\gamma, r+1, \alpha)].
$$

Hence

$$
\begin{aligned}
S &= -\left(\frac{\partial F}{\partial T}\right)_{V,N} \\
&= -A_r\Gamma(r+1)k_B^{r+2}(r+2)T^{r+1}[\gamma I(\gamma, r, \alpha) - I(\gamma, r+1, \alpha)] \\
&\quad -A_r\Gamma(r+1)(k_B T)^{r+2}[I(\gamma, r, \alpha) + \gamma I(\gamma, r-1, \alpha) - I(\gamma, r, \alpha)] \\
&\quad \times\left(-\frac{(r+1)}{T}\frac{I(\gamma, r, \alpha)}{I(\gamma, r-1, \alpha)}\right) \\
&= -A_r\Gamma(r+1)k_B^{r+2}T^{r+1}\left\{(r+2)[\gamma I(\gamma, r, \alpha) - I(\gamma, r+1, \alpha)]\right. \\
&\quad \left. - \gamma(r+1)I(\gamma, r, \alpha)\right\} \\
&= -A_r\Gamma(r+1)k_B(k_B T)^{r+1}I(\gamma, r, \alpha)\left\{\gamma - (r+2)\frac{I(\gamma, r+1, \alpha)}{I(\gamma, r, \alpha)}\right\}
\end{aligned}
$$

and the answer follows using Equation (3.74). $\square$

# Chapter 4
# Density of states and applications—1

The Fermi-Dirac and Bose-Einstein distributions which were derived in Chapter 3 will now be used to describe a number of physical situations, including blackbody radiation, classical and quantum aspect of specific heat, Bose-Einstein condensation, thermionic emission, and a brief introduction to semiconductor statistics.

## 4.1 Electron number and energy densities

The number of single-particle quantum states per unit volume in the energy range $(E, E + dE)$ is $A_r E^r dE$ where $r$ is a constant, and $A_r$ depends on volume $V$ and not on particle temperature $T$. From now on, we will deal only with fermions. The Fermi-Dirac occupancy probability $f(E)$ is given by $\langle n(E) \rangle = g(E) f(E)$. Then, from Equation (3.71) with $\alpha = +1$, it follows that

$$f(E) = \frac{1}{1 + e^{\frac{1}{k_B T}(E - E_F)}} \tag{4.1}$$

where the chemical potential $\mu$ is replaced by the *Fermi level* $E_F$. In the case of electrons, energies are measured from the bottom of the conduction band $E_c$, which will be described in Sect. 4.7. Hence if $\phi(E)$ is any function of the energy, its average value will be

$$\langle \phi \rangle = A_r \int_{E_c}^{\infty} \frac{\phi(E - E_c)(E - E_c)^r dE}{1 + e^{\frac{1}{k_B T}(E - E_F)}} \tag{4.2}$$

$$= A_r (k_B T)^{r+1} \int_0^{\infty} \frac{\phi(k_B T y) y^r dy}{1 + e^{y - \frac{1}{k_B T}(E_F - E_c)}}, \tag{4.3}$$

where the substitution $y \equiv (E - E_c)/(k_B T)$ has been used. The integrals $I(\gamma, r, +1)$ which were introduced in Sect. 3.6 are denoted by $F_r(\gamma)$; from Equations (3.75) and (3.76), they are

E.A.B. Cole, *Mathematical and Numerical Modelling of Heterostructure Semiconductor Devices: From Theory to Programming*, DOI 10.1007/978-1-84882-937-4_4, © Springer-Verlag London Limited 2009

$$\Gamma(r+1) \equiv \int_0^\infty x^r e^{-x} dx, \tag{4.4}$$

$$F_r(\gamma) \equiv \frac{1}{\Gamma(r+1)} \int_0^\infty \frac{x^r dx}{e^{x-\gamma}+1}, \tag{4.5}$$

and, from Equations (3.77)–(3.79), they have the properties

$$\lim_{\gamma \to -\infty} F_r(\gamma) = e^\gamma \quad \text{for all } r, \tag{4.6}$$

$$\frac{d}{d\gamma} F_r(\gamma) = F_{r-1}(\gamma) \quad \text{for all } r > 0, \tag{4.7}$$

$$\Gamma(r+1) = r\Gamma(r). \tag{4.8}$$

In particular, in the specific case of $\phi(E) \equiv 1$, the electron density is given by

$$n = A_r(k_B T)^{r+1} \int_0^\infty \frac{y^r dy}{1 + e^{y - \frac{1}{k_B T}(E_F - E_c)}}$$

$$= A_r(k_B T)^{r+1} \Gamma(r+1) F_r \left( \frac{E_F - E_c)}{k_B T} \right)$$

$$= N_{cr} F_r \left( \frac{E_F - E_c)}{k_B T} \right) \quad \text{where} \tag{4.9}$$

$$N_{cr} \equiv A_r(k_B T)^{r+1} \Gamma(r+1). \tag{4.10}$$

Similarly, taking the special case $\phi(E) \equiv E$, the total energy, denoted by $W$ in the case of electrons, is

$$W = A_r(k_B T)^{r+2} \int_0^\infty \frac{y^{r+1} dy}{1 + e^{y - \frac{1}{k_B T}(E_F - E_c)}}$$

$$= A_r(k_B T)^{r+2} \Gamma(r+2) F_{r+1} \left( \frac{E_F - E_c)}{k_B T} \right)$$

$$= W_{cr} F_{r+1} \left( \frac{E_F - E_c)}{k_B T} \right), \quad \text{where} \tag{4.11}$$

$$W_{cr} \equiv A_r(k_B T)^{r+2} \Gamma(r+2). \tag{4.12}$$

### 4.1.1 Density of states—general

Using the de Broglie relation $p\lambda = h$, the energy of a single free particle with mass $m$ is

$$E = \frac{p^2}{2m} = \frac{h^2}{2m} \frac{1}{\lambda^2}.$$

Hence

$$dE = -\frac{h^2}{2m}\frac{2}{\lambda^3}d\lambda = -\frac{h^2}{m}\frac{1}{\lambda^3}d\lambda,$$

or

$$d\lambda = -\frac{m\lambda^3}{h^2}dE. \tag{4.13}$$

Therefore the number of single-particle quantum states per unit volume in the energy range $(E, E + dE)$, corresponding to wavelength $\lambda$, is

$$A_r E^r dE = dN(\lambda) = \frac{dN}{d\lambda}d\lambda = \frac{dN}{d\lambda}\left(-\frac{m\lambda^3}{h^2}dE\right),$$

from which it follows that

$$A_r E^r = -\frac{m\lambda^3}{h^2}\frac{dN}{d\lambda}. \tag{4.14}$$

The quantity $N(\lambda_0)$ is the number of modes of vibration which corresponds to a wavelength between $\lambda = \infty$ and $\lambda = \lambda_0$. In order to be able to use this result, expressions are needed for $N(\lambda)$ and $dN/d\lambda$ in terms of $E$. These expressions will depend on the number of spatial dimensions in which the electrons are free.

### 4.1.2 Density of states—particles free in three dimensions

For particles in zero potential, the TISE and its solution are

$$-\frac{\hbar^2}{2m}\nabla^2 u + Uu = Eu, \qquad u(x, y, z) = e^{ik_1 x}e^{ik_2 y}e^{ik_3 z}$$

where the quantities $k_1$, $k_2$ and $k_3$ are real. The values of $k_1$, $k_2$ and $k_3$ are found using the process of *box normalisation*, in which it is assumed that the solution is periodic in a sufficiently large box with sides $D_1$, $D_2$ and $D_3$, and with volume $V = D_1 D_2 D_3$. It follows that $k_i D_i = n_i \pi$, for $i = 1, 2$, and 3, where the $n_i$ are integers. Then

$$\frac{h^2}{2m}\frac{1}{\lambda^2} = E = \frac{\hbar^2}{2m}(k_1^2 + k_2^2 + k_3^2) = \frac{\hbar^2 \pi^2}{2m}\left(\frac{n_1^2}{D_1^2} + \frac{n_2^2}{D_2^2} + \frac{n_3^2}{D_3^2}\right),$$

giving

$$\frac{1}{\lambda^2} = \frac{n_1^2}{4D_1^2} + \frac{n_2^2}{4D_2^2} + \frac{n_3^2}{4D_3^2}.$$

Let $N(\lambda_0)$ be the number of modes of vibration $(n_1, n_2, n_3)$ which give a wavelength between $\lambda = \infty$ and $\lambda = \lambda_0$. Then this number of modes must satisfy

$$0 \le \frac{n_1^2}{4D_1^2} + \frac{n_2^2}{4D_2^2} + \frac{n_3^2}{4D_3^2} \le \frac{1}{\lambda_0^2}$$

or

$$\left(\frac{n_1}{a_1}\right)^2 + \left(\frac{n_2}{a_2}\right)^2 + \left(\frac{n_3}{a_3}\right)^2 \leq 1 \quad \text{where } a_i \equiv \frac{2D_i}{\lambda_0}, \quad (i = 1, 2, 3).$$

It is only necessary to specify the positive values of the $n_i$, since the corresponding negative values give the same solution. The required points will be contained in the positive octant of the above ellipsoid; since they occur at integer points, the number of modes will be equal to the volume of the octant:

$$N(\lambda_0) = \frac{1}{8} \text{ (volume of octant)} = \frac{1}{8}\frac{4\pi}{3}a_1a_2a_3 = \frac{1}{8}\frac{4\pi}{3}\frac{8D_1D_2D_3}{\lambda_0^3}$$

$$= \frac{4\pi}{3}\frac{V}{\lambda_0^3}.$$

Hence

$$\frac{dN}{d\lambda} = -\frac{4\pi V}{\lambda^4}. \tag{4.15}$$

Denoting the degeneracy by $g$, which has the value $g = 2$ for two degrees of polarisation, the number of states per unit volume is therefore

$$A_r E^r = -\frac{m\lambda^3}{h^2}\left(-\frac{4\pi}{\lambda^4}\right)g = \frac{4\pi m}{h^2\lambda}g = \frac{4\pi m}{h^2}\frac{\sqrt{2mE}}{h}g$$

$$= \frac{8\pi m\sqrt{2mE}}{h^3}. \tag{4.16}$$

Therefore, in this case, it follows that

$$r = \frac{1}{2}, \qquad A_{\frac{1}{2}} = \frac{8\pi m\sqrt{2m}}{h^3}, \tag{4.17}$$

with

$$n = 2\left(\frac{2\pi mk_BT}{h^2}\right)^{\frac{3}{2}}F_{\frac{1}{2}}\left(\frac{1}{k_BT}(E_F - E_c)\right), \tag{4.18}$$

$$W = 3k_BT\left(\frac{2\pi mk_BT}{h^2}\right)^{\frac{3}{2}}F_{\frac{3}{2}}\left(\frac{1}{k_BT}(E_F - E_c)\right). \tag{4.19}$$

Before leaving this section, it will be useful to derive the density of states in terms of the momentum $\mathbf{p}$. Let $dz(p)$ be the number of single-particle states in the range $(p, p + dp)$ of the momentum magnitude. Then since $E = p^2/(2m)$, we have

$$dz(p) = \frac{8\pi V}{h^3}\sqrt{2mE}d(mE) = \frac{8\pi V}{h^3}p\frac{1}{2}d(p^2)$$

$$= \frac{8\pi V}{h^3}p^2dp. \tag{4.20}$$

Hence the number of states in the vector momentum range $(\mathbf{p}, \mathbf{p} + d\mathbf{p})$ is

$$dz(\mathbf{p}) = \frac{\text{volume of element } dp_x dp_y dp_z}{\text{volume of shell } (p, p + dp)} \times dz(p)$$

$$= \frac{dp_x dp_y dp_z}{4\pi p^2 dp} \times \frac{8\pi V p^2 dp}{h^3}$$

$$= \frac{2V}{h^3} dp_x dp_y dp_z. \tag{4.21}$$

### 4.1.3 Density of states—particles free in two dimensions

For particles which are free to move in the $x$- and $z$-directions, and are constrained by the potential $U \equiv U(y)$ in the $y$-direction, the TISE and its solution are

$$-\frac{\hbar^2}{2m}\nabla^2 u + Uu = Eu, \qquad u(x, y, z) = e^{ik_1 x} e^{ik_3 z} u'(y)$$

for some function $u'(y)$. Therefore

$$-\frac{\hbar^2}{2m}(-k_1{}^2 - k_3{}^2 + d_{yy})u' + U(y)u' = Eu'(y)$$

where the notation $d_{yy} \equiv d^2/dy^2$ has been used. This result can be written

$$-\frac{\hbar^2}{2m}d_{yy}u' + U(y)u' = \left(E - \frac{\hbar^2}{2m}(k_1{}^2 + k_3{}^2)\right)u'(y) \equiv (E - E')u'(y)$$

where $E' \equiv -\hbar^2(k_1{}^2 + k_3{}^2)/(2m)$. Again, using the process of box normalisation, the solution is periodic in a box with sides $D_1$ and $D_3$, with area $V = D_1 D_3$. Therefore $k_i D_i = n_i \pi$ for $(i = 1, 3)$ and the $n_i$ are integers. It follows that

$$\frac{\hbar^2}{2m}\frac{1}{\lambda'^2} = E' = \frac{\hbar^2}{2m}(k_1{}^2 + k_3{}^2) = \frac{\hbar^2 \pi^2}{2m}\left(\frac{n_1^2}{D_1^2} + \frac{n_3^2}{D_3^2}\right),$$

giving

$$\frac{1}{\lambda'^2} = \frac{n_1^2}{4D_1^2} + \frac{n_3^2}{4D_3^2}.$$

Let $N(\lambda'_0)$ be the number of modes of vibration $(n_1, n_3)$ which give a wavelength between $\lambda' = \infty$ and $\lambda' = \lambda'_0$. Then this number of modes must satisfy

$$0 \le \frac{n_1^2}{4D_1^2} + \frac{n_3^2}{4D_3^2} \le \frac{1}{\lambda'^2_0},$$

or

$$\left(\frac{n_1}{a_1}\right)^2 + \left(\frac{n_3}{a_3}\right)^2 \le 1 \quad \text{where } a_i \equiv \frac{2D_i}{\lambda_0'}, \ (i = 1, 3).$$

Again, only the positive values of the $n_i$ are needed, since the corresponding negative values correspond to the same solution. The required points will be contained in the positive quadrant of the above ellipse; since they occur at integer points, the number of modes will be equal to the volume of the quadrant:

$$N(\lambda_0) = \frac{1}{4} \text{ (area of ellipse)} = \frac{1}{4}\pi a_1 a_3 = \frac{1}{4}\pi \frac{4 D_1 D_3}{\lambda_0^2}$$

$$= \frac{\pi V}{\lambda_0^2}.$$

Hence

$$\frac{dN}{d\lambda} = -\frac{2\pi V}{\lambda^3}.$$

Again taking the degeneracy by $g = 2$, the number of states per unit volume will be given by

$$A_r E^r = -\frac{m\lambda^3}{h^2}\left(-\frac{2\pi}{\lambda^3}\right)g = \frac{2\pi m}{h^2}g$$

$$= \frac{4\pi m}{h^2}. \tag{4.22}$$

Hence in this case, it follows that

$$r = 0, \qquad A_0 = \frac{4\pi m}{h^2}, \tag{4.23}$$

with

$$n = 2\left(\frac{2\pi m k_B T}{h^2}\right) F_0\left(\frac{1}{k_B T}(E_F - E_c)\right)$$

$$= 2\left(\frac{2\pi m k_B T}{h^2}\right) \ln\left(1 + e^{\frac{1}{k_B T}(E_F - E_c)}\right), \tag{4.24}$$

$$W = 2k_B T\left(\frac{2\pi m k_B T}{h^2}\right) F_1\left(\frac{1}{k_B T}(E_F - E_c)\right). \tag{4.25}$$

### 4.1.4 Density of states—particles free in one dimension

For particles which are free to move in the $z$-direction and are constrained by potential $U \equiv U(x, y)$ in the $x$- and $y$-directions, the TISE and its solution are

$$-\frac{\hbar^2}{2m}\nabla^2 u + Uu = Eu, \qquad u(x, y, z) = e^{ik_3 z} u'(x, y)$$

for some function $u'(x, y)$. Therefore

$$-\frac{\hbar^2}{2m}(-k_3{}^2 + \partial_{xx} + \partial_{yy})u' + U(x, y)u' = Eu'(x, y),$$

or

$$-\frac{\hbar^2}{2m}(\partial_{xx} + \partial_{yy})u' + U(x, y)u' = \left(E - \frac{\hbar^2}{2m}k_3{}^2\right)u'(x, y) \equiv (E - E')u'(x, y).$$

Again, using the process of box normalisation, the solution is periodic on a line with length $D_3$, with length $V = D_3$. Therefore $k_3 D_3 = n_3 \pi$ where $n_3$ is an integer. It follows that

$$\frac{h^2}{2m}\frac{1}{\lambda'^2} = E' = \frac{\hbar^2}{2m}k_3{}^2 = \frac{\hbar^2 \pi^2}{2m}\frac{n_3^2}{D_3^2},$$

giving

$$\frac{1}{\lambda'^2} = \frac{n_3^2}{4D_3^2}.$$

Let $N(\lambda_0')$ be the number of modes of vibration $(n_3)$ which give a wavelength between $\lambda' = \infty$ and $\lambda' = \lambda_0'$. Then this number of modes must satisfy

$$0 \le \frac{n_3^2}{4D_3^2} \le \frac{1}{\lambda_0'^2}$$

or

$$\left(\frac{n_3}{a_3}\right)^2 \le 1 \quad \text{where } a_3 \equiv \frac{2D_3}{\lambda_0'}.$$

Again, only the positive values of $n_3$ need be considered since the corresponding negative values give the same solution. The required points will be contained in the positive section of the above line; since they occur at integer points, the number of modes will be equal to the length of this positive section:

$$N(\lambda_0) = (\text{length of section}) = a_3 = \frac{2D_3}{\lambda_0} = \frac{2V}{\lambda_0},$$

giving

$$\frac{dN}{d\lambda} = -\frac{2V}{\lambda^2}.$$

Taking the degeneracy $g = 2$ as before, the number of states per unit volume will then be

$$A_r E^r = -\frac{m\lambda^3}{h^2}\left(-\frac{2}{\lambda^2}\right)g = \frac{2m\lambda}{h^2}g = \frac{4m\lambda}{h^2} = \frac{4m}{h^2}\frac{h}{\sqrt{2mE}}$$

$$= \frac{2\sqrt{2m}}{h} E^{-\frac{1}{2}}. \tag{4.26}$$

Hence in this case, it follows that

$$r = -\frac{1}{2}, \qquad A_{-\frac{1}{2}} = \frac{2\sqrt{2m}}{h}, \tag{4.27}$$

with

$$n = 2 \left( \frac{2\pi m k_B T}{h^2} \right)^{\frac{1}{2}} F_{-\frac{1}{2}} \left( \frac{1}{k_B T} (E_F - E_c) \right) \tag{4.28}$$

$$W = k_B T \left( \frac{2\pi m k_B T}{h^2} \right)^{\frac{1}{2}} F_{\frac{1}{2}} \left( \frac{1}{k_B T} (E_F - E_c) \right). \tag{4.29}$$

## 4.2 Blackbody radiation

A *blackbody* is one that absorbs all incident radiation. We wish to calculate the spectral distribution of the radiation in a cavity of volume $V$, with perfectly reflecting walls, which contains a blackbody at temperature $T$. The photon gas will be treated as bosons ($\alpha = -1$). The fact that they have zero mass means that we do not need to keep the particle number constant in deriving the statistical entropy $S$; this is formally equivalent to taking $\mu = 0$.

In order to take the appropriate density of states relating to radiation of frequency $\nu$ (or wavelength $\lambda$), note that the energy is given by $E = h\nu = hc/\lambda$. Hence $dE = -hc/\lambda^2 \, d\lambda$, and Equation (4.15) then gives

$$A_r E^r dE = dN(\lambda) = -\frac{4\pi V}{\lambda^4} \left( -\lambda^2 \frac{dE}{hc} \right) g = \frac{8\pi V}{h^3 c^3} E^2 dE \tag{4.30}$$

where the degeneracy has been taken as $g = 2$ for two degrees of polarisation. Hence

$$r = 2, \quad \text{with } A_2 = \frac{8\pi V}{h^3 c^3}. \tag{4.31}$$

On defining a constant $D$ by

$$D \equiv \frac{k_B^4 A_2}{V} \Gamma(3) I(0, 3, -1). \tag{4.32}$$

Equations (3.74) and (3.83)–(3.85) become

$$p = DT^4, \tag{4.33}$$

$$N = \frac{DV}{k_B} \frac{I(0, 2, -1)}{I(0, 3, -1)} T^3, \tag{4.34}$$

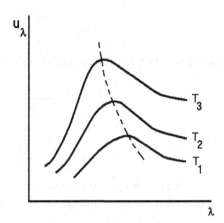

**Fig. 4.1** Plot of the blackbody energy density for various values of $T$, where $T_1 < T_2 < T_3$. The peaks of maximum intensity (dashed line) shift to the left for higher values of $T$ (Wien's law).

$$S = 4DVT^3, \tag{4.35}$$

$$U = 3DVT^4. \tag{4.36}$$

Let $u_\nu d\nu$ be the energy per unit volume lying in the frequency range $(\nu, \nu + d\nu)$. Then the total energy is

$$V \int_0^\infty u_\nu d\nu = U = \int_0^\infty E \cdot A_2 E^2 \langle n(E) \rangle dE$$

$$= \int_0^\infty A_2 E^3 \frac{1}{e^{\frac{E}{k_B T}} - 1} dE = \int_0^\infty A_2 (h\nu)^3 \frac{d(h\nu)}{e^{\frac{h\nu}{k_B T}} - 1}$$

$$= \int_0^\infty \frac{A_2 h^4 \nu^3}{e^{\frac{h\nu}{k_B T}} - 1} d\nu.$$

Hence

$$u_\nu = \frac{8\pi h}{c^3} \frac{\nu^3}{e^{\frac{h\nu}{k_B T}} - 1}. \tag{4.37}$$

This energy density may be written in terms of the wavelength $\lambda = c/\nu$ as

$$u_\nu d\nu = \frac{8\pi h}{c^3} \frac{\nu^3}{e^{\frac{h\nu}{k_B T}} - 1} d\nu = \frac{8\pi hc}{\lambda^5 (1 - e^{\frac{h\nu}{k_B T}})} d\lambda$$

$$\equiv u_\lambda d\lambda. \tag{4.38}$$

Fig. 4.1 shows a plot of $u_\lambda$ for various values of the temperature $T$. It shows how the maximum intensity for larger temperatures shifts towards smaller wavelengths—this is *Wien's Law*. See Problem 4.3 for a proof of this result.

By taking the classical limit of this result, we see why Planck found it necessary to introduce the formula $E = h\nu$. First, in the limit $h\nu \gg k_B T$, then

$$u_\nu \approx \frac{8\pi h}{c^3}\nu^3 e^{-\frac{h\nu}{k_B T}} = a\nu^3 e^{-\frac{b\nu}{T}},$$

agreeing with Wien's semi-empirical formula of 1896. Second, for large $T$,

$$\frac{1}{e^{\frac{h\nu}{k_B T}} - 1} \approx \frac{k_B T}{h\nu},$$

and then

$$u_\nu \approx \frac{8\pi h}{c^3}\nu^3\frac{k_B T}{h\nu} = \frac{8\pi k_B T}{c^3}\nu^2 \equiv u_{\nu c}.$$

This is the formula *for all* $\nu$ derived using classical physics only (note that the Planck constant $h$ is not involved in the expression for $u_{\nu c}$). Using this value of $u_{\nu c}$, then

$$U = V\int_0^\infty u_{\nu c}d\nu = \frac{8\pi k_B T V}{c^3}\int_0^\infty \nu^2 d\nu,$$

and this integral diverges with large $\nu$ (or small $\lambda$). The problem was known as the *ultra-violet catastrophe*, and Planck's solution to the problem was one of the major factors which led to the birth of quantum mechanics.

## 4.3 Classical aspects of specific heat

The classical discussion of statistical mechanics based on the Maxwell-Boltzmann distribution is conveniently accessed through the *Equipartition of Energy* theorem. The result of this theorem is extremely useful in determining expressions for the macroscopic variables such as the internal energy $U$ and specific heat $C_V$ when the number of degrees of freedom of the molecules of a gas is specified.

### 4.3.1 The Equipartition of Energy theorem

**Theorem 4.1.** *A generalised position or momentum coordinate which appears in the expression for $E$ only as a square term contributes an energy $k_B T/2$ to the mean energy of the system.*

*Proof.* Let the $L$ generalised coordinates and momenta describing the classical system be $q_1, \ldots, q_L$ and $p_1, \ldots, p_L$, and suppose that the energy can be written in the form

$$E = ap_1^2 + b \tag{4.39}$$

where $a$ and $b$ are independent of $p_1$ but may depend on the remaining coordinates and momenta. The contribution to $U = \langle E \rangle$ only from the first term is, where all integrals are evaluated between 0 and $\infty$,

$$\langle a p_1{}^2 \rangle = \frac{\int dp_1 \dots \int dp_L \int dq_1 \dots \int dq_L a p_1{}^2 e^{-\frac{E}{k_B T}}}{\int dp_1 \dots \int dp_L \int dq_1 \dots \int dq_L e^{-\frac{E}{k_B T}}}$$

$$= \frac{\int dp_2 \dots \int dp_L \int dq_1 \dots \int dq_L e^{-\beta b} \int a p_1{}^2 e^{-\alpha \beta p_1{}^2} dp_1}{\int dp_2 \dots \int dp_L \int dq_1 \dots \int dq_L e^{-\beta b} \int e^{-\alpha \beta p_1'{}^2} dp_1'}$$

$$= \frac{\int dp_2 \dots \int dp_L \int dq_1 \dots \int dq_L e^{-\beta b}[-\frac{d}{d\beta} \ln \int e^{-\alpha \beta p_1{}^2} dp_1]}{\int dp_2 \dots \int dp_L \int dq_1 \dots \int dq_L e^{-\beta b}}$$

$$= \frac{\int dp_2 \dots \int dp_L \int dq_1 \dots \int dq_L e^{-\beta b}[-\frac{d}{d\beta} \ln \sqrt{\frac{\pi}{\alpha \beta}}]}{\int dp_2 \dots \int dp_L \int dq_1 \dots \int dq_L e^{-\beta b}}$$

$$= \frac{\int dp_2 \dots \int dp_L \int dq_1 \dots \int dq_L e^{-\beta b}[\frac{1}{2} k_B T]}{\int dp_2 \dots \int dp_L \int dq_1 \dots \int dq_L e^{-\beta b}}$$

$$= \frac{1}{2} k_B T.$$

This completes the proof. □

## 4.3.2 Examples on the Equipartition of Energy theorem

### Example 1

A monatomic gas with $N$ single structureless non-interacting particles, each with mass $m$, has energy

$$E = \sum_{i=1}^{N} \frac{1}{2m}(p_{ix}{}^2 + p_{iy}{}^2 + p_{iz}{}^2).$$

Then $E$ contains a total of $3N$ square terms, and so the total mean energy is

$$U = 3N \times \frac{1}{2} k_B T = \frac{3}{2} N k_B T, \tag{4.40}$$

$$C_V = \frac{\partial U}{\partial T} = \frac{3}{2} N k_B. \tag{4.41}$$

### Example 2

A gas of freely moving rigid diatomic molecules consists of $N$ non-interacting molecules. Each molecule is composed of two particles with masses $m_1$ and $m_2$ at a fixed separation $r_0$. The *translational* component of the kinetic energy of each molecule is

$$(KE)_t = \frac{p^2}{2M} = \frac{1}{2M}(p_x{}^2 + p_y{}^2 + p_z{}^2)$$

where $M = m_1 + m_2$ is the total mass. To find the *rotational* component $(KE)_r$ of the kinetic energy, note that the square of the angular velocity about an axis perpendicular to the molecule length through the centre of mass using spherical polar coordinates $(r, \theta, \phi)$ is $\dot{\theta}^2 + \sin^2\theta\dot{\phi}^2$, and the moment of inertia about this axis is

$$m_1\left(\frac{m_2 r_0}{m_1 + m_2}\right)^2 + m_2\left(\frac{m_1 r_0}{m_1 + m_2}\right)^2 = \left(\frac{m_1 m_2}{m_1 + m_2}\right) r_0^2 \equiv \mu_0 r_0^2$$

where $\mu_0$ is the reduced mass. Hence the total kinetic energy is $K = (KE)_t + (KE)_r$, and the Lagrangian for the system is

$$L = \frac{1}{2M}p^2 + \frac{1}{2}\mu_0 r_0^2(\dot{\theta}^2 + \sin^2\theta\dot{\phi}^2).$$

The generalised momenta are

$$p_\theta \equiv \frac{\partial L}{\partial\dot{\theta}} = \mu_0 r_0^2\dot{\theta}, \qquad p_\phi \equiv \frac{\partial L}{\partial\dot{\phi}} = \mu_0 r_0^2\sin^2\theta\dot{\phi}.$$

Hence the energy of each molecule is

$$E_{mol} = \dot{x}p_x + \dot{y}p_y + \dot{z}p_z + \dot{\theta}p_\theta + \dot{\phi}p_{phi}$$
$$= \frac{1}{2M}(p_x^2 + p_y^2 + p_z^2) + \frac{1}{2\mu_0 r_0^2}\left(p_\theta^2 + \frac{p_\phi^2}{\sin^2\theta}\right),$$

and the total energy for $N$ molecules is then

$$E_{gas} = \sum_{i=1}^{N}\frac{1}{2M}(p_{ix}^2 + p_{iy}^2 + p_{iz}^2) + \frac{1}{2\mu_0 r_0^2}\sum_{i=1}^{N}\left(p_{i\theta}^2 + \frac{p_{i\phi}^2}{\sin^2\theta_i}\right).$$

This expression contains a total of $5N$ square terms, and hence

$$U = 5N \times \frac{1}{2}k_B T = \frac{5}{2}Nk_B T, \tag{4.42}$$

$$C_V = \frac{\partial U}{\partial T} = \frac{5}{2}Nk_B. \tag{4.43}$$

*Example 3*

A gas of freely moving vibrating diatomic molecules. The two particles of each molecule now oscillate about their mean distance $r_0$. If $r$ is the instantaneous separation, then the components of the energy are

$$(KE)_t = \frac{1}{2M}(p_x^2 + p_y^2 + p_z^2),$$

$$(KE)_r = \frac{1}{2\mu_0 r^2}\left(p_\theta{}^2 + \frac{p_\phi{}^2}{\sin^2\theta}\right),$$

$$(KE)_{vib} = \frac{1}{2}m_2\left(\frac{m_1\dot{r}}{m_1+m_2}\right)^2 + \frac{1}{2}m_1\left(\frac{m_2\dot{r}}{m_1+m_2}\right)^2 = \frac{1}{2}\mu_0\dot{r}^2$$

$$(E)_{pot} = u(r) = u(r_0) + \frac{1}{2}(r - r_0)^2\left(\frac{d^2u}{dr^2}\right)_{r=r_0} + \cdots$$

$$\approx u(r_0) + \frac{1}{2}\gamma(r - r_0)^2$$

to the second order, where $\gamma \equiv (d^2u/dr^2)_{r=r_0}$. Hence for each molecule there are seven square terms in this energy, or $7N$ such terms for the whole gas. Hence

$$U = 7N \times \frac{1}{2}k_BT = \frac{7}{2}Nk_BT, \qquad (4.44)$$

$$C_V = \frac{\partial U}{\partial T} = \frac{7}{2}Nk_B. \qquad (4.45)$$

On this classical basis only, this should be the model for $O_2$ and $N_2$, but experimentally it is found that $C_V = 5Nk_B/2$ at room temperature, indicating that only the translational and rotational degrees of freedom are important. A proper treatment of this phenomena must involve a quantum approach.

## 4.4 Quantum aspects of specific heat

The Equipartition of Energy theorem applies only to classical systems. A more complete approach involving quantum statistics must be used to obtain the corresponding results for quantum particles. In this Section, we will consider the quantum contributions which both vibrational and rotational properties of atoms make to the calculation of specific heat.

### 4.4.1 Quantum vibrational aspects: the Einstein solid

A number $N$ of atoms perform simple harmonic motion about fixed positions, each moving independently in three dimensions with a common circular frequency $\omega_c$. This will be equivalent to having a total of $3N$ linear distinguishable oscillators. For one oscillator, the allowed (non-degenerate) energies are

$$e_j = \left(j + \frac{1}{2}\right)\hbar\omega_c, \quad (j = 0, 1, \ldots). \qquad (4.46)$$

Hence defining the *vibrational characteristic temperature* to be $\theta_v \equiv \hbar\omega/k_B$, the partition function for one particle is

$$Z_1 = \sum_{j=0}^{\infty} e^{-\frac{e_j}{k_B T}} = \sum_{j=0}^{\infty} e^{-(j+\frac{1}{2})\frac{\theta_v}{T}} = e^{-\frac{\theta_v}{2T}} \sum_{j=0}^{\infty} e^{-j\frac{\theta_v}{T}}$$

$$= \frac{e^{-\frac{\theta_v}{2T}}}{1 - e^{-\frac{\theta_v}{T}}}. \tag{4.47}$$

Hence the free energy $F_N$, and other thermodynamic quantities, for the system of $N$ particles will be

$$F_N = 3N F_1 = 3N(-k_B T \ln Z_1)$$
$$= 3N k_B T \ln(1 - e^{-\frac{\theta_v}{T}}) + \frac{3}{2} N k_B \theta_v, \tag{4.48}$$

$$S = -\left(\frac{\partial F_N}{\partial T}\right)_V = 3N k_B T \frac{\theta_v}{T^2} \frac{e^{-\frac{\theta_v}{T}}}{1 - e^{-\frac{\theta_v}{T}}} - 3N k_B \ln(1 - e^{-\frac{\theta_v}{T}})$$
$$= 3N k_B \frac{\theta_v}{T}(e^{\frac{\theta_v}{T}} - 1)^{-1} - 3N k_B \ln(1 - e^{-\frac{\theta_v}{T}}), \tag{4.49}$$

$$U = F_N + TS$$
$$= \frac{3}{2} N k_B \theta_v + 3N k_B \theta_v (e^{\frac{\theta_v}{T}} - 1)^{-1}, \tag{4.50}$$

$$C_V = \left(\frac{\partial U}{\partial T}\right)_V$$
$$= 3N k_B \left(\frac{\theta_v}{T}\right)^2 (e^{\frac{\theta_v}{T}} - 1)^{-2} e^{\frac{\theta_v}{T}}. \tag{4.51}$$

In the expression for $U$ in Equation (4.50), the first term on the right hand side is called the *zero-point energy* (being the value of $U$ as $T \to 0$), while the second term is the *thermal energy*. Note that $C_V$ depends on the particular solid through the quantity $\theta_v$, but a plot of $C_V$ against $T/\theta_v$ should be the same for all solids. This plot is shown in Fig. 4.2. For large temperatures, that is, for $T \gg \theta_v$, then $C_V \to 3N k_B$, which is the law of Dulong and Petit. The fall-off in $C_V$ could not be explained using classical physics, and was explained satisfactorily by Einstein in 1907.

### 4.4.2 *Quantum rotational aspects*

The classical energy of rotation for a rotor with moment of inertia $I$ about an axis is given by

$$E = \frac{1}{2} I \omega^2 = \frac{(I\omega)^2}{2I} = \frac{L^2}{2I}$$

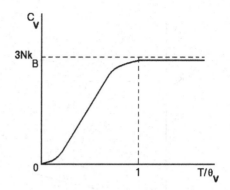

**Fig. 4.2** The fall-off in $C_V$ predicted by Einstein in 1907. The quantity $\theta_v$ is the characteristic vibrational temperature.

when its angular speed is $\omega$, and where $L$ is the angular momentum about the axis. In the quantum treatment of angular momentum, it has been found that the eigenvalues of $L^2$ are $l(l+1)\hbar^2$, $(l = 0, 1, \ldots)$, and this eigenvalue is $(2l+1)$-fold degenerate. Hence the rotational energy eigenvalues are $\hbar^2 l(l+1)/(2I)$, and the partition function for one particle becomes

$$Z_1 = \sum_{l=0}^{\infty} (2l+1) e^{-\frac{\hbar^2}{2Ik_BT} l(l+1)} = \sum_{l=0}^{\infty} (2l+1) e^{-l(l+1)\frac{\theta_r}{T}} \tag{4.52}$$

where $\theta_r \equiv \hbar^2/(2Ik_B)$ is the *rotational characteristic temperature*. The high temperature limit is obtained by noting that the quantised differences in the energy are small compared with $k_BT$, and we may then make the replacement

$$Z_1 \to \int_0^{\infty} (2l+1) e^{-l(l+1)\frac{\theta_r}{T}} dl = \frac{T}{\theta_r}$$

giving, for a gas of $N$ molecules,

$$U = Nk_BT^2 \frac{d}{dt} \ln Z_1 = Nk_BT,$$

$$C_V = Nk_B.$$

However, generally when the high-temperature approximation is not assumed, a numerical evaluation of $U$ and $C_V$ must be performed for the $N$-particle case using Equation (3.56) in the form

$$U = Nk_BT^2 \frac{d}{dT} \ln Z_1,$$

$$C_V = \left(\frac{\partial U}{\partial T}\right)_V. \tag{4.53}$$

These values are plotted in Fig. 4.3.

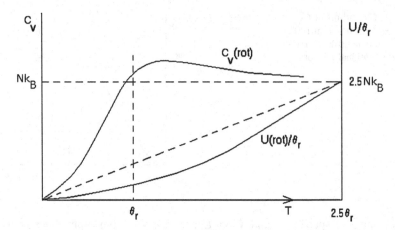

**Fig. 4.3** Rotational aspects of specific heat $C$ against $T$. The quantity $\theta_r$ is the rotational characteristic temperature.

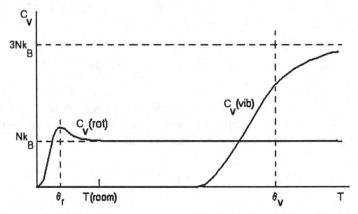

**Fig. 4.4** Combined rotational and vibrational aspects of specific heat, with $\theta_r \approx 10^0$ and $\theta_v \approx 10^3$. The rotational aspect dominates at room temperature $T$ (room).

**Table 4.1** The rotational characteristic temperature $\theta_r$ and the vibrational characteristic temperature $\theta_v$ for various gases.

|              | $H_2$   | $N_2$   | $O_2$   | CO      |
| ------------ | ------- | ------- | ------- | ------- |
| $\theta_r$ (K) | 85.38   | 2.863   | 2.096   | 2.766   |
| $\theta_v$ (K) | 5987.0  | 3352.0  | 2239.3  | 3083.7  |

Table 4.1 shows various values of $\theta_r$ and $\theta_v$, and indicates the large variation in the relative values.

Fig. 4.4 pieces together both the rotational and vibrational contributions to $C_V$. Note how the rotational aspect dominates at room temperature.

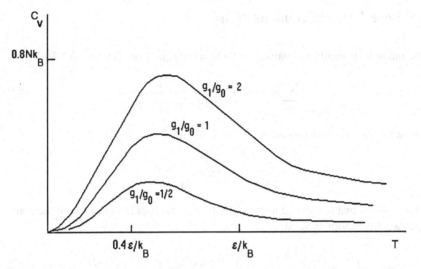

**Fig. 4.5** Schottky peaks in the plot of $C_V$ against $T$, plotted for various ratios $g_1/g_0$ of the degeneracy values.

### 4.4.3 Schottky peaks

For most gases, electronic excited states lie at energies greater than 1 eV above the ground state, and these are hardly occupied at room temperature. But certain gases, for example $O_2$ and NO, have their first excited states not far above the ground state, and this gives rise to some unusual $C_V$–$T$ curves, involving *Schottky peaks*.

Let the energies be measured from the ground state, with the first excited state having energy $\varepsilon$. Let $g_0$ and $g_1$ be the degeneracies of these levels respectively. Then the partition function for one molecule, expanded to the first two significant terms, is

$$Z_1 = \sum_E g(E)e^{-\frac{E}{k_B T}} = g_0 + g_1 e^{-\frac{\varepsilon}{k_B T}}.$$ (4.54)

Hence for a gas of $N$ non-interacting particles, the internal energy is given by

$$U = Nk_B T^2 \frac{d}{dT} \ln Z_1 = N\varepsilon \left(1 + \frac{g_0}{g_1} e^{\frac{\varepsilon}{k_B T}}\right)^{-1},$$ (4.55)

and the specific heat is given by

$$C_V = \left(\frac{\partial U}{\partial T}\right)_V = \frac{N\varepsilon^2}{k_B T^2} \frac{g_0}{g_1} e^{\frac{\varepsilon}{k_B T}} \left(1 + \frac{g_0}{g_1} e^{\frac{\varepsilon}{k_B T}}\right)^{-2}.$$ (4.56)

The structure of peaks in the plot of $C_V$ against $T$ is shown in Fig. 4.5, which is plotted for various degeneracy ratios. This phenomenon is seen in some paramagnetic salts and rare earth metals.

## 4.5 Bose-Einstein condensation

The mean total number of particles of a boson gas is, from Equation (3.70),

$$N = \sum_j \langle n_j \rangle = \sum_j \frac{1}{\exp\{\frac{1}{k_B T}(e_j - \mu)\} - 1}. \tag{4.57}$$

Again take the groundstate energy $e_0 = 0$. Then

$$\langle n_0 \rangle = \frac{1}{e^{-\frac{\mu}{k_B T}} - 1} \geq 0,$$

which means that $\mu$ must be negative. Now assume for the moment that the continuous approximation is valid:

$$N \to \int_0^\infty \frac{A_{\frac{1}{2}} E^{\frac{1}{2}} dE}{e^{\frac{1}{k_B T}(E-\mu)} - 1} = \frac{4\pi V g m \sqrt{2m}}{h^3} \int_0^\infty \frac{\sqrt{E} dE}{e^{\frac{1}{k_B T}(E-\mu)} - 1}. \tag{4.58}$$

Then the boson density $\rho$ is

$$\rho \equiv \frac{N}{V} = \frac{4\pi g m \sqrt{2m}}{h^3} \int_0^\infty \frac{\sqrt{E} dE}{e^{\frac{1}{k_B T}(E-\mu)} - 1}. \tag{4.59}$$

Now lower the temperature $T$ while keeping this density constant. Define $T_{\rho c}$ as that temperature required to give $\mu = 0$ for constant density:

$$\rho = \frac{4\pi g m \sqrt{2m}}{h^3} \int_0^\infty \frac{\sqrt{E} dE}{e^{\frac{1}{k_B T_{\rho c}} E} - 1}$$

$$= g \left( \frac{2\pi m k_B T_{\rho c}}{h^2} \right)^{\frac{3}{2}} \left[ \frac{2}{\sqrt{\pi}} \int_0^\infty \frac{\sqrt{x} dx}{e^x - 1} \right] \tag{4.60}$$

$$= g \left( \frac{2\pi m k_B T_{\rho c}}{h^2} \right)^{\frac{3}{2}} \times 2.61. \tag{4.61}$$

Hence

$$T_{\rho c} = \frac{h^2}{g^{\frac{2}{3}} 2\pi m k_B} \left( \frac{\rho}{2.61} \right)^{\frac{2}{3}}. \tag{4.62}$$

The consequence of this result is that, for this non-zero value of $T_{\rho c}$, it is not possible to reduce the temperature below $T_{\rho c}$ while keeping the density constant, for then $\mu$ would become positive. This conclusion is physically unrealistic: the mistake has been to assume that the continuous approximation in Equation (4.58) is valid. Instead, the contribution for $E = 0$ must be explicitly split off by writing

$$N = \langle n_0 \rangle + N_{E>0}$$

$$= \frac{1}{e^{-\frac{\mu}{k_B T}} - 1} + \frac{4\pi V gm\sqrt{2m}}{h^3} \int_0^\infty \frac{\sqrt{E}\, dE}{e^{\frac{1}{k_B T}(E-\mu)} - 1}. \tag{4.63}$$

The two regimes $T > T_{\rho c}$ and $T < T_{\rho c}$ must then be considered separately:

1. For $T > T_{\rho c}$, the term $(e^{-\frac{\mu}{k_B T}} - 1)^{-1}$ is negligible, and we are left with the last term. In this case, the fraction of particles in the ground state is approximately zero.

2. For $T < T_{\rho c}$, put $\mu = 0$ into the integral term in Equation (4.63). Using Equation (4.60), we have

$$\begin{aligned}
N_{E>0} &= \frac{4\pi V gm\sqrt{2m}}{h^3} \int_0^\infty \frac{\sqrt{E}\, dE}{e^{\frac{1}{k_B T}E} - 1} \\
&= \frac{4\pi V gm\sqrt{2m}}{h^3} (k_B T)^{\frac{3}{2}} \frac{\sqrt{\pi}}{2} \left[ \frac{2}{\sqrt{\pi}} \int_0^\infty \frac{\sqrt{x}\, dx}{e^x - 1} \right] \\
&= N \left( \frac{T}{T_{\rho c}} \right)^{\frac{3}{2}}.
\end{aligned} \tag{4.64}$$

Hence the fraction of particles in the ground state is

$$\frac{\langle n_0 \rangle}{N} = 1 - \frac{N_{E>0}}{N} = 1 - \left( \frac{T}{T_{\rho c}} \right)^{\frac{3}{2}}. \tag{4.65}$$

This process of concentrating particles into the groundstate for temperatures less than $T_{\rho c}$ is called *Bose-Einstein Condensation*. The quantity $T_{\rho c}$ is called the *condensation temperature* . This model is partly applicable to liquid helium, for which $T_{\rho c} = 3.1$ K.

The internal energy in this case is

$$U = \frac{4\pi V gm\sqrt{2m}}{h^3} \int_0^\infty \frac{E^{\frac{3}{2}} dE}{e^{\frac{1}{k_B T}(E-\mu)} - 1}, \tag{4.66}$$

and $C_V = (\partial U/\partial T)_{V,N}$ can be calculated for the two separate cases $T > T_{\rho c}$ and $T < T_{\rho c}$ (for which $\mu = 0$). The result is shown in Fig. 4.6.

We will return to this topic again in Chapter 14 when the process of *simulated annealing* based on Fermi-Dirac statistics is introduced.

## 4.6 Thermionic emission

In the process of thermionic emission, electrons are ejected at temperature $T$ from a metal through a surface which is perpendicular to the $x$-direction. The electrons

**Fig. 4.6** Plot of the specific heat $C_V$ against $T$ for Bose-Einstein condensation. $T_{\rho c}$ is the condensation temperature.

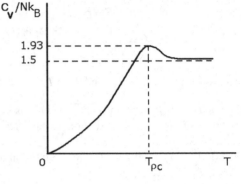

**Fig. 4.7** The energy structure for thermionic emission. The electrons must have minimum energy $W_0$ in order to escape the surface. $\phi$ is the work function.

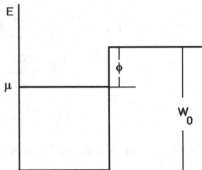

must have a minimum energy $W_0$ to escape the surface. Fig. 4.7 shows the energy structure. The quantity $\phi \equiv W_0 - \mu$ is called the *work function* for thermionic emission.

For electrons with two degrees of spin, we take the degeneracy $g = 2$. The density of states which is given by Equation (4.21), and the energy, are

$$dz(\mathbf{p}) = \frac{2V}{h^3}dp_x dp_y dp_z, \tag{4.67}$$

$$E = \frac{1}{2m}(p_x{}^2 + p_y{}^2 + p_z{}^2). \tag{4.68}$$

It follows that the number of electrons per unit volume with momenta in the range $(\mathbf{p}, \mathbf{p} + d\mathbf{p})$ is

$$\frac{1}{V} \times (\text{number of states in range}) \times (\text{mean number of electrons per state})$$

$$= \frac{1}{V} \times \frac{2V}{h^3}dp_x dp_y dp_z \times \frac{1}{e^{\frac{E-\mu}{k_B T}} + 1}$$

$$= \frac{2}{h^3} \frac{dp_x dp_y dp_z}{e^{\frac{E-\mu}{k_B T}} + 1}. \tag{4.69}$$

Hence the number of electrons crossing the unit area perpendicular to the $x$-axis per unit time with momenta in the range $(p_x, p_x + dp_x)$ is $dI$, which is given by

$$
dI = \frac{2}{h^3}\left(\frac{p_x}{m}\right)dp_x \int_{-\infty}^{\infty}\int_{-\infty}^{\infty}\frac{dp_y dp_z}{e^{\frac{E-\mu}{k_BT}}+1}
$$

$$
= \frac{2}{h^3}dW \int_{-\infty}^{\infty}\int_{-\infty}^{\infty}\frac{dp_y dp_z}{e^{\frac{W-\mu}{k_BT}}e^{\frac{p_x^2+p_y^2}{2mk_BT}}+1}, \tag{4.70}
$$

where the substitution $W = p_x^2/(2m)$ has been used. The integral is evaluated using plane polar coordinates $p_y = \rho\cos\theta$, $p_z = \rho\sin\theta$. Then with $\partial(p_y, p_z)/\partial(\rho,\theta) = \rho$, Equation (4.70) becomes

$$
dI = \frac{2}{h^3}dW \int_0^{2\pi}d\theta \int_0^{\infty}\frac{\rho\,d\rho}{e^{\frac{W-\mu}{k_BT}}e^{\frac{\rho^2}{2mk_BT}}+1}
$$

$$
= \frac{4\pi}{h^3}dW \int_0^{\infty}\frac{\rho e^{\frac{-\rho^2}{2mk_BT}}\,d\rho}{e^{\frac{W-\mu}{k_BT}}+e^{\frac{-\rho^2}{2mk_BT}}}
$$

$$
= -\frac{4\pi}{h^3}dW(mk_BT)\left[\ln\left(e^{\frac{W-\mu}{k_BT}}+e^{\frac{-\rho^2}{2mk_BT}}\right)\right]_0^{\infty}
$$

$$
= \frac{4\pi mk_BT}{h^3}\ln\left(1+e^{\frac{\mu-W}{k_BT}}\right)dW \tag{4.71}
$$

$$
= N_S(T,\mu,W)dW \tag{4.72}
$$

where $N_S(T,\mu,W)$ is called the *supply function* for thermionic emission, defined by

$$
N_S(T,\mu,W) \equiv \frac{4\pi mk_BT}{h^3}\ln\left(1+e^{\frac{\mu-W}{k_BT}}\right). \tag{4.73}
$$

For emission to take place, the condition $W_0 \le E = W + p_y^2 + p_z^2$ must hold. This implies that $W \ge W_0$, and hence the current density in the $x$-direction is

$$
j = e\int_{W_0}^{\infty}N_S(T,\mu,W)dW
$$

$$
= \frac{4\pi emk_BT}{h^3}\int_{W_0}^{\infty}\ln\left(1+e^{\frac{\mu-W}{k_BT}}\right)dW
$$

$$
= \frac{4\pi em(k_BT)^2}{h^3}\int_0^{\infty}\ln\left(1+e^{-x+\frac{\mu-W_0}{k_BT}}\right)dx
$$

$$
= \frac{4\pi em(k_BT)^2}{h^3}\int_0^{\infty}\frac{x\,dx}{e^{x+\frac{\phi}{k_BT}}+1}
$$

$$
= \frac{4\pi em(k_BT)^2}{h^3}I\left(-\frac{\phi}{k_BT},1,1\right). \tag{4.74}
$$

If the temperature $T$ is small enough so that the approximation $\phi \gg k_B T$ holds—for example, $\phi \approx 100 k_B T$ at room temperature—then

$$j = \frac{4\pi em(k_B T)^2}{h^3} e^{-\frac{\phi}{k_B T}} = AT^2 e^{-\frac{\phi}{k_B T}} \quad \text{where}$$

$$A \equiv \frac{4\pi emk_B{}^2}{h^3}.$$

(4.75)

These are the *Dushman-Richardson equations*. These results must be modified if the metals under consideration have dirty surfaces, for then there may be reflection at the surface. The equations will then be modified to the form

$$j = e \int_{W_0}^{\infty} N_S(T, \mu, W)[1 - r(W)] dW$$

(4.76)

where $r(W)$ is the reflection coefficient. A further complication may arise if the work function $\phi$ varies with temperature.

The unmodified Dushman-Richardson equations may be written as

$$\ln \left( \frac{j}{T^2} \right) = \ln A - \frac{\phi}{k_B T},$$

and a plot of $\ln(j/T^2)$ against $1/(k_B T)$ will give a straight line with slope $(-\phi)$. A smaller value of $\phi$ will give a larger value of $j$ for a given value of $T$; the work function takes the value $\phi = 4.1$ eV for tungsten, while for a mixture of tungsten and caesium it takes the value $\phi = 3.0$ eV. This mixture will therefore give a bigger current at the same temperature.

## 4.7 Semiconductor statistics

A brief introduction to the general physical behaviour of electrons and holes in semiconductors will be given, with particular reference to allowed and forbidden energy bands, carrier effective mass, and the non-degenerate approximation.

### 4.7.1 Allowed and forbidden bands

The electron energies in a single atom take certain discrete values. When atoms are brought together to form a crystal lattice, these energy levels are perturbed to produce bands of allowed and forbidden energies. It is found that the solution of the TISE with a periodic potential gives allowed values of the energy $E$ where

$$E_0 \le E \le E_1, \qquad E_2 \le E \le E_3, \quad \text{etc.}$$

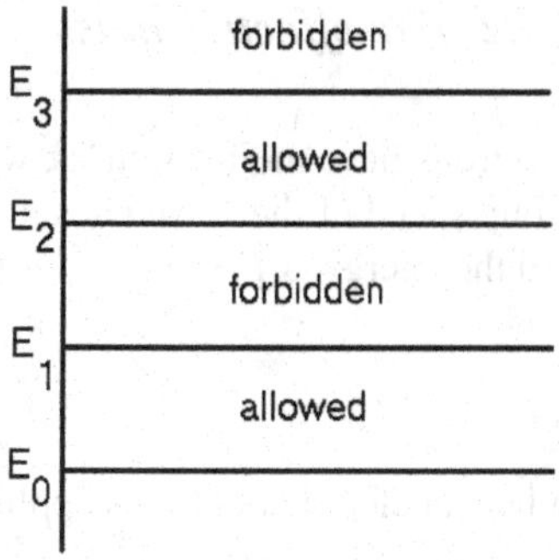

**Fig. 4.8** The energy band structure, showing the allowed and forbidden bands.

**Fig. 4.9** The conduction and valence bands, showing the top of the valence band $E_v$ and the bottom of the conduction band $E_c$. The band gap is $E_G = E_c - E_v$.

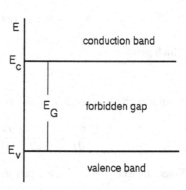

Fig. 4.8 shows the structure of these bands. Two main classifications arise in the situation at absolute zero temperature $T = 0$:

1. if an allowed band is only partly full, *any* added energy can move an electron into a higher state: this situation applies to a conductor;
2. if electrons completely fill a band, with none left over for a higher band, then an energy input above a certain threshold is needed to excite electrons into the next band: this situation applies to a semiconductor or an insulator.

In general, semiconductors have smaller forbidden gaps $E_G$ than insulators—approximately 1 eV as opposed to 10 eV. For example, Ge has a forbidden gap of 0.72 eV, silicon has a forbidden gap of 1.12 eV, while diamond has a gap of 7 eV. If impurities are present, then these cause isolated allowed energies in the forbidden gap of the pure, or *intrinsic*, semiconductor. At ordinary temperatures, a *valence* band with highest energy $E_v$ and a *conduction* band with lowest energy $E_c$ is formed, as shown in Fig. 4.9. The valence band is almost full, and the electrons at these energies are bound to lattice sites. The positively charged holes which remain are isolated to energy states situated mainly near the top of the valence band. The conduction band is almost empty, and electrons at these energies are free. When an electric field is applied across the material, a current is carried by both the negatively charged electrons in the conduction band and by the positively charged holes in the valence band.

## 4.7.2 The effective mass

Electrons have momentum $\hbar\mathbf{k}$ where $\mathbf{k}$ is the wave vector, and they move with the group speed of the wave packet. This group speed is $v = d\omega/dk$ where $\omega = 2\pi v$, and the energy is $E = hv = h\omega/(2\pi) = \hbar\omega$. Hence the group speed is

$$v = \frac{1}{\hbar}\frac{dE}{dk}.$$

When an electric field $F$ is applied, the energy gain in time $dt$ is therefore

$$dE = (eF)(v\,dt) = \frac{eF}{\hbar}\frac{dE}{dk}dt,$$

which gives

$$\frac{dk}{dt} = \frac{eF}{\hbar}.$$

Therefore the acceleration of the electron is

$$f = \frac{dv}{dt} = \frac{1}{\hbar}\frac{d}{dt}\left(\frac{dE}{dk}\right) = \frac{1}{\hbar}\frac{d^2E}{dk^2}\frac{dk}{dt} = \frac{1}{\hbar^2}(eF)\frac{d^2E}{dk^2}.$$

Since the total force on the electron is

$$eF = \hbar^2\left(\frac{d^2E}{dk^2}\right)^{-1}f,$$

the electron therefore has an *effective mass* $m_e$ given by

$$m_e \equiv \hbar^2\left(\frac{d^2E}{dk^2}\right)^{-1}. \tag{4.77}$$

Hence the effective mass of the charge carries depends on its position on the energy scale. It can be shown (Landsberg 1969) that this second derivative near the bottom of the conduction band is positive, while it is negative near the top of the valence band. This means that, when an electric field is applied, the electrons in the conduction band and valence band move in opposite directions. This somewhat inconvenient picture of electrons with negative effective mass is replaced by one in which we think of the empty states at the top of the valence band as positively charged *holes* with a *positive* effective mass.

## 4.7.3 Electron and hole densities

Let $m_e$ be the effective mass of an electron in the conduction band near $E_c$. Again taking the degeneracy $g = 2$, the density of states is

$$N(E)dE = \frac{8\pi V}{h^3} m_e \sqrt{2m_e} \sqrt{E - E_c} dE. \qquad (4.78)$$

We must now perform an integration over the energy in the conduction band, with the lower limit being $E_c$ and the top limit being the top of this band. However, the occupation number at this top limit is so small that a good approximation will be to replace the top limit by an infinite limit. With this in mind, the electron density in the conduction band is

$$
\begin{aligned}
n &= \frac{1}{V} \int_{E_c}^{\infty} \frac{8\pi V}{h^3} m_e \sqrt{2m_e} \sqrt{E - E_c} \frac{dE}{e^{\frac{E-\mu}{k_B T}} + 1} \\
&= 4\pi \left(\frac{2m_e}{h^2}\right)^{\frac{3}{2}} \int_{E_c}^{\infty} \frac{\sqrt{E - E_c}}{e^{\frac{E-\mu}{k_B T}} + 1} dE \\
&= 4\pi \left(\frac{2k_B T m_e}{h^2}\right)^{\frac{3}{2}} \int_{0}^{\infty} \frac{\sqrt{x} dx}{e^{x - \frac{\mu-E}{k_B T}} + 1} dx \\
&= 4\pi \left(\frac{2k_B T m_e}{h^2}\right)^{\frac{3}{2}} \Gamma\left(\frac{3}{2}\right) I\left(\frac{\mu - E_c}{k_B T}, \frac{1}{2}, +1\right) \\
&= 2 \left(\frac{2\pi k_B T m_e}{h^2}\right)^{\frac{3}{2}} I\left(\frac{\mu - E_c}{k_B T}, \frac{1}{2}, +1\right) \\
&= N_e I\left(\frac{\mu - E_c}{k_B T}, \frac{1}{2}, +1\right) \qquad (4.79)
\end{aligned}
$$

where

$$N_e \equiv 2 \left(\frac{2\pi k_B T m_e}{h^2}\right)^{\frac{3}{2}}. \qquad (4.80)$$

A similar calculation applies to the hole density in the valence band. At energy $E$ in the valence band,

$$
\begin{aligned}
\text{mean number of holes} &= 1 - (\text{mean number of electrons}) \\
&= 1 - \frac{1}{e^{\frac{E-\mu}{k_B T}} + 1} \\
&= \frac{1}{e^{\frac{\mu-E}{k_B T}} + 1}, \qquad (4.81)
\end{aligned}
$$

and the density of states below $E_v$ is

$$\frac{8\pi V}{h^3} |m_h| \sqrt{2|m_h|} \sqrt{E_v - E} dE$$

where $m_h$ is the effective mass of a hole near $E_v$. Hence the hole density in the valence band is

$$p = \frac{1}{V} \int_{-\infty}^{E_v} \frac{8\pi V}{h^3} |m_h| \sqrt{2|m_h|} \sqrt{E_v - E} \frac{dE}{e^{\frac{\mu - E}{k_B T}} + 1}$$

$$= N_h I \left( \frac{E_v - \mu}{k_B T}, \frac{1}{2}, +1 \right) \tag{4.82}$$

where

$$N_h \equiv 2 \left( \frac{2\pi k_B T |m_h|}{h^2} \right)^{\frac{3}{2}}. \tag{4.83}$$

These results are summarised as

$$n = N_e I \left( \frac{\mu - E_c}{k_B T}, \frac{1}{2}, +1 \right) \quad \text{where } N_e \equiv 2 \left( \frac{2\pi k_B T m_e}{h^2} \right)^{\frac{3}{2}}, \tag{4.84}$$

$$p = N_h I \left( \frac{E_v - \mu}{k_B T}, \frac{1}{2}, +1 \right) \quad \text{where } N_h \equiv 2 \left( \frac{2\pi k_B T |m_h|}{h^2} \right)^{\frac{3}{2}}. \tag{4.85}$$

These quantities are more commonly written in terms of the Fermi integrals defined in Equation (4.5):

$$n = N_e F_{\frac{1}{2}} \left( \frac{\mu - E_c}{k_B T} \right) \quad \text{where } N_e \equiv 2 \left( \frac{2\pi k_B T m_e}{h^2} \right)^{\frac{3}{2}}, \tag{4.86}$$

$$p = N_h F_{\frac{1}{2}} \left( \frac{E_v - \mu}{k_B T} \right) \quad \text{where } N_h \equiv 2 \left( \frac{2\pi k_B T |m_h|}{h^2} \right)^{\frac{3}{2}}. \tag{4.87}$$

### 4.7.4 The non-degenerate approximation

These results are simplified when the Fermi level $\mu$ lies almost half way inside the forbidden gap. In this special case, the quantities $(\mu - E_c)/(k_B T)$ and $(E_v - \mu)/(k_B T)$ are negative and have large magnitudes, so that the approximation of Equation (4.6) is valid:

$$n \approx N_e e^{\frac{\mu - E_c}{k_B T}}, \tag{4.88}$$

$$p \approx N_h e^{\frac{E_v - \mu}{k_B T}}. \tag{4.89}$$

This is the *non-degenerate approximation*.

If all of the conduction electrons have come from the valence band, then $n = p$. In that case,

$$N_e e^{\frac{\mu - E_c}{k_B T}} = N_h e^{\frac{E_v - \mu}{k_B T}}, \quad \text{or} \quad e^{\frac{2\mu}{k_B T}} = \frac{N_h}{N_e} e^{\frac{E_v + E_c}{k_B T}}.$$

Hence the temperature dependence of $\mu$ is given by

$$\mu(T) = \frac{1}{2}(E_v + E_c) + \frac{1}{2}k_BT \ln\left(\frac{N_h}{N_e}\right). \tag{4.90}$$

Now the quantity $N_h/N_e$ is independent of $T$. Hence, at $T = 0$, it follows that

$$\mu_{(T=0)} \equiv \mu(0) = \frac{1}{2}(E_v + E_c). \tag{4.91}$$

This shows that $\mu_{(T=0)}$ lies exactly in the middle of the forbidden gap, and then Equation (4.90) becomes

$$\mu(T) = \mu_{(T=0)} + \frac{1}{2}k_BT \ln\left(\frac{N_h}{N_e}\right). \tag{4.92}$$

Further, it follows from Equations (4.88) and (4.89) that

$$n^2 = np = N_eN_h e^{\frac{\mu-E_c+E_v-\mu}{k_BT}} = N_eN_h e^{\frac{E_v-E_c}{k_BT}}$$

$$= N_eN_h e^{-\frac{E_G}{k_BT}}. \tag{4.93}$$

Hence

$$n = \sqrt{N_eN_h} e^{-\frac{E_G}{2k_BT}} = 2\left(\frac{2\pi k_B\sqrt{m_e|m_h|}}{h^2}\right)^{\frac{3}{2}} T^{\frac{3}{2}} e^{-\frac{E_G}{2k_BT}}. \tag{4.94}$$

It follows that the temperature dependencies of both $n$ and $p$ in the non-degenerate approximation are given by

$$n = p = AT^{\frac{3}{2}} e^{-\frac{E_G}{2k_BT}} \tag{4.95}$$

where $A$ is a constant.

## Problems

**Problem 4.1.** The canonical distribution for a system at temperature $T$ is

$$P(E) = Ae^{-\frac{E}{k_BT}}$$

where $A$ is a constant. Suppose that the energy of the system can be expressed in terms of the generalised coordinates $q_i$ and momenta $p_i$ $(i = 1, \ldots, r)$ as

$$E = K(p_1, \ldots, p_r) + F(q_1, \ldots, q_r)$$

where the functions $K$ and $F$ are the kinetic and potential energies respectively. If $P_p dp_1 \cdots dp_r$ is the probability of finding the momenta in the ranges $(p_i, p_i+dp_i)$, show that $P_P = Be^{-\frac{K}{k_BT}}$ where $B$ is a constant for given values of $E$ and $T$.

**Problem 4.2.** In the case of a gas consisting of $N$ weakly interacting particles with mass $m$, we can write $K = \sum_{i=1}^{3N} p_i^2/(2m)$. If the values of $p_i$ are unbounded, find the quantity $B$ defined in Problem 4.1.

**Problem 4.3.** In the case of blackbody radiation, the energy per unit volume lying in the frequency range $(\nu, \nu + d\nu)$ is given by

$$u_\nu = \frac{8\pi h}{c^3} \frac{\nu^3}{e^{\frac{h\nu}{k_B T}} - 1}.$$

Prove Wien's Law.
*Solution.* Define $x \equiv h\nu/(k_B T)$. Then

$$u_\nu = \frac{8\pi h}{c^3} \frac{\nu^3}{e^{\frac{h\nu}{k_B T}} - 1} = \frac{8\pi k_B^3 T^3}{c^3 h^3} \times \frac{x^3}{e^x - 1}.$$

For a fixed value of $T$, the maximum value of $u_\nu$ is given by $du/dx = 0$ at $x = x_0$, or

$$(3 - x_0)e^{x_0} = 0.$$

The solution of this equation is independent of $\nu_0$ and $T$ separately, and hence the value $\nu_0$ of $\nu$ at which the peak occurs for a given value of $T$ is proportional to $T$. Hence as $T$ increases, $\nu_0$ increases, or equivalently, the value of $\lambda$ for this maximum decreases.  □

# Chapter 5
# Density of states and applications—2

The basic results of quantum mechanics and statistical mechanics which were derived in Chapter 3 and Chapter 4 will be targeted to relate more closely to the processes of device modelling. Particular reference will be made to periodic potentials, the dependence of effective mass on position, the effective mass approximation, and quantum wells in semiconducting devices.

## 5.1 Periodic potential: the Bloch theorem

The Bloch theorem applies to the solutions of the Schrödinger equation when the potential is periodic (Dekker 1963; Landsberg 1969). Its statement and proof will be given for one spatial dimension, and the statement will be given for the full three dimensional case later.

Suppose that a constant effective mass $m_e$ is assumed, and that the potential $V(x)$ has a period $a$:

$$V(x + a) = V(x) \quad \text{for all } x. \tag{5.1}$$

The TISE for the system will have the form

$$-\frac{\hbar^2}{2m_e}\phi''(x) + V(x)\phi(x) = E\phi(x). \tag{5.2}$$

**Theorem 5.1.** *The Bloch theorem. When the potential has period a, there exist solutions of Equation (5.2) of the form*

$$\phi_k(x) = e^{\pm ikx}u_k(x) \quad \text{where } u_k(x + a) = u_k(x). \tag{5.3}$$

*Proof.* In fact, Bloch's theorem is a special case of Floquet's theorem (Jordan and Smith 1979), but we will give a simple proof of Bloch's theorem which follows that outlined by Dekker (1963). Using the periodicity of the potential, Equation (5.2) gives

E.A.B. Cole, *Mathematical and Numerical Modelling of Heterostructure Semiconductor Devices: From Theory to Programming,* DOI 10.1007/978-1-84882-937-4_5, © Springer-Verlag London Limited 2009

$$-\frac{\hbar^2}{2m_e}\phi''(x+a) + V(x)\phi(x+a) = E\phi(x+a). \tag{5.4}$$

Suppose that $\phi_r(x)$ and $\phi_s(x)$ are two real independent solutions of Equation (5.2). Then any linear combination of these two solutions is also a solution. Further, any solution can be written as a linear combination, since the differential equation is second order. Then two such combinations which are solutions of Equation (5.4) are

$$\phi_r(x+a) = r_1\phi_r(x) + s_1\phi_s(x) \tag{5.5}$$
$$\phi_s(x+a) = r_2\phi_r(x) + s_2\phi_s(x)$$

where $r_1$ and $s_1$ are constants. The general solution of Equation (5.2) will be

$$\phi(x) = A_r\phi_r(x) + A_s\phi_s(x).$$

Hence

$$\phi(x+a) = A_r(r_1\phi_r(x) + s_1\phi_s(x)) + A_s(r_2\phi_r(x) + s_2\phi_s(x))$$
$$= (A_r r_1 + A_s r_2)\phi_r(x) + (A_r s_1 + A_s s_2)\phi_s(x).$$

Since the quantities $A_r$, $A_s$, $r_1$, $r_2$, $s_1$ and $s_2$ are arbitrary, consider the special case

$$A_r r_1 + A_s r_2 = QA_r$$
$$A_r s_1 + A_s s_2 = QA_s \tag{5.6}$$

where $Q$ is a constant to be determined. Then

$$\phi(x+a) = QA_r\phi_r(x) + QA_s\phi_s(x) = Q\phi(x). \tag{5.7}$$

Equations (5.6) can be rearranged to give

$$(r_1 - Q)A_r + r_2 A_s = 0$$
$$s_1 A_r + (s_2 - Q)A_s = 0$$

which have non-trivial solutions only if

$$(r_1 - Q)(s_2 - Q) - s_1 r_2 = 0,$$

that is,

$$Q^2 - (r_1 + s_2)Q + (r_1 s_2 - s_1 r_2) = 0. \tag{5.8}$$

Applying Equation (5.2) to both $\phi_r$ and $\phi_s$, we have

$$-\frac{\hbar^2}{2m_e}\phi_r''(x) + V(x)\phi_r(x) = E\phi_r(x) \quad \text{and}$$

$$-\frac{\hbar^2}{2m_e}\phi_s''(x) + V(x)\phi_s(x) = E\phi_s(x).$$

On multiplying the first by $\phi_s$ and the second by $\phi_r$ and subtracting, it follows that

$$0 = \phi_s\phi_r'' - \phi_r\phi_s'' = (\phi_s\phi_r' - \phi_r\phi_s')',$$

or

$$\phi_s(x)\phi_r'(x) - \phi_r(x)\phi_s'(x) = C \tag{5.9}$$

where $C$ is a constant independent of $x$. The value of $C$ is not zero in general, since $\phi_r$ and $\phi_s$ are two independent solutions of a second order differential equation. On making the replacement $x \to x + a$, Equation (5.9) becomes

$$\begin{aligned}
C &= \phi_s(x+a)\phi_r'(x+a) - \phi_r(x+a)\phi_s'(x+a) \\
&= (r_2\phi_r(x) + s_2\phi_s(x))(r_1\phi_r'(x) + s_1\phi_s'(x)) \\
&\quad - (r_1\phi_r(x) + s_1\phi_s(x))(r_2\phi_r'(x) + s_2\phi_s'(x)) \\
&= (r_1 s_2 - r_2 s_1)(\phi_s(x)\phi_r'(x) - \phi_r(x)\phi_s'(x)) \\
&= (r_1 s_2 - r_2 s_1)C.
\end{aligned}$$

Since $C \neq 0$, it follows that

$$r_1 s_2 - r_2 s_1 = 1,$$

and Equation (5.8) then gives

$$Q^2 - (r_1 + s_2)Q + 1 = 0$$

which has solutions

$$Q = \frac{1}{2}\left(r_1 + s_2 \pm \sqrt{(r_1 + s_2)^2 - 4}\right).$$

The roots of this equation are reciprocal, and hence have the form $e^{ika}$ for some $k = k_1$ and $k = -k_1$. Now define the function $u_k(x) \equiv \phi(x)e^{-ikx}$. Then

$$u_k(x+a) = \phi(x+a)e^{-ik(x+a)} = Q\phi(x)e^{-ika}e^{-ikx} = u_k(x).$$

It follows that the Schrödinger equation possesses solutions

$$\phi_k(x) = e^{ikx}u_k(x), \qquad \phi_k(x+a) = e^{ika}\phi_k(x), \qquad u_k(x+a) = u_k(x). \tag{5.10}$$

This completes the proof. $\square$

The quantity $k$ must be real, since an imaginary value would give solutions $\phi_k(x) = e^{\pm|k|x}u_k(x)$ which correspond to unbound solutions. The quantity $k$ is called the *Bloch wave number*, and $\hbar k$ is called the *crystal momentum*. The Bloch function consists of two terms: the term $e^{ikx}$ corresponds to a particle unlocalised in space, while the function $u_k(x)$ includes the periodicity of the crystal.

The value of $k$ is undetermined, and it may be written as

$$k = n\left(\frac{2\pi}{a}\right) + k', \quad \left(-\frac{\pi}{a} < k' < \frac{\pi}{a}\right) \tag{5.11}$$

where $n$ is an integer called the *band index*. The interval $(-\pi/a < k' < \pi/a)$ is called the *first Brillouin zone*. Then

$$\begin{aligned}\phi_k(x) &= u_k(x)e^{i\left(\frac{2n\pi}{a}+k'\right)x} = u_k(x)e^{i\frac{2n\pi}{a}x}e^{ik'x}\\ &= \left[u_{n\left(\frac{2\pi}{a}\right)+k'}(x)e^{i\frac{2n\pi}{a}x}\right]e^{ik'x}.\end{aligned}$$

Since the term in square brackets also has period $a$, it follows that on dropping the prime,

$$\phi \equiv \phi_{n,k}(x) = u_{n,k}(x)e^{ikx}, \quad \left(-\frac{\pi}{a} < k < \frac{\pi}{a}\right), \tag{5.12}$$

where

$$u_{n,k}(x + a) = u_{n,k}(x). \tag{5.13}$$

This result may be generalised to the full three dimensional case in the following way. Suppose that axes are taken along the crystal lines, and that the lattice repeats after the principal (minimum) lengths $a_1$, $a_2$ and $a_3$ along each axis. Then a spatial displacement of the form $\mathbf{R} = (l_1a_1, l_2a_2, l_3a_3)$, where $l_1, l_2$ and $l_3$ are integers, will cause a shift to an equivalent point in the lattice. Such a vector $\mathbf{R}$ will be called a *lattice displacement*. The periodicity of the potential is then written

$$V(\mathbf{r} + \mathbf{R}) = V(\mathbf{r}) \quad \text{for all } \mathbf{r}. \tag{5.14}$$

If $H_0$ represents the hamiltonian operator for an electron in the crystal lattice with this symmetry, then there exist solutions of the TISE in the form

$$\phi_{\mathbf{n},\mathbf{k}}(\mathbf{r}) = u_{\mathbf{n},\mathbf{k}}(\mathbf{r})e^{i\mathbf{k}\cdot\mathbf{r}}, \quad \left(-\frac{\pi}{a_i} < k_i < \frac{\pi}{a_i}\right). \tag{5.15}$$

$$\phi_{\mathbf{n},\mathbf{k}}(\mathbf{r} + \mathbf{R}) = e^{i\mathbf{k}\cdot\mathbf{R}}\phi_{\mathbf{n},\mathbf{k}}(\mathbf{r}), \quad \text{where} \tag{5.16}$$

$$u_{\mathbf{n},\mathbf{k}}(\mathbf{r} + \mathbf{R}) = u_{\mathbf{n},\mathbf{k}}(\mathbf{r}), \quad \text{and} \tag{5.17}$$

$$H_0\phi_{\mathbf{n},\mathbf{k}}(\mathbf{r}) = E_{\mathbf{n}}(\mathbf{k})\phi_{\mathbf{n},\mathbf{k}}(\mathbf{r}). \tag{5.18}$$

It will now be demonstrated how the Bloch theorem is linked to the Floquet theorem. Let $\mathbf{y}(x)$ be an $N$-component vector. Then the Floquet theorem states that the regular system $\mathbf{y}' = \mathbf{P}(x)\mathbf{y}$, where $\mathbf{P}(x)$ is an $N \times N$ matrix function with principal period $a$, has at least one non-trivial solution $\mathbf{y} = \mathbf{Y}(x)$ such that

$$\mathbf{Y}(x + a) = Q\mathbf{Y}, \quad -\infty < x < \infty$$

where $Q$ is a constant. The TISE Equation (5.2) may be re-written in the following way. Define $y_1 \equiv \phi$ and $y_2 \equiv y_1'$. Then the equation becomes

$$-\frac{\hbar^2}{2m_e}y_2' + V(x)y_1 = Ey_1.$$

Hence the second order TISE is equivalent to the two first order equations

$$y_1' = y_2, \quad y_2' = \frac{2m_e}{\hbar^2}(V(x) - E)y_1,$$

or $\mathbf{y}' = \mathbf{Py}$ where $\mathbf{P}$ is the matrix

$$\mathbf{P} \equiv \begin{pmatrix} 0 & 1 \\ \frac{2m_e}{\hbar^2}(V(x) - E) & 0 \end{pmatrix}.$$

The matrix $\mathbf{P}$ has period $a$ because the potential $V$ has this period, and so the conditions for the Floquet theorem with $N = 2$ are satisfied. In fact, TISE Equation (5.2) is an example of *Hill's equation* (Jordan and Smith 1979); the fact that the matrix $\mathbf{P}$ has zero trace means that the two possible values of $Q$ are reciprocal, as is the case in the Bloch theorem.

## 5.2 Heterostructures: position-dependent mass

In quantising the classical energy equation, the kinetic energy operator $K$ has been replaced by $p^2/(2m_e)$ where $m_e$ is the effective mass. To ease the notation throughout this section, we write $\mu \equiv 1/m_e$: this should not be confused with the mobility $\mu$ introduced in later chapters, or the chemical potential introduced in earlier chapters. Since the momentum operator $\mathbf{p}$ is hermitian and $\mu$ is real, it follows from Equation (2.86) that the adjoint operator is $K^\times = \frac{1}{2}p^2\mu$, and this is equal to $K = \frac{1}{2}\mu p^2$ only if $[p^2, \mu] = 0$. But it is easily shown that $[p, \mu] = -i\hbar(\nabla\mu)$, so that $p$ and $\mu$ comute only if $\mu$ is independent of position. This is not the case in semiconductor heterostructures, and hence the form of $K$ introduced above is not generally hermitian. Since operators which represent observables will be taken to be hermitian, an hermitian form of $K$ must be found which reduces to the one above in the case of constant effective mass, or constant $\mu$.

In order to obtain a suitable hermitian form of $K$, define the operator

$$K_{ab} \equiv \frac{1}{2}\mu^a p\mu^{1-a-b}p\mu^b \tag{5.19}$$

for any real constants $a$ and $b$. This operator has the property that $K_{ab}^\times = K_{ba}$, and it becomes the usual kinetic energy operator when $\mu$ is independent of position. This operator is not necessarily hermitian, but the operator

$$K(a, b) \equiv \frac{1}{2}(K_{ab} + K_{ba}) = \frac{1}{4}(\mu^a p\mu^{1-a-b}p\mu^b + \mu^b p\mu^{1-b-a}p\mu^a) \tag{5.20}$$

is hermitian (see Problem 5.1). This is the form that will be taken for the kinetic energy operator. In fact, the more general operator

$$K \equiv \sum_a \sum_b A(a,b) K(a,b), \qquad A(a,b) \text{ real}, \qquad \sum_a \sum_b A(a,b) = 1 \quad (5.21)$$

will also be an hermitian operator which reduces to the usual form of the kinetic energy operator.

The time-dependent Schrödinger equation and its complex conjugate are

$$i\hbar \frac{\partial \Psi}{\partial t} = K\Psi + V\Psi$$

$$-i\hbar \frac{\partial \overline{\Psi}}{\partial t} = K\overline{\Psi} + V\overline{\Psi}$$

since the potential $V$ is real. Hence it follows that (see Problem 5.2)

$$i\hbar \frac{\partial}{\partial t} |\Psi|^2 = i\hbar \left( \Psi \frac{\partial \overline{\Psi}}{\partial t} + \overline{\Psi} \frac{\partial \Psi}{\partial t} \right) = \overline{\Psi} K\Psi - \Psi K\overline{\Psi}. \qquad (5.22)$$

To further ease the notation, restrict the calculation to the one dimensional case: the three-dimensional case will follow as a straightforward generalisation. Since the momentum operator $p$ is the derivative operator $p = -i\hbar\partial_x$, it follows that $(p\mu^c) = c\mu^{c-1}(p\mu)$ and $(p(fg)) = f(pg) + ((pf)g)$. Using Equation (5.19), we have

$$2\overline{\Psi} K_{ab}\Psi = \overline{\Psi}\mu^a p[\mu^{1-a-b} p(\mu^b \Psi)]$$
$$= \overline{\Psi}\mu^a p[\mu^{1-a-b}(\mu^b p\Psi + \Psi b\mu^{b-1} p\mu)]$$
$$= \overline{\Psi}\mu^a p[\mu^{1-a}(p\Psi) + \Psi b\mu^{-a}(p\mu)]$$
$$= \overline{\Psi}[(1-a+b)(p\mu)(p\Psi) + \mu(p^2\Psi) - ab\mu^{-1}\Psi(p\mu)^2 + b\Psi(p^2\mu)]$$
$$= \overline{\Psi}[(1-a+b)(p\mu)(p\Psi) + \mu(p^2\Psi)] + |\Psi|^2[b(p^2\mu) - ab\mu^{-1}(p\mu)^2],$$
$$2\overline{\Psi} K_{ba}\Psi = \overline{\Psi}[(1-b+a)(p\mu)(p\Psi) + \mu(p^2\Psi)] + |\Psi|^2[a(p^2\mu) - ba\mu^{-1}(p\mu)^2],$$
$$2\Psi K_{ab}\overline{\Psi} = \Psi[(1-a+b)(p\mu)(p\overline{\Psi}) + \mu(p^2\overline{\Psi})] + |\Psi|^2[b(p^2\mu) - ab\mu^{-1}(p\mu)^2],$$
$$2\Psi K_{ba}\overline{\Psi} = \Psi[(1-b+a)(p\mu)(p\overline{\Psi}) + \mu(p^2\overline{\Psi})] + |\Psi|^2[a(p^2\mu) - ba\mu^{-1}(p\mu)^2].$$

It follows from Equation (5.20) that

$$4(\overline{\Psi} K\Psi - \Psi K\overline{\Psi})$$
$$= 2\overline{\Psi} K_{ab}\Psi + 2\overline{\Psi} K_{ba}\Psi - 2\Psi K_{ab}\overline{\Psi} - 2\Psi K_{ba}\overline{\Psi}$$
$$= 2\overline{\Psi}(p\mu)(p\Psi) + 2\overline{\Psi}\mu(p^2\Psi) - 2\Psi(p\mu)(p\overline{\Psi}) - 2\Psi\mu(p^2\overline{\Psi})$$
$$= 2p[\overline{\Psi}(\mu p\Psi) - \Psi(\mu p\overline{\Psi})]$$
$$= -2\hbar^2 \partial_x[\overline{\Psi}(\mu\partial_x\Psi) - \Psi(\mu\partial_x\overline{\Psi})].$$

Substitution of this result into Equation (5.22) gives

$$\frac{\partial}{\partial t}|\Psi|^2 = \frac{1}{i\hbar}(\overline{\Psi} K\Psi - \Psi K\overline{\Psi})$$

$$= -\frac{\hbar}{2i}\partial_x[\overline{\Psi}(\mu\partial_x\Psi) - \Psi(\mu\partial_x\overline{\Psi})].$$

On reinstating the full three dimensional scheme, and remembering that $\mu = 1/m_e$, this result gives

$$\frac{\partial}{\partial t}|\Psi|^2 + \nabla \cdot \mathbf{s} = 0, \qquad (5.23)$$

$$\mathbf{s} \equiv \frac{\hbar}{2im_e}(\overline{\Psi}\nabla\Psi - \Psi\nabla\overline{\Psi}), \qquad (5.24)$$

where $\mathbf{s}$ is the probability current density which has already been defined in Equation (2.21). Note the remarkable fact that results (5.23) and (5.24) are independent of the values of $a$ and $b$. These results will therefore apply when the more general form of $K$ defined in Equation (5.21) is used.

There is disagreement in the literature concerning which form of $K$ to take. Gora and Williams (1969) and Bastard (1981) use

$$K = \frac{1}{2}(K_{10} + K_{01}) = \frac{1}{4}\left(p^2\frac{1}{m_e} + \frac{1}{m_e}p^2\right),$$

while Zhu and Kroemer (1983) use

$$K = \frac{1}{2}(K_{\frac{1}{2}\frac{1}{2}} + K_{\frac{1}{2}\frac{1}{2}}) = \frac{1}{2}\sqrt{\frac{1}{m_e}}p^2\sqrt{\frac{1}{m_e}}.$$

Neither author gives justification for their choice, apart from the hermitian nature of the operator. Morrow and Brownstein (1984) use

$$K = \frac{1}{2}(K_{00} + K_{00}) = \frac{1}{2}p\frac{1}{m_e}p \qquad (5.25)$$

and demonstrate that this is the correct form to use when there is a discontinuity in the value of $m_e$ at material boundaries in heterojunctions. This form is also used by Davies (1998), Ferry and Goodnick (1997) and von Roos (1983), and will be the form adopted in later chapters.

Whichever form is taken, it can be seen from Equation (5.24) that both $\Psi$ and $(\nabla\Psi)/m_e$ must be continuous at the interfaces.

## 5.3 Heterostructures: the effective mass approximation

Let $\mathbf{R} = (l_1a_1, l_2a_2, l_3a_3)$ be a lattice displacement, where $l_1$, $l_2$ and $l_3$ are integers. Suppose that the hamiltonian for an electron in a perfect crystal is $H_0 = K + V_0$

where the potential $V_0$ has the periodicity of the lattice:

$$V_0(\mathbf{r} + \mathbf{R}) = V_0(\mathbf{r}) \quad \text{for all } \mathbf{r}. \tag{5.26}$$

Bloch's theorem will apply to the solutions of this unperturbed system.

Suppose that a perturbation, for example, a quantum well, is added to the pure crystal structure. If this perturbation supplies a potential $V_p(\mathbf{r})$ to the hamiltonian, then the TISE is

$$H\Psi(\mathbf{r}) = E\Psi(\mathbf{r}) \tag{5.27}$$

where the total hamiltonian is

$$H = H_0 + V_p(\mathbf{r}). \tag{5.28}$$

The complexity of the full system is such that it is impossible to solve the full Equation (5.27), since $H_0$ involves the potential of the periodic lattice. In the *effective mass approximation* (EMA), Equation (5.27) is replaced with a simpler equation for an envelope function which is assumed to be slowly varying over the span of the crystal unit cell. We first need the result of *Wannier's theorem*.

**Theorem 5.2.** *Wannier's theorem. Let the function $f(\mathbf{k})$ be periodic in $\mathbf{k}$, with*

$$f(\mathbf{k}) = \sum_{\mathbf{R}} f_{\mathbf{R}} e^{i\mathbf{k}\cdot\mathbf{R}}$$

*where the summation is taken over all lattice points $\mathbf{R}$ in a given region. Then*

$$f(-i\nabla)\phi_{\mathbf{n},\mathbf{k}}(\mathbf{r}) = f(\mathbf{k})\phi_{\mathbf{n},\mathbf{k}}(\mathbf{r}). \tag{5.29}$$

*Proof.* Let $v(\mathbf{r})$ be a differentiable function of position space. Then for any vector $\mathbf{s}$, it follows that

$$e^{\mathbf{s}\cdot\nabla}v(\mathbf{r}) = \left(1 + \mathbf{s}\cdot\nabla + \frac{1}{2}(\mathbf{s}\cdot\nabla)^2 + \cdots\right)v(\mathbf{r})$$

$$= v(\mathbf{r}) + \mathbf{s}\cdots\nabla v(\mathbf{r}) + \frac{1}{2}\sum_{i,j=1}^{3} s_i s_j \partial_{x_i}\partial_{x_j} v(\mathbf{r}) + \cdots$$

$$= v(\mathbf{r} + \mathbf{s}).$$

Hence

$$e^{\mathbf{s}\cdot\nabla}v(\mathbf{r}) = v(\mathbf{r} + \mathbf{s}). \tag{5.30}$$

Applying result (5.30) to the Bloch function and using Equation (5.16), it follows that

$$f(-i\nabla)\phi_{\mathbf{n},\mathbf{k}}(\mathbf{r}) = \left(\sum_{\mathbf{R}} f_{\mathbf{R}} e^{\mathbf{R}\cdot\nabla}\right)\phi_{\mathbf{n},\mathbf{k}}(\mathbf{r})$$

$$= \sum_{\mathbf{R}} f_{\mathbf{R}} \phi_{n,k}(\mathbf{r} + \mathbf{R})$$

$$= \sum_{\mathbf{R}} f_{\mathbf{R}} e^{i\mathbf{k}.\mathbf{R}} \phi_{n,k}(\mathbf{r})$$

$$= f(\mathbf{k}) \phi_{n,k}(\mathbf{r}).$$

This completes the proof. $\square$

The set of Bloch functions forms a complete set, and hence the solution $\Psi$ of Equation (5.27) must be a linear superposition of these functions:

$$\Psi(\mathbf{r}) = \sum_{n} \sum_{k} g_{n,k} \phi_{n,k}(\mathbf{r}). \tag{5.31}$$

Then

$$E\Psi = H\Psi = (H_0 + V_p) \sum_{n} \sum_{k} g_{n,k} \phi_{n,k}(\mathbf{r})$$

$$= \sum_{n} \sum_{k} g_{n,k} \left( H_0 \phi_{n,k}(\mathbf{r}) + V_p \phi_{n,k}(\mathbf{r}) \right)$$

$$= \sum_{n} \sum_{k} g_{n,k} \left( E_n(\mathbf{k}) \phi_{n,k}(\mathbf{r}) + V_p \phi_{n,k}(\mathbf{r}) \right)$$

$$= \sum_{n} \sum_{k} g_{n,k} \left( E_n(-i\nabla) \phi_{n,k}(\mathbf{r}) + V_p \phi_{n,k}(\mathbf{r}) \right)$$

$$= \sum_{n} \left( E_n(-i\nabla) + V_p \right) \sum_{k} g_{n,k} \phi_{n,k}(\mathbf{r}), \tag{5.32}$$

where the result of Wannier's theorem (5.29) has been used.

In order to proceed further, two simplifying assumptions must be made.

1. The summation over the index $\mathbf{n}$ in Equation (5.32) is taken over all bands. It will now be assumed that only one band makes a significant contribution to this summation. For example, if a donor atom has energy levels in the forbidden gap close to the bottom of the conduction band, then the only significant contribution to the sum will come from the bottom of the conduction band. In this case, Equation (5.32) will replaced with

$$\left( E_n(-i\nabla) + V_p \right) \sum_{k} g_{n,k} \phi_{n,k}(\mathbf{r}) = E\Psi. \tag{5.33}$$

2. It can be assumed that the constant energy surfaces in $\mathbf{k}$ space are spheres (that is, the three effective masses which contribute to the ellipsoidal energy surfaces are equal). This means that the energies can be written

$$E_n(\mathbf{k}) = E_{n0} + \sum_{j=1}^{3} \frac{\hbar^2 k_j^2}{2m_e}. \tag{5.34}$$

Remembering to maintain the hermitian character of the kinetic energy for an effective mass which depends on position, then

$$E_n(-i\nabla) = E_{n0} - \frac{\hbar^2}{2}\nabla \cdot \frac{1}{m_e}\nabla. \tag{5.35}$$

Hence Equation (5.33) can be written

$$-\frac{\hbar^2}{2}\nabla \cdot \left(\frac{1}{m_e}\nabla\Psi\right) + V_p\Psi = (E - E_{n0})\Psi. \tag{5.36}$$

In this way, Equation (5.27) has been replaced by a simpler equation in which the periodic potential has been absorbed into an effective mass $m_e$.

## 5.4 Quantum wells

When two materials such as GaAs and AlGaAs are brought together to form an abrupt interface, a quantum well forms at the interface. This comes about because there is a mismatch in the $E_c$ levels of the two materials: this mismatch will be quantified fully in Sect. 6.7. The discrete energy levels formed in these wells will contribute to the density of states at these positions, and must be combined in a consistent manner with the continuum density of states which was derived in Sect. 4.1. In particular, care must be taken to ensure that states inside the wells are not double counted.

### 5.4.1 The general structure of a quantum well

The potential applied to the Schrödinger equation will take the form

$$V(\mathbf{r}) = V_{xc} + E_h - q\psi(\mathbf{r}) = V_{xc} + E_c(\mathbf{r})$$

where $E_h$ is the band offset factor relating to the abrupt band mismatch, and $\psi(\mathbf{r})$ is the electrostatic potential which is the solution of the Poisson equation. The quantity $V_{xc}$ is the exchange correlation potential which will be described in Sect. 6.8. Fig. 5.1 shows the situation which arises in the simple case when there is only one interface, with the normal to this interface lying along the $y$ direction. The interface lies at the point $y = Y$. The figure shows the plot of $E_c$ superimposed on the function $(-\psi)$: it can be seen how a quantum well is formed which gives rise to discrete energy eigenvalues $\lambda_0, \lambda_1, \ldots$ in the well. In what follows, it will be assumed that the eigensolutions are ordered with $\lambda_0 < \lambda_1 < \lambda_2 < \ldots$. The first $L$ of eigenvalues $\lambda_0, \ldots, \lambda_{L-1}$ are superimposed on the plot.

**Fig. 5.1** The quantum well formed at the abrupt interface $y = Y$ of two materials, showing the discontinuity in the conduction band $E_c$, the negative of the electrostatic potential $\psi(y)$, and the first $L$ energy eigenvalues $\lambda_0, \dots, \lambda_{L-1}$. The term *inside* the well relates to the section $Y < y < Y_L$.

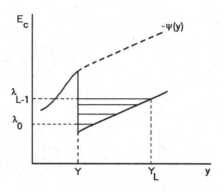

## 5.4.2 Density of states in quantum wells

In order to obtain a working model of the quantum well structure, it is necessary to choose a maximum number of eigensolutions with which to work. One of the most time consuming parts of a full numerical device simulation arises in the solution of the Schrödinger equation. Hence the number of eigensolutions which must be generated in this process should be kept to a minimum, subject to the accuracy of the solution. One method of limiting this number is to stop generating these eigensolutions when the difference between consecutive energy eigenvalues falls below the thermal energy $k_B T/q$ (Drury and Snowden 1995). However, this method could possibly give a false sense of security in the situation in which a double well produces pairs of very close eigenvalues. It has been shown (Cole et al. 1997) that there is very little to be gained in going beyond a maximum of $L = 10$ eigensolutions, and that there is relatively little contribution even after $L = 3$ eigensolutions.

Having decided on a maximum number $L$ of eigensolutions, we must now make sure that states inside the well are not counted twice when producing expressions for the electron density $n$ and the energy density $W$. We must first decide what we mean by *inside* and outside the well. Referring to Fig. 5.1, let $Y_L$ be that value of $y$ at which $E_c = \lambda_{L-1}$. Then the term *inside* the well will relate to those values of $y$ for which $Y < y < Y_L$, and *outside* the well relates to all other values of $y$ (in the case of a single well). Then outside the well, the integral expressions for $n$ and $W$ derived in Chapter 4 will be valid. Inside the well, however, the lower limit of the integrals must be $\lambda_{L-1}$ rather than $E_c$. The contribution to the densities calculated using this new lower limit must then be added in a consistent manner to the contributions from the first $L$ eigensolutions $\lambda_0, \dots, \lambda_{L-1}$.

Let $\xi_j$ be the normalised energy eigenfunction corresponding to the eigenvalue $\lambda_j$ for $j = 0, \dots, L - 1$. Let $\phi(E)$ be any function of the energy $E$. Outside the well, the particles will be free to move in three dimensions, and the number of states per unit volume will be given by $A_{1/2}\sqrt{E}$. Inside the well, the discrete states will produce the number of states per unit volume as $A_r E^r$, where the value of $r$ will depend on the number of dimensions in which they are free to move. Then

*Outside the well:*

$$\langle \phi \rangle = A_{\frac{1}{2}} \int_{E_c}^{\infty} \frac{\phi(E - E_c)(E - Ec)^{\frac{1}{2}}}{1 + e^{\frac{1}{k_B T}(E - E_F)}} dE.$$

*Inside the well:*

$$\langle \phi \rangle = A_{\frac{1}{2}} \int_{\lambda_L}^{\infty} \frac{\phi(E - E_c)(E - Ec)^{\frac{1}{2}}}{1 + e^{\frac{1}{k_B T}(E - E_F)}} dE$$

$$+ \sum_{j=0}^{L-1} |\xi_j|^2 A_r \int_{\lambda_j}^{\infty} \frac{\phi(E - E_c)(E - Ec)^r}{1 + e^{\frac{1}{k_B T}(E - E_F)}} dE.$$

In particular, if $\phi(E - E_c) = (E - E_c)^p$ for some real constant $p$, then

*Outside the well:*

$$\langle (E - E_c)^p \rangle = A_{\frac{1}{2}} \int_{E_c}^{\infty} \frac{(E - Ec)^{p+\frac{1}{2}}}{1 + e^{\frac{1}{k_B T}(E - E_F)}} dE. \tag{5.37}$$

*Inside the well:*

$$\langle (E - E_c)^p \rangle = A_{\frac{1}{2}} \int_{\lambda_L}^{\infty} \frac{(E - Ec)^{p+\frac{1}{2}}}{1 + e^{\frac{1}{k_B T}(E - E_F)}} dE$$

$$+ \sum_{j=0}^{L-1} |\xi_j|^2 A_r \int_{\lambda_j}^{\infty} \frac{(E - Ec)^{p+r}}{1 + e^{\frac{1}{k_B T}(E - E_F)}} dE. \tag{5.38}$$

The above integrals can be unified in a consistent manner by defining the integrals

$$I_r(a, b) \equiv \frac{1}{\Gamma(r + 1)} \int_b^{\infty} \frac{x^r}{1 + e^{x - (a+b)}} dx \tag{5.39}$$

$$= \frac{1}{\Gamma(r + 1)} \int_0^{\infty} \frac{(x + b)^r}{1 + e^{x - a}} dx. \tag{5.40}$$

See Problem 5.3 for a list of the properties of this function. Using the substitution $y = E/(k_B T)$, it follows that

$$\int_{\lambda}^{\infty} \frac{(E - Ec)^q}{1 + e^{\frac{1}{k_B T}(E - E_F)}} dE = (k_B T)^{q+1} \int_{\frac{\lambda}{k_B T}}^{\infty} \frac{(y - \frac{E_c}{k_B T})^q}{1 + e^{y - \frac{1}{k_B T} E_F}}$$

$$= (k_B T)^{q+1} \Gamma(q+1)$$
$$\times I_q \left( \frac{1}{k_B T}(E_F - \lambda), \frac{1}{k_B T}(\lambda - E_c) \right). \quad (5.41)$$

It follows that

*Outside the well:*

$$\langle (E - E_c)^p \rangle = A_{\frac{1}{2}} \Gamma \left( p + \frac{3}{2} \right)(k_B T)^{p+\frac{3}{2}} F_{p+\frac{1}{2}} \left( \frac{1}{k_B T}(E_F - E_c) \right) \quad (5.42)$$

*Inside the well:*

$$\langle (E - E_c)^p \rangle = A_{\frac{1}{2}} \Gamma \left( p + \frac{3}{2} \right)(k_B T)^{p+\frac{3}{2}} I_{p+\frac{1}{2}}$$
$$\times \left( \frac{1}{k_B T}(E_F - \lambda_L), \frac{1}{k_B T}(\lambda_L - E_c) \right)$$
$$+ A_r \sum_{j=0}^{L-1} |\xi_j|^2 \Gamma(r + p + 1)(k_B T)^{r+p+1}$$
$$\times I_{r+p} \left( \frac{1}{k_B T}(E_F - \lambda_j), \frac{1}{k_B T}(\lambda_j - E_c) \right). \quad (5.43)$$

In particular, taking $p = 0$ in order to calculate the electron density $n$, and $p = 1$ in order to calculate the electron energy density $W$, it follows that

*Outside the well:*

$$n = \langle 1 \rangle = A_{\frac{1}{2}} \Gamma \left( \frac{3}{2} \right)(k_B T)^{\frac{3}{2}} F_{\frac{1}{2}} \left( \frac{1}{k_B T}(E_F - E_c) \right), \quad (5.44)$$
$$W = \langle E - E_c \rangle = A_{\frac{1}{2}} \Gamma \left( \frac{5}{2} \right)(k_B T)^{\frac{5}{2}} F_{\frac{3}{2}} \left( \frac{1}{k_B T}(E_F - E_c) \right). \quad (5.45)$$

*Inside the well:*

$$n = \langle 1 \rangle = A_{\frac{1}{2}} \Gamma \left( \frac{3}{2} \right)(k_B T)^{\frac{3}{2}} I_{\frac{1}{2}} \left( \frac{1}{k_B T}(E_F - \lambda_L), \frac{1}{k_B T}(\lambda_L - E_c) \right)$$
$$+ A_r \sum_{j=0}^{L-1} |\xi_j|^2 \Gamma(r + 1)(k_B T)^{r+1}$$
$$\times I_r \left( \frac{1}{k_B T}(E_F - \lambda_j), \frac{1}{k_B T}(\lambda_j - E_c) \right), \quad (5.46)$$

$$W = \langle E - E_c \rangle = A_{\frac{1}{2}} \Gamma\left(\frac{5}{2}\right)(k_B T)^{\frac{5}{2}} I_{\frac{3}{2}}\left(\frac{1}{k_B T}(E_F - \lambda_L), \frac{1}{k_B T}(\lambda_L - E_c)\right)$$

$$+ A_r \sum_{j=0}^{L-1} |\xi_j|^2 \Gamma(r+2)(k_B T)^{r+2}$$

$$\times I_{r+1}\left(\frac{1}{k_B T}(E_F - \lambda_j), \frac{1}{k_B T}(\lambda_j - E_c)\right). \tag{5.47}$$

The value of $r$ will depend on whether we take the particles in the well to be free in two or three spatial dimensions.

### 5.4.3 Quantum wells—particles free in two dimensions

If the particles are free to move in two spatial dimensions, then it has been shown in Sect. 4.1.3 that

$$r = 0, \qquad A_0 = \frac{4\pi m_e}{h^2}.$$

It follows that

*Outside the well:*

$$n = 2\left(\frac{2\pi m_e k_B}{h^2}\right)^{\frac{3}{2}} T^{\frac{3}{2}} F_{\frac{1}{2}}\left(\frac{1}{k_B T}(E_F - E_c)\right), \tag{5.48}$$

$$W = 3\left(\frac{2\pi m_e k_B}{h^2}\right)^{\frac{3}{2}} k T^{\frac{5}{2}} F_{\frac{3}{2}}\left(\frac{1}{k_B T}(E_F - E_c)\right). \tag{5.49}$$

*Inside the well:*

$$n = 2\left(\frac{2\pi m_e k_B}{h^2}\right)^{\frac{3}{2}} T^{\frac{3}{2}} I_{\frac{1}{2}}\left(\frac{1}{k_B T}(E_F - \lambda_L), \frac{1}{k_B T}(\lambda_L - E_c)\right)$$

$$+ 2\left(\frac{2\pi m_e k_B}{h^2}\right) T \sum_{j=0}^{L-1} |\xi_j|^2 I_0\left(\frac{1}{k_B T}(E_F - \lambda_j), \frac{1}{k_B T}(\lambda_j - E_c)\right)$$

$$= 2\left(\frac{2\pi m_e k_B}{h^2}\right)^{\frac{3}{2}} T^{\frac{3}{2}} I_{\frac{1}{2}}\left(\frac{1}{k_B T}(E_F - \lambda_L), \frac{1}{k_B T}(\lambda_L - E_c)\right)$$

$$+ 2\left(\frac{2\pi m_e k_B}{h^2}\right) T \sum_{j=0}^{L-1} |\xi_j|^2 \ln\left(1 + e^{\frac{1}{k_B T}(E_F - \lambda_j)}\right), \tag{5.50}$$

$$W = 3\left(\frac{2\pi m_e k_B}{h^2}\right)^{\frac{3}{2}} k_B T^{\frac{5}{2}} I_{\frac{3}{2}} \left(\frac{1}{k_B T}(E_F - \lambda_L), \frac{1}{k_B T}(\lambda_L - E_c)\right)$$

$$+ 2\left(\frac{2\pi m_e k_B}{h^2}\right) k_B T^2 \sum_{j=0}^{L-1} |\xi_j|^2 I_1 \left(\frac{1}{k_B T}(E_F - \lambda_j), \frac{1}{k_B T}(\lambda_j - E_c)\right)$$

$$= 3\left(\frac{2\pi m_e k_B}{h^2}\right)^{\frac{3}{2}} k_B T^{\frac{5}{2}} I_{\frac{3}{2}} \left(\frac{1}{k_B T}(E_F - \lambda_L), \frac{1}{k_B T}(\lambda_L - E_c)\right)$$

$$+ 2\left(\frac{2\pi m_e k_B}{h^2}\right) k_B T^2 \sum_{j=0}^{L-1} |\xi_j|^2 \left[F_1\left(\frac{1}{k_B T}(E_F - \lambda_j)\right)\right.$$

$$\left. + \left(\frac{1}{k_B T}(\lambda_j - E_c)\right) \ln\left(1 + e^{\frac{1}{k_B T}(E_F - \lambda_j)}\right)\right]. \tag{5.51}$$

### 5.4.4 Quantum wells—particles free in one dimension

If the particles are free to move in one spatial dimension, then it has been shown in Sect. 4.1.4 that

$$r = -\frac{1}{2}, \qquad A_{-\frac{1}{2}} = \frac{2\sqrt{2m_e}}{h}.$$

It follows that

*Outside the well:*

$$n = 2\left(\frac{2\pi m_e k_B}{h^2}\right)^{\frac{3}{2}} T^{\frac{3}{2}} F_{\frac{1}{2}} \left(\frac{1}{k_B T}(E_F - E_c)\right), \tag{5.52}$$

$$W = 3\left(\frac{2\pi m_e k_B}{h^2}\right)^{\frac{3}{2}} k T^{\frac{5}{2}} F_{\frac{3}{2}} \left(\frac{1}{k_B T}(E_F - E_c)\right). \tag{5.53}$$

*Inside the well:*

$$n = 2\left(\frac{2\pi m_e k_B}{h^2}\right)^{\frac{3}{2}} T^{\frac{3}{2}} I_{\frac{1}{2}} \left(\frac{1}{k_B T}(E_F - \lambda_L), \frac{1}{k_B T}(\lambda_L - E_c)\right)$$

$$+ 2\left(\frac{2\pi m_e k_B}{h^2}\right)^{\frac{1}{2}} T^{\frac{1}{2}}$$

$$\times \sum_{j=0}^{L-1} |\xi_j|^2 I_{-\frac{1}{2}} \left(\frac{1}{k_B T}(E_F - \lambda_j), \frac{1}{k_B T}(\lambda_j - E_c)\right), \tag{5.54}$$

$$W = 3 \left( \frac{2\pi m_e k_B}{h^2} \right)^{\frac{3}{2}} k_B T^{\frac{5}{2}} I_{\frac{3}{2}} \left( \frac{1}{k_B T}(E_F - \lambda_L), \frac{1}{k_B T}(\lambda_L - E_c) \right)$$

$$+ \left( \frac{2\pi m_e k_B}{h^2} \right)^{\frac{1}{2}} k_B T^{\frac{3}{2}}$$

$$\times \sum_{j=0}^{L-1} |\xi_j|^2 I_{\frac{1}{2}} \left( \frac{1}{k_B T}(E_F - \lambda_j), \frac{1}{k_B T}(\lambda_j - E_c) \right). \tag{5.55}$$

In order to be of practical use in device modelling, the integrals $I_r(a, b)$ will need to be evaluated numerically, and as rapidly as possible. It will be shown in Chapter 14 how suitable polynomial approximations can be made to these integrals: these approximations will provide a simple method of evaluating the integrals, while at the same time reducing approximation errors to a minimum.

## Problems

**Problem 5.1.** Let $a$ and $b$ be real constants, let $\mu \equiv 1/m_e$, and let $p$ be the one-dimensional momentum operator. Show that the operator

$$K(a, b) \equiv \frac{1}{4}(\mu^a p \mu^{1-a-b} p \mu^b + \mu^b p \mu^{1-b-a} p \mu^a)$$

is hermitian.

Verify that the operator

$$K \equiv \sum_a \sum_b A(a, b) K(a, b), \qquad A(a, b) \text{ real}$$

reduces to the usual form of the kinetic operator in the case of constant mass provided that

$$\sum_a \sum_b A(a, b) = 1.$$

**Problem 5.2.** Writing the time-dependent Schrödinger equation as

$$i\hbar \frac{\partial \Psi}{\partial t} = K\Psi + V\Psi,$$

show that

$$i\hbar \frac{\partial}{\partial t} |\Psi|^2.$$

**Problem 5.3.** The associated Fermi integral is defined as

$$I_r(a, b) \equiv \frac{1}{\Gamma(r+1)} \int_b^\infty \frac{x^r}{1 + e^{x-(a+b)}} dx.$$

Show that it has the properties

$$I_r(a, 0) = F_r(a),$$

$$I_0(a, b) = F_0(a) = \ln(1 + e^a),$$

$$I_1(a, b) = F_1(a) + b \ln(1 + e^a),$$

$$I_n(a, b) = F_n(a) + b \sum_{r=1}^{n} \frac{1}{r!} b^{r-1} F_{n-r}(a), \quad n \text{ integer } \geq 1,$$

$$\partial_b I_r(a, b) = I_{r-1}(a, b),$$

$$\partial_a I_r(a, b) = \frac{1}{\Gamma(r+1)} \frac{b^r}{1 + e^{-a}} + I_{r-1}(a, b).$$

# Chapter 6
# The transport equations and the device equations

The process of device modelling begins by obtaining a set of working equations which are distilled from the underlying physical principles. This set consists of two main subsets:

- a subset consisting of equations—most usually differential equations—for the main physical variables which have been chosen to describe the system;
- a subset of subsidiary equations linking the equations of the first subset. These equations will normally provide expressions for physical data based on material properties, and will also provide linking equations for any intermediate variables used in the main set of differential equations.

The modelling equations of many current semiconductor devices are based on the classical Boltzmann transport equation (BTE), with several quantum effects bolted on to provide a *semiclassical* approach. The equations derived in the first part of this Chapter will be based on the BTE, which is a classical integro-differential equation for a distribution function $f(\mathbf{r}, \mathbf{p}, t)$, for which the carrier's position $\mathbf{r}$ and momentum $\mathbf{p}$ are specified at time $t$. There are several ways in which the BTE can be used when modelling devices. One main method is to use a Monte Carlo approach to finding the distribution function when all of the carrier scattering processes have been specified (Kelsall 1998; Lugli 1993; Mietzner et al. 2001). The main physical quantities can then be calculated once this distribution function has been found. Generally, this method is computationally expensive. A second approach, and one which will be adopted here, is to obtain a set of differential equations based on the first three moments of the BTE. This method provides the time-dependent equations for the carrier concentration, momentum conservation, and energy transport.

Since the Boltzmann distribution function $f$ assumes that both the position and momentum of a particle can be specified simultaneously and precisely, this approach is not strictly applicable to the charge carriers, which are quantum particles. The use of the Wigner distribution function, which is defined in terms of the wave function satisfied by the Schrödinger equation, presents a satisfactory method of ammending the approach based on the BTE.

E.A.B. Cole, *Mathematical and Numerical Modelling of Heterostructure Semiconductor Devices: From Theory to Programming*, DOI 10.1007/978-1-84882-937-4_6, © Springer-Verlag London Limited 2009

## 6.1 The Boltzmann transport equation

The following derivation of the BTE for a distribution function $f$ is based on the assumption that carrier collision times are a very small fraction of their lifetimes. This implies that only binary collisions will be taken into account. The BTE is an integro-differential equation for the distribution function $f$, and there is no general solution of this equation. However, it can be transformed into a useful differential equation using the *relaxation time approximation* when certain simplifying assumptions are made regarding carrier collisions. These assumptions will depend on the particular applications to which the equation is put (Chapman and Cowling 1970; Landsberg 1969).

### 6.1.1 Derivation of the BTE

Suppose that a classical particle with constant effective mass $m_e$ has position $\mathbf{r}$, velocity $\mathbf{v}$, and momentum $\mathbf{p}$ given by

$$\mathbf{p} = m_e\mathbf{v} = \hbar\mathbf{k} \tag{6.1}$$

at time $t$. If it is acted on by an external force $\mathbf{F}$, then

$$\frac{d\mathbf{r}}{dt} = \mathbf{v}(\mathbf{k}) \tag{6.2}$$

$$\frac{d\mathbf{p}}{dt} = \hbar\frac{d\mathbf{k}}{dt} = \mathbf{F} = -q\mathbf{E} \tag{6.3}$$

where $\mathbf{E}$ is the electric field and $q$ is the magnitude of the electron charge. Note that Equation (6.2) does not constitute a definition of $\mathbf{v}$ as $d\mathbf{r}/dt$, since $\mathbf{v}$ is the group velocity of the carrier, and depends only on $\mathbf{k}$: $\mathbf{v} \equiv \mathbf{v}(\mathbf{k})$. Let

$$f(\mathbf{r}, \mathbf{p}, t)d\mathbf{r}d\mathbf{p}$$

be defined as the number of carriers at time $t$ in the region of space $d\mathbf{r}$ centred on position $\mathbf{r}$ and the region of momentum space $d\mathbf{p}$ centred on momentum $\mathbf{p}$. Then in the following short interval of time $dt$, the net gain of carriers to the region $d\mathbf{r}d\mathbf{p}$ surrounding the position $\mathbf{r} + d\mathbf{r}$ and momentum $\mathbf{p} + d\mathbf{p}$ is $df\,d\mathbf{r}\,d\mathbf{p}$ where

$$\frac{df}{dt} \approx \frac{\partial f}{\partial t} + \frac{d\mathbf{p}}{dt} \cdot \nabla_{\mathbf{p}}f + \frac{d\mathbf{r}}{dt} \cdot \nabla_{\mathbf{r}}f$$

$$= \frac{\partial f}{\partial t} + \mathbf{F} \cdot \nabla_{\mathbf{p}}f + \mathbf{v} \cdot \nabla_{\mathbf{r}}f, \tag{6.4}$$

where $\nabla_{\mathbf{r}}$ and $\nabla_{\mathbf{p}}$ are the gradient operators with respect to the position and momentum coordinates respectively. In the absence of collisions, the term $df/dt$ would be zero since particles would drift without loss into the new region, and none would drift into the new region which were not originally in the old region. However, this number will change if some of the particles are deflected by collisions, and hence we must write

$$\frac{\partial f}{\partial t} + \mathbf{F} \cdot \nabla_{\mathbf{p}} f + \mathbf{v} \cdot \nabla_{\mathbf{r}} f = \left(\frac{\partial f}{\partial t}\right)_c \tag{6.5}$$

where the right hand side $(\partial f/\partial t)_c$ of this equation simply denotes the rate at which these collisions take place. For this equation to be of any use, an expression must be found for this collision term.

The spatial density of particles at position $\mathbf{r}$ at time $t$, irrespective of momentum, is

$$n(\mathbf{r}, t) = \int f(\mathbf{r}, \mathbf{p}', t)d\mathbf{p}'. \tag{6.6}$$

Hence the proportion of carriers at position $\mathbf{r}$ at time $t$ which are not in the momentum state $\mathbf{p}$ is

$$1 - \frac{f(\mathbf{r}, \mathbf{p}, t)}{n(\mathbf{r}, t)}.$$

Let $P(\mathbf{r}', \mathbf{p}', t'; \mathbf{r}, \mathbf{p}, t)$ be the probability that a carrier is scattered from the state $(\mathbf{r}', \mathbf{p}', t')$ to the state $(\mathbf{r}, \mathbf{p}, t)$. Then the net increase in the collision term due to carriers being scattered both into and out of the state $(\mathbf{r}, \mathbf{p}, t)$ will be

$$\left(\frac{\partial f}{\partial t}\right)_c = \int \int \int f(\mathbf{r}', \mathbf{p}', t') \left(1 - \frac{f(\mathbf{r}, \mathbf{p}, t)}{n(\mathbf{r}, t)}\right) P(\mathbf{r}', \mathbf{p}', t'; \mathbf{r}, \mathbf{p}, t)d\mathbf{r}' \, d\mathbf{p}' \, dt'$$
$$- \int \int \int f(\mathbf{r}, \mathbf{p}, t) \left(1 - \frac{f(\mathbf{r}', \mathbf{p}', t')}{n(\mathbf{r}', t')}\right) P(\mathbf{r}, \mathbf{p}, t; \mathbf{r}', \mathbf{p}', t')d\mathbf{r}' \, d\mathbf{p}' \, dt'. \tag{6.7}$$

## 6.1.2 The relaxation-time approximation

When the collision term in Equation (6.7) is used on the right hand side of Equation (6.5), the integro-differential equation nature of the BTE becomes apparent. The relaxation-time approximation involves several assumptions which are made to enable the integral terms to be written in a simplified way.

It will first be assumed that collisions take place over a very short time, and that the scattering is localised. This enables us to write

$$P(\mathbf{r}', \mathbf{p}', t'; \mathbf{r}, \mathbf{p}, t) = \delta(t - t')\delta(\mathbf{r} - \mathbf{r}')\Pi(\mathbf{p}'; \mathbf{r}, \mathbf{p}, t) \tag{6.8}$$

for some function $\Pi(\mathbf{p}'; \mathbf{r}, \mathbf{p}, t)$. Then Equation (6.7) can be integrated to give

$$\left(\frac{\partial f}{\partial t}\right)_c = \int f(\mathbf{r}, \mathbf{p}', t)\left(1 - \frac{f(\mathbf{r}, \mathbf{p}, t)}{n(\mathbf{r}, t)}\right)\Pi(\mathbf{p}'; \mathbf{r}, \mathbf{p}, t), d\mathbf{p}'$$
$$- \int f(\mathbf{r}, \mathbf{p}, t)\left(1 - \frac{f(\mathbf{r}, \mathbf{p}', t)}{n(\mathbf{r}, t)}\right)\Pi(\mathbf{p}; \mathbf{r}, \mathbf{p}', t)d\mathbf{p}'. \qquad (6.9)$$

The second assumption to be made is that detailed balance occurs in the form

$$\Pi(\mathbf{p}'; \mathbf{r}, \mathbf{p}, t) = \Pi(\mathbf{p}; \mathbf{r}, \mathbf{p}', t).$$

Using this assumption, it can then be shown that Equation (6.9) becomes

$$\left(\frac{\partial f}{\partial t}\right)_c = \int \Pi(\mathbf{p}'; \mathbf{r}, \mathbf{p}, t)\{f(\mathbf{r}, \mathbf{p}', t) - f(\mathbf{r}, \mathbf{p}, t)\}d\mathbf{p}'. \qquad (6.10)$$

Finally, on defining the function $f_0(\mathbf{r}, \mathbf{p}, t)$ by

$$f_0(\mathbf{r}, \mathbf{p}, t) \equiv \frac{\int \Pi(\mathbf{p}'; \mathbf{r}, \mathbf{p}, t)f(\mathbf{r}, \mathbf{p}', t)d\mathbf{p}'}{\int \Pi(\mathbf{p}'; \mathbf{r}, \mathbf{p}, t)d\mathbf{p}'}, \qquad (6.11)$$

the collision term of Equation (6.10) becomes

$$\left(\frac{\partial f}{\partial t}\right)_c = -\frac{f - f_0}{\tau} \qquad (6.12)$$

where the *relaxation time* $\tau$ is defined as

$$\tau \equiv \tau(\mathbf{r}, \mathbf{p}, t) \equiv \frac{1}{\int \Pi(\mathbf{p}'; \mathbf{r}, \mathbf{p}, t)d\mathbf{p}'}. \qquad (6.13)$$

Hence, under these assumptions, the BTE becomes

$$\frac{\partial f}{\partial t} + \mathbf{F} \cdot \nabla_{\mathbf{p}} f + \mathbf{v} \cdot \nabla_{\mathbf{r}} f = -\frac{f - f_0}{\tau} \qquad (6.14)$$

with $\mathbf{v} \equiv \mathbf{v}(\mathbf{p})$, $\mathbf{F} \equiv \mathbf{F}(\mathbf{r}, t)$ and $f \equiv f(\mathbf{r}, \mathbf{p}, t)$.

## 6.2 The moments of the BTE

The BTE Equation (6.5) can be transformed into a number of useful device equations by taking *moments* of the equation. These moments are found by multiplying the equation by appropriate functions of $\mathbf{p}$ and integrating over all $\mathbf{p}$-space. The first three moments yield the equations for the carrier concentration, momentum conservation, and energy transport.

## 6.2.1 The general moment

Let $A(\mathbf{p})$ be any function of $\mathbf{p}$, which could be either a scalar or a vector quantity. We will first consider the case in which $A$ is a scalar quantity. Multiply Equation (6.5) by $A(\mathbf{p})$ and integrate over all $\mathbf{p}$-space:

$$\int \frac{\partial f}{\partial t} A(\mathbf{p}) d\mathbf{p} + \mathbf{F} \cdot \int A(\mathbf{p}) \nabla_{\mathbf{p}} f d\mathbf{p} + \int A(\mathbf{p}) \mathbf{v} \cdot \nabla_{\mathbf{r}} f d\mathbf{p} = \left( \frac{\partial (\int A(\mathbf{p}) f d\mathbf{p})}{\partial t} \right)_c .$$
(6.15)

Now, for any function $B(\mathbf{p})$ of $\mathbf{p}$, define its average over $\mathbf{p}$-space as

$$\langle B \rangle \equiv \frac{\int B(\mathbf{p}) f d\mathbf{p}}{\int f d\mathbf{p}} = \frac{1}{n} \int B(\mathbf{p}) f \, d\mathbf{p}.$$

Hence

$$\int B(\mathbf{p}) f \, d\mathbf{p} = n \langle B \rangle .$$
(6.16)

Before proceeding, it is important to remind ourselves about the functional dependencies of the various quantities which have been introduced. The distribution function $f$ is a function of position $\mathbf{r}$, momentum $\mathbf{p}$, and time $t$: $f \equiv f(\mathbf{r}, \mathbf{p}, t)$. The average of the function $B(\mathbf{p})$ is a function of $\mathbf{r}$ and $t$: $\langle B \rangle \equiv \langle B \rangle(\mathbf{r}, t)$. The carrier velocity is a function of $\mathbf{p}$: $\mathbf{v} \equiv \mathbf{v}(\mathbf{p})$, but the average velocity $\langle \mathbf{v} \rangle$ is a function of $\mathbf{r}$ and $t$: $\langle \mathbf{v} \rangle \equiv \langle \mathbf{v} \rangle(\mathbf{r}, t)$. The carrier density $n$ is a function of $\mathbf{r}$ and $t$: $n \equiv n(\mathbf{r}, t)$. The average momentum density $\langle \mathbf{p} \rangle \equiv n m_e \langle \mathbf{v} \rangle$ is also a function of $\mathbf{r}$ and $t$: $\langle \mathbf{p} \rangle \equiv \langle \mathbf{p} \rangle(\mathbf{r}, t)$.

A comment should also be made at this stage about the dimensionality of the position space we are using. The carriers may be restricted to move in a physical space whose dimension is less than three. With this in mind, let $d$ be the dimensionality of the space, and let the partial derivatives with respect to the space coordinates be denoted by $\partial_i$ ($i = 1, \ldots, d$).

The first term in Equation (6.15) can then be written as

$$\frac{\partial}{\partial t} (n \langle A \rangle).$$

The integral in the second term can be written as

$$\int A(\mathbf{p}) \nabla_{\mathbf{p}} f \, d\mathbf{p} = \int \nabla_{\mathbf{p}} (Af) \, d\mathbf{p} - \int f \nabla_{\mathbf{p}} A \, d\mathbf{p}.$$

Using the divergence theorem, the first term on the right hand side of this expression is zero because $f$ is assumed to be zero on the boundary of the region, while the second term on the right hand side is $-n \langle \nabla_{\mathbf{p}} A \rangle$. Hence the second term in Equation (6.15) can be written as

$$-n \mathbf{F} \cdot \langle \nabla_{\mathbf{p}} A \rangle.$$

If $A$ represents a vector quantity, then this expression applies to each component of $A$ separately. In this case, $\mathbf{F} \cdot \langle \nabla_\mathbf{p} A \rangle$ will itself represent a vector quantity such that its $i$th component is

$$(\mathbf{F} \cdot \langle \nabla_\mathbf{p} A \rangle)_i \equiv \mathbf{F} \cdot \langle \nabla_\mathbf{p} A_i \rangle.$$

The third term in Equation (6.15) is

$$\int A(\mathbf{p})\mathbf{v} \cdot \nabla_\mathbf{r} f = \sum_{j=1}^{d} \int A(\mathbf{p}) v_j \partial_j f \, d\mathbf{p}$$

$$= \sum_{j=1}^{d} \partial_j \int A(\mathbf{p}) v_j f \, d\mathbf{p} = \sum_{j=1}^{d} \partial_j (n \langle A v_j \rangle)$$

$$= \nabla_\mathbf{r} \cdot (n \langle A\mathbf{v} \rangle).$$

Again, if $A$ represents a vector quantity, then this result applies to each component of $A$, in the sense that

$$(\nabla_\mathbf{r} \cdot (n \langle A\mathbf{v} \rangle))_i \equiv \nabla_\mathbf{r} \cdot (n \langle A_i \mathbf{v} \rangle) \quad (i = 1, \ldots, d).$$

Hence the general moment Equation (6.15) can be written as

$$\frac{\partial}{\partial t} L + L_K + L_R = \left( \frac{\partial L}{\partial t} \right)_c \tag{6.17}$$

where

$$L \equiv n \langle A \rangle \tag{6.18}$$

$$L_K \equiv -n\mathbf{F} \cdot \langle \nabla_\mathbf{p} A \rangle \tag{6.19}$$

$$L_R \equiv \nabla_\mathbf{r} \cdot (n \langle A\mathbf{v} \rangle). \tag{6.20}$$

The first three moments of the BTE are then found by taking $A(\mathbf{p}) = 1$, $A(\mathbf{p}) = \mathbf{p}$, and $A(\mathbf{p}) = m_e \mathbf{v}^2 / 2$ in turn.

### 6.2.2 First moment: carrier concentration equation

Take $A(\mathbf{p}) = 1$. Then Equations (6.18)–(6.20) become

$$L = n, \qquad L_K = 0, \qquad L_R = \nabla \cdot (n \langle \mathbf{v} \rangle),$$

and Equation (6.17) becomes

$$\frac{\partial n}{\partial t} + \nabla \cdot (n \langle \mathbf{v} \rangle) = \left( \frac{\partial n}{\partial t} \right)_c. \tag{6.21}$$

In this case, the collision term is

$$\left(\frac{\partial n}{\partial t}\right)_c = G$$

where $G$ is the net carrier generation and recombination rate. The carrier concentration equation then becomes

$$\frac{\partial n}{\partial t} + \nabla \cdot (n\langle \mathbf{v}\rangle) = G. \tag{6.22}$$

### 6.2.3 Second moment: momentum conservation equation

Take $A(\mathbf{p}) = \mathbf{p}$. In this case, the quantities $L$, $L_K$ and $L_R$ are vector quantities. Then Equations (6.18)–(6.20) become

$$L = nm_e\langle\mathbf{v}\rangle = \langle\mathbf{p}\rangle,$$
$$L_K = -n\mathbf{F}\cdot\nabla_{\mathbf{p}}(\mathbf{p}) = -n\mathbf{F}$$
$$= nq\mathbf{E},$$
$$L_R = \sum_{j=1}^{d} \partial_j(nm_e\langle\mathbf{v}v_j\rangle),$$

and the $i$th component of $L_R$ will be

$$(L_R)_i = m_e \sum_{j=1}^{d} \partial_j(n\langle v_i v_j\rangle)$$
$$= m_e \sum_{j=1}^{d} \partial_j\left(n\langle(v_i - \langle v_i\rangle)(v_j - \langle v_j\rangle)\rangle\right) + m_e \sum_{j=1}^{d} \partial_j(n\langle v_i\rangle\langle v_j\rangle). \tag{6.23}$$

In order to proceed further, it will now be assumed that the deviations in the velocity components are uncorrelated, in the sense that

$$\langle(v_i - \langle v_i\rangle)(v_j - \langle v_j\rangle)\rangle = 0 \quad \text{if} \quad i \neq j,$$

and that spatial symmetry holds in the sense that

$$\langle(v_i - \langle v_i\rangle)^2\rangle = \frac{1}{d}\langle(v - \langle v\rangle)^2\rangle \quad (i = 1,\dots,d).$$

Again, the carrier temperature $T_e$ satisfies

$$d \cdot \frac{1}{2}k_B T_e = \frac{1}{2}m_e\langle(v - \langle v\rangle)^2\rangle$$

by virtue of the theorem on equipartition of energy which was introduced in Sect. 4.3.1, which provides an energy $k_B T_e/2$ for each degree of freedom. Then Equation (6.23) becomes

$$
(L_R)_i = m_e \sum_{j=1}^{d} \partial_j \left( n \left\langle (v_i - \langle v_i \rangle)^2 \right\rangle \right) \delta_{ij} + m_e \sum_{j=1}^{d} \left( \langle v_j \rangle \partial_j (n \langle v_i \rangle) + n \langle v_i \rangle \partial_j \langle v_j \rangle \right)
$$
$$
= \partial_i (n k_B T_e) + ((\langle \mathbf{v} \rangle \cdot \nabla) \langle \mathbf{p} \rangle + \langle \mathbf{p} \rangle \nabla \cdot \langle \mathbf{v} \rangle)_i ,
$$

from which it follows that

$$
L_R = \nabla (n k_B T_e) + (\langle \mathbf{v} \rangle \cdot \nabla) \langle \mathbf{p} \rangle + \langle \mathbf{p} \rangle \nabla \cdot \langle \mathbf{v} \rangle.
$$

Equation (6.17) then becomes

$$
\frac{\partial \langle \mathbf{p} \rangle}{\partial t} + \nabla (n k_B T_e) + (\langle \mathbf{v} \rangle \cdot \nabla) \langle \mathbf{p} \rangle + \langle \mathbf{p} \rangle \nabla \cdot \langle \mathbf{v} \rangle + n q \mathbf{E} = \left( \frac{\partial \langle \mathbf{p} \rangle}{\partial t} \right)_c , \qquad (6.24)
$$

which is the momentum conservation equation.

## 6.2.4 Third moment: energy transport equation

Let $\xi \equiv \frac{1}{2} m_e v^2$ be the single carrier energy, and let $W \equiv n\xi = \frac{1}{2} n m_e v^2$ be the total energy of $n$ such carriers. Take $A(\mathbf{p}) = \frac{1}{2} m_e v^2$. Then Equations (6.18)–(6.20) become

$$
L = \frac{1}{2} n m_e \langle v^2 \rangle \equiv \langle W \rangle
$$
$$
L_K = -n \mathbf{F} \cdot \left\langle \nabla_{\mathbf{p}} \left( \frac{1}{2} m_e v^2 \right) \right\rangle
$$
$$
= -m_e n v \mathbf{F} \cdot \langle \nabla_{\mathbf{p}} v \rangle = -m_e n v \mathbf{F} \cdot \left\langle \nabla_{\mathbf{p}} \left( \frac{1}{m_e} p \right) \right\rangle
$$
$$
= -n \mathbf{F} \cdot \langle \mathbf{v} \rangle = q n \mathbf{E} \cdot \langle \mathbf{v} \rangle
$$
$$
L_R = \frac{1}{2} m_e \sum_{i=1}^{d} \partial_i (n \langle v^2 v_i \rangle)
$$
$$
= \nabla_{\mathbf{r}} (n \langle W \mathbf{v} \rangle),
$$

and Equation (6.17) becomes

$$
\frac{\partial \langle W \rangle}{\partial t} + q n \mathbf{E} \cdot \langle \mathbf{v} \rangle + \nabla_{\mathbf{r}} (n \langle W \mathbf{v} \rangle) = \left( \frac{\partial \langle W \rangle}{\partial t} \right)_c . \qquad (6.25)
$$

This is the energy transport equation.

It can easily be seen that the energy $W$ consists of the sum of a kinetic energy component and a thermal energy, since

$$\langle W \rangle = \frac{1}{2} n m_e \langle v^2 \rangle = \frac{1}{2} n m_e \left( \langle v \rangle^2 + \left\langle (v - \langle v \rangle)^2 \right\rangle \right) = n \frac{1}{2} m_e \langle v \rangle^2 + n \frac{d}{2} k_B T_e.$$

Hence

$$\langle W \rangle = n \langle \xi \rangle, \quad \text{where } \langle \xi \rangle \equiv \frac{1}{2} m_e \langle v \rangle^2 + \frac{d}{2} k_B T_e \qquad (6.26)$$

is the average single carrier energy.

## 6.3 Models based on the BTE moments

Up to this point, we have been careful to distinguish the dependencies of various quantities on the space coordinates and time through the use of the average $\langle . \rangle$ notation. In order to ease this notation, we will now drop the bracket average notation, while bearing in mind that quantities such as $\mathbf{v}$, $\mathbf{p}$, $W$ will now represent average values which will be functions of the space coordinates and time.

The equations based on the first three moments of the BTE have been found in Sect. 6.2 to be

*Carrier concentration equation:*

$$\frac{\partial n}{\partial t} + \nabla \cdot (n\mathbf{v}) = G. \qquad (6.27)$$

*Momentum conservation equation:*

$$\frac{\partial \mathbf{p}}{\partial t} + \nabla(n k_B T_e) + (\mathbf{v} \cdot \nabla)\mathbf{p} + \mathbf{p}\nabla \cdot \mathbf{v} + nq\mathbf{E} = \left( \frac{\partial \mathbf{p}}{\partial t} \right)_c. \qquad (6.28)$$

*Energy transport equation:*

$$\frac{\partial W}{\partial t} + qn\mathbf{E} \cdot \mathbf{v} + \nabla_\mathbf{r}(nW\mathbf{v}) = \left( \frac{\partial W}{\partial t} \right)_c. \qquad (6.29)$$

Two important approximations to these equations will now be described—the simplified energy transport model and the drift diffusion model. Many devices which are modelled can be assumed to be unipolar, and this will further reduce the degree of complexity. Consequently, we will only be concerned here with unipolar modelling. Discussions of bipolar modelling can be found elsewhere (Graaff and Klaassen 1990; Liou 1992; Tsai et al. 2002). Again, we will only be concerned with one level of carrier, and not consider carriers in multivalley structures (Blotekjaer 1970; Cheng and Chennupati 1995; Sandborn et al. 1989).

The above equations must be supplemented with the Poisson equation, which is a differential equation for the electrostatic potential $\psi$.

## 6.3.1 The Poisson equation

The Maxwell electromagnetic equations are

$$\nabla_{\mathbf{r}} \times \mathbf{E} = -\partial_t \mathbf{B}$$
$$\nabla_{\mathbf{r}} \times \mathbf{H} = \mathbf{J} + \partial_t \mathbf{D}$$
$$\nabla_{\mathbf{r}} \cdot \mathbf{D} = \rho$$
$$\nabla_{\mathbf{r}} \cdot \mathbf{B} = 0$$

where $\mathbf{H}$ is the magnetic field, $\mathbf{B}$ is the magnetic induction, $\mathbf{E}$ is the electric field, $\mathbf{D}$ is the electric displacement, $\mathbf{J}$ is the current density, and $\rho$ is the charge density.

It follows from the last of these equations that $\mathbf{B} = \nabla_{\mathbf{r}} \times \mathbf{A}$ for some *magnetic potential* $\mathbf{A}$. The first of these equations will then become

$$\nabla_{\mathbf{r}} \times \mathbf{E} = -\nabla_{\mathbf{r}} \times \partial_t \mathbf{A}, \quad \text{or} \quad \nabla_{\mathbf{r}} \times (\mathbf{E} + \partial_t \mathbf{A}) = \mathbf{0}.$$

Hence $\mathbf{E} + \partial_t \mathbf{A} = -\nabla_{\mathbf{r}}\psi$ for some scalar potential $\psi$. In the present situation in which semiconductor devices are being modelled, the term involving the time derivative turns out to be much smaller in magnitude than the remaining terms, and we can then write

$$\mathbf{E} = -\nabla_{\mathbf{r}}\psi.$$

In order to derive the Poisson equation, it is assumed that the permittivity is independent of time, and that polarisation due to mechanical forces is negligible. We can then write $\mathbf{D} = \epsilon_0 \epsilon_r \mathbf{E}$ where

$$\epsilon_0 = permittivity\ of\ free\ space = 8.854 \times 10^{-12}\ \mathrm{F\,m^{-1}},$$
$$\epsilon_r = relative\ permittivity\ \text{of the semiconductor.}$$

It follows that

$$\nabla_{\mathbf{r}} \cdot (\epsilon_0 \epsilon_r \nabla_{\mathbf{r}}\psi) = -\rho,$$

or

$$\nabla_{\mathbf{r}} \cdot (\epsilon_0 \epsilon_r \nabla_{\mathbf{r}}\psi) = -q(N_D^+ - n - T^-) \tag{6.30}$$

where $N_D^+$ is the ionised donor trap density, and $T^-$ is the ionised filled trap density. As we have already seen in Sect. 5.4, the expression for $n$ to be used in Equation (6.30) is given by

$$n = n_{2d} + n_{3d}$$

where

$$n_{3d} \equiv 2A^{\frac{3}{2}} F_{\frac{1}{2}} \left( \frac{1}{k_B T}(E_F - E_c) \right), \tag{6.31}$$

$$n_{2d} \equiv 2A \sum_k |\xi_k(y)|^2 F_0 \left( \frac{1}{k_B T}(E_F - q\lambda_k) \right) \quad \text{for electrons free in 2d}$$
$$\tag{6.32}$$

$$n_{2d} \equiv 2A^{\frac{1}{2}} \sum_k |\xi_k(x, y)|^2 F_{-\frac{1}{2}} \left( \frac{1}{k_B T}(E_F - q\lambda_k) \right) \quad \text{for electrons free in 1d}$$

and $A \equiv 2\pi m_e k_B T_e / h^2$.

It follows from the second of the Maxwell equations that

$$0 = \nabla_r \cdot (\nabla_r \times \mathbf{H}) = \nabla_r \cdot (\mathbf{J} + \partial_t \mathbf{D}) = \nabla_r \cdot \mathbf{J} + \partial_t \nabla_r \cdot \mathbf{D} = \nabla_r \cdot \mathbf{J} + \partial_t \rho.$$

Hence

$$\nabla_r \cdot \mathbf{J} = -\partial_t \rho = -q \partial_t (N_D^+ - n_{2d} - n_{3d} - T^-).$$

If all but the electron density $n$ are independent of $t$, then we have the *continuity equation*

$$\frac{\partial n}{\partial t} = \frac{1}{q} \nabla_r \cdot \mathbf{J}. \tag{6.33}$$

Normally, terms involving generation and recombination rates for charge carriers are present on the right hand side of this equation. However, they are neglected here for the unipolar MESFET and HEMT models.

## 6.3.2 The simplified energy transport model

A simplified model may be produced from Equations (6.27)–(6.29) by ignoring the kinetic energy term in Equation (6.26), and by ignoring the first, third and fourth terms on the left hand side of Equation (6.28). With the mobility $\mu$ defined as

$$\mathbf{J} = -qn\mathbf{v} \equiv \mu \left( \frac{\partial \mathbf{p}}{\partial t} \right)_c,$$

then the remaining terms in Equation (6.28) give

$$\mathbf{J} = \mu \nabla_r (n k_B T_e) + \mu n q \mathbf{E}. \tag{6.34}$$

Using the phenomenological approach (Cook and Frey 1982; Gardner 1991), the collision term in Equation (6.29) can be written

$$\left( \frac{\partial \mathbf{W}}{\partial t} \right)_c = -\frac{W - W_0}{\tau_W}$$

where $\tau_W$ is the effective *energy relaxation time* and $W_0$ is the average of $W$ with respect to the function $f_0$ which was defined in Equation (6.11). The energy transport equation (6.29) then becomes

$$\frac{\partial W}{\partial t} = -qn\mathbf{E} \cdot \mathbf{v} - \nabla_{\mathbf{r}} \cdot \mathbf{s} - \frac{W - W_0}{\tau_W} \tag{6.35}$$

where

$$\mathbf{s} \equiv nW\mathbf{v} = -W\mathbf{E} - \frac{k_B}{q}\nabla_{\mathbf{r}}(\mu W T_e). \tag{6.36}$$

In fact, the expression for the current density given in Equation (6.34) has been generalised by several authors to the form

$$\mathbf{J} = -qn\mu\nabla_{\mathbf{r}}\psi + k_B\mu T_e\nabla_{\mathbf{r}}n + ak_B\mu n\nabla_{\mathbf{r}}T_e + bk_B T_e n\nabla_{\mathbf{r}}\mu \tag{6.37}$$

where $a$ and $b$ are constants. Tang (1984) takes ($a = 1, b = 1$), McAndrew et al. (1985) and Selberherr (1984) take ($a = \frac{1}{2}, b = 0$), and Snowden and Loret (1987) and Feng and Hintz (1988) take ($a = 1, b = 0$).

### 6.3.3 The drift-diffusion model

In this model, we assume that the electron temperature $T_e$ is constant throughout the device. In this case, $\nabla_{\mathbf{r}}T_e = \mathbf{0}$, and then Equation (6.34) becomes

$$\begin{aligned}\mathbf{J} &= q\mu n\mathbf{E} + \mu k_B T_e \nabla_{\mathbf{r}}n \\ &= q\mu n\mathbf{E} + qD\nabla_{\mathbf{r}}n \end{aligned} \tag{6.38}$$

where $D \equiv \mu k_B T_e/q$ is the diffusion coefficient. This assumption is not a good one when HEMTs are being modelled, since the presence of very high electric fields at some hotspots give rise to a nonuniform electron temperature distribution.

## 6.4 The Wigner distribution function

The BTE is a classical transport equation in which the particle positions $\mathbf{r}$ and momenta $\mathbf{p}$ are defined simultaneously and precisely. Yet the carriers in any semiconductor device are quantum particles, for which this description does not apply. The Wigner function and its associated transport differential equation replace the Boltzmann distribution and the BTE, to produce a full quantum description of the transport (Bordone et al. 2001; Grubin and Buggeln 2001; Wigner 1932).

In what follows, we will use the following notation in order to ease the reading of the equations. As before, the dimensionality of the physical space will be $d$. Then the gradient operators with respect to position $\mathbf{r}$ and momentum $\mathbf{p}$ will be denoted by

$$\nabla_{\mathbf{r}} = \left(\frac{\partial}{\partial x_1}, \ldots, \frac{\partial}{\partial x_d}\right) = (\partial_{r1}, \ldots, \partial_{rd}), \quad \text{and}$$

$$\nabla_{\mathbf{p}} = \left(\frac{\partial}{\partial p_1}, \ldots, \frac{\partial}{\partial p_d}\right) = (\partial_{p1}, \ldots, \partial_{pd}).$$

### 6.4.1 Definition of the Wigner function

Let $\Psi(\mathbf{r}, t)$ be the pure state satisfying the time-dependent Schrödinger equation

$$i\hbar \frac{\partial \Psi}{\partial t} = -\frac{\hbar^2}{2}\nabla_{\mathbf{r}} \cdot \left(\frac{1}{m_e}\nabla_{\mathbf{r}}\Psi\right) + V(\mathbf{r})\Psi \tag{6.39}$$

where $m_e$ is the effective carrier mass and $V(\mathbf{r})$ is the (real) potential energy. Then the Wigner function is defined as (Wigner 1932)

$$f_W(\mathbf{r}, \mathbf{p}, t) \equiv \frac{1}{(\pi\hbar)^d} \int_{-\infty}^{\infty} d\mathbf{y}\,\overline{\Psi(\mathbf{r}+\mathbf{y}, t)}\Psi(\mathbf{r}-\mathbf{y}, t)e^{2i\mathbf{p}\cdot\mathbf{y}/\hbar} \tag{6.40}$$

where $\int_{-\infty}^{\infty} d\mathbf{y}$ denotes the integral $\int_{-\infty}^{\infty} dy_1 \cdots \int_{-\infty}^{\infty} dy_d$. As Wigner pointed out, this function cannot be regarded as the simultaneous probability of finding position $\mathbf{r}$ and momentum $\mathbf{p}$, and it is not even always positive. It is not even unique in producing the desirable properties which are listed in the next subsection, but it has the simplest form.

### 6.4.2 Properties of the Wigner function

In what follows, we will drop the explicit mention of the time $t$ in both the wave function and the Wigner function, writing them simply as $\Psi(\mathbf{r})$ and $f_W(\mathbf{r}, \mathbf{p})$. However, it must be remembered that both are really functions of $t$. Further, in deriving the following properties, we use the result

$$\int_{-\infty}^{\infty} d\mathbf{p}\, e^{2i\mathbf{p}\cdot\mathbf{a}/\hbar} = (\pi\hbar)^d \delta(\mathbf{a}), \tag{6.41}$$

where $\delta(\mathbf{a})$ is the Dirac $\delta$-function.

1. The Wigner function defined in Equation (6.40) is real, since

$$\overline{f_W(\mathbf{r}, \mathbf{p})} = \frac{1}{(\pi\hbar)^d} \int_{-\infty}^{\infty} d\mathbf{y}\,\Psi(\mathbf{r}+\mathbf{y})\overline{\Psi(\mathbf{r}-\mathbf{y})}e^{-2i\mathbf{p}\cdot\mathbf{y}/\hbar}$$

$$= \frac{1}{(\pi\hbar)^d} \int_{\infty}^{-\infty} (-d\mathbf{z})\,\Psi(\mathbf{r}-\mathbf{z})\overline{\Psi(\mathbf{r}+\mathbf{z})}e^{2i\mathbf{p}\cdot\mathbf{z}/\hbar} \quad (\mathbf{z} \equiv -\mathbf{y})$$

$$= \frac{1}{(\pi\hbar)^d} \int_{-\infty}^{\infty} dz\, \overline{\Psi(\mathbf{r}+\mathbf{z})}\Psi(\mathbf{r}-\mathbf{z})e^{2i\mathbf{p}\cdot\mathbf{z}/\hbar}$$

$$= f_W(\mathbf{r},\mathbf{p}).$$

2. Integration of $f_W(\mathbf{r},\mathbf{p})$ over all $\mathbf{p}$ produces the correct spatial probability density, since

$$\int_{-\infty}^{\infty} d\mathbf{p}\, f_W(\mathbf{r},\mathbf{p}) = \frac{1}{(\pi\hbar)^d}\int_{-\infty}^{\infty} d\mathbf{y}\,\overline{\Psi(\mathbf{r}+\mathbf{y})}\Psi(\mathbf{r}-\mathbf{y})\int_{-\infty}^{\infty} d\mathbf{p}\, e^{2i\mathbf{p}\cdot\mathbf{y}/\hbar}$$

$$= \frac{1}{(\pi\hbar)^d}\int_{-\infty}^{\infty} d\mathbf{y}\,\overline{\Psi(\mathbf{r}+\mathbf{y})}\Psi(\mathbf{r}-\mathbf{y})\cdot(\pi\hbar)^d\delta(\mathbf{y})$$

$$= \overline{\Psi(\mathbf{r})}\Psi(\mathbf{r}) = |\Psi(\mathbf{r})|^2.$$

3. Integration of $f_W(\mathbf{r},\mathbf{p})$ over all $\mathbf{r}$ will produce the correct momentum probability density. This integral is

$$I \equiv \int_{-\infty}^{\infty} d\mathbf{r}\, f_W(\mathbf{r},\mathbf{p}) = \frac{1}{(\pi\hbar)^d}\int_{-\infty}^{\infty} d\mathbf{r}\int_{-\infty}^{\infty} d\mathbf{y}\,\overline{\Psi(\mathbf{r}+\mathbf{y})}\Psi(\mathbf{r}-\mathbf{y})e^{2i\mathbf{p}\cdot\mathbf{y}/\hbar}.$$

In order to proceed with the evaluation of this integral, the substitutions $\mathbf{u} \equiv \mathbf{r}+\mathbf{y}$ and $\mathbf{v} \equiv \mathbf{r} - \mathbf{y}$ are made. Then

$$I = \frac{1}{(\pi\hbar)^d}\int_{-\infty}^{\infty} d\mathbf{u}\int_{-\infty}^{\infty} d\mathbf{v}\,\overline{\Psi(\mathbf{u})}\Psi(\mathbf{v})e^{i\mathbf{p}\cdot(\mathbf{u}-\mathbf{v})/\hbar}$$

$$= \frac{1}{(\pi\hbar)^d}\int_{-\infty}^{\infty} d\mathbf{u}\,\overline{\Psi(\mathbf{u})}e^{i\mathbf{p}\cdot\mathbf{u}/\hbar}\int_{-\infty}^{\infty} d\mathbf{v}\,\Psi(\mathbf{v})e^{-i\mathbf{p}\cdot\mathbf{v}/\hbar}$$

$$= \frac{1}{(\pi\hbar)^d}\left|\int_{-\infty}^{\infty} d\mathbf{v}\,\Psi(\mathbf{v})e^{-i\mathbf{p}\cdot\mathbf{v}/\hbar}\right|^2$$

which is the correct momentum probability distribution.

4. For a system in a mixed state $\{\Psi_k(\mathbf{r}) : k = 0, 1, \ldots\}$, the Wigner function is generalised to

$$f_W(\mathbf{r},\mathbf{p},t) \equiv \frac{1}{(\pi\hbar)^d}\sum_k a_k\int_{-\infty}^{\infty} d\mathbf{y}\,\overline{\Psi_k(\mathbf{r}+\mathbf{y})}\Psi_k(\mathbf{r}-\mathbf{y})e^{2i\mathbf{p}\cdot\mathbf{y}/\hbar}. \qquad (6.42)$$

## 6.5 The Wigner transport equation

The Wigner transport equation is the equivalent of the BTE when full quantum effects are taken into account. As in the case of the BTE, moments of the equation can be taken to provide useful transport equations. It is found that in the classical limit $\hbar \to 0$, the Wigner equation reduces to the BTE.

### 6.5.1 Derivation of the Wigner equation

The Wigner equation is found by taking the time derivative of the Wigner function which was defined in Equation (6.40):

$$\frac{\partial f_W}{\partial t} = \frac{1}{(\pi\hbar)^d} \int_{-\infty}^{\infty} d\mathbf{y} \left[ \frac{\partial \overline{\Psi(\mathbf{r}+\mathbf{y})}}{\partial t} \Psi(\mathbf{r}-\mathbf{y}) + \overline{\Psi(\mathbf{r}+\mathbf{y})}\frac{\partial \Psi(\mathbf{r}-\mathbf{y})}{\partial t} \right] e^{2i\mathbf{p}\cdot\mathbf{y}/\hbar}$$

(6.43)

where

$$i\hbar\frac{\partial \Psi}{\partial t} = -\frac{\hbar^2}{2}\nabla_{\mathbf{r}} \cdot \left(\frac{1}{m_e}\nabla_{\mathbf{r}}\Psi\right) + V(\mathbf{r})\Psi \quad \text{and}$$

$$i\hbar\frac{\partial \overline{\Psi}}{\partial t} = \frac{\hbar^2}{2}\nabla_{\mathbf{r}} \cdot \left(\frac{1}{m_e}\nabla_{\mathbf{r}}\overline{\Psi}\right) - V(\mathbf{r})\overline{\Psi},$$

since both $V$ and $m_e$ are real. Remember that the effective mass $m_e$ is generally a function of position: $m_e \equiv m_{e(\mathbf{r})}$. Substitution of these expressions for $\partial\Psi/\partial t$ and $\partial\overline{\Psi}/\partial t$ into Equation (6.43) gives

$$\frac{\partial f_W}{\partial t} = A + B$$

(6.44)

where

$$A \equiv -\frac{i\hbar}{2}\frac{1}{(\pi\hbar)^d} \int_{-\infty}^{\infty} d\mathbf{y}\, e^{2i\mathbf{p}\cdot\mathbf{y}/\hbar} \left[ \nabla_{\mathbf{r}} \cdot \left(\frac{1}{m_{e(\mathbf{r}+\mathbf{y})}}\nabla_{\mathbf{r}}\overline{\Psi(\mathbf{r}+\mathbf{y})}\right) \Psi(\mathbf{r}-\mathbf{y}) \right.$$
$$\left. - \overline{\Psi(\mathbf{r}+\mathbf{y})}\nabla_{\mathbf{r}} \cdot \left(\frac{1}{m_{e(\mathbf{r}-\mathbf{y})}}\nabla_{\mathbf{r}}\Psi(\mathbf{r}-\mathbf{y})\right) \right],$$

(6.45)

$$B \equiv -\frac{i}{\hbar}\frac{1}{(\pi\hbar)^d} \int_{-\infty}^{\infty} d\mathbf{y}\, e^{2i\mathbf{p}\cdot\mathbf{y}/\hbar} \left(V(\mathbf{r}+\mathbf{y}) - V(\mathbf{r}-\mathbf{y})\right) \overline{\Psi(\mathbf{r}+\mathbf{y})}\Psi(\mathbf{r}-\mathbf{y}).$$

(6.46)

The quantity $A$ is evaluated by first expanding the gradient operator, and then evaluating some of the integrals by making the replacement $\mathbf{r} \rightarrow \mathbf{y}$ (and making the reverse replacement once the integrals have been evaluated). Care must be taken in ensuring that the correct signs are taken in this replacement—for example, in the case $\nabla_{\mathbf{r}}\Psi(\mathbf{r}-\mathbf{y}) = -\nabla_{\mathbf{y}}\Psi(\mathbf{r}-\mathbf{y})$. The result can be shown to be (see Problem 6.1)

$$A = -\frac{1}{(\pi\hbar)^d} \int_{-\infty}^{\infty} d\mathbf{y}\, e^{2i\mathbf{p}\cdot\mathbf{y}/\hbar} \mathbf{p} \cdot \left(\frac{1}{m_{e(\mathbf{x}+\mathbf{y})}}\nabla_{\mathbf{r}}\overline{\Psi(\mathbf{r}+\mathbf{y})}\Psi(\mathbf{r}-\mathbf{y})\right.$$
$$\left. + \frac{1}{m_{e(\mathbf{x}-\mathbf{y})}}\overline{\Psi(\mathbf{r}+\mathbf{y})}\nabla_{\mathbf{r}}\Psi(\mathbf{r}-\mathbf{y})\right)$$
$$- \frac{i\hbar}{2}\frac{1}{(\pi\hbar)^d} \int_{-\infty}^{\infty} d\mathbf{y}\, e^{2i\mathbf{p}\cdot\mathbf{y}/\hbar} \left(\frac{1}{m_{e(\mathbf{x}-\mathbf{y})}} - \frac{1}{m_{e(\mathbf{x}+\mathbf{y})}}\right) \nabla_{\mathbf{r}}\overline{\Psi(\mathbf{r}+\mathbf{y})}\nabla_{\mathbf{r}}\Psi(\mathbf{r}-\mathbf{y}).$$

(6.47)

In the special case in which the effective mass is independent of position: $m_{e(\mathbf{r})} \equiv m_e$, then this result will reduce to

$$A = -\frac{\mathbf{p}}{m_e} \cdot \nabla_{\mathbf{r}} f_W. \tag{6.48}$$

To evaluate the quantity $B$, we start by taking a step which seems to make things more complicated:

$$
\begin{aligned}
V(\mathbf{r}+\mathbf{y}) - V(\mathbf{r}-\mathbf{y}) &= \int_{-\infty}^{\infty} d\mathbf{q}\{V(\mathbf{r}+\mathbf{q}) - V(\mathbf{r}-\mathbf{q})\}\delta(\mathbf{q}-\mathbf{y}) \\
&= \frac{1}{(\pi\hbar)^d} \int_{-\infty}^{\infty} d\mathbf{q}\{V(\mathbf{r}+\mathbf{q}) - V(\mathbf{r}-\mathbf{q})\} \\
&\quad \times \int_{-\infty}^{\infty} d\mathbf{w}\, e^{2i\mathbf{w}\cdot(\mathbf{q}-\mathbf{y}/\hbar)} \\
&= \frac{1}{(\pi\hbar)^d} \int_{-\infty}^{\infty} d\mathbf{w}\, V_W(\mathbf{r},\mathbf{w})e^{-2i\mathbf{w}\cdot\mathbf{y}/\hbar}
\end{aligned}
\tag{6.49}
$$

where the function $V_W(\mathbf{r}, \mathbf{w})$, called the *Wigner potential*, is defined by

$$V_W(\mathbf{r}, \mathbf{w}) \equiv -\frac{i}{\hbar} \int_{-\infty}^{\infty} d\mathbf{q}\, e^{2i\mathbf{w}\cdot\mathbf{q}/\hbar}\{V(\mathbf{r}+\mathbf{q}) - V(\mathbf{r}-\mathbf{q})\}. \tag{6.50}$$

On substituting this result for $V(\mathbf{r}+\mathbf{y}) - V(\mathbf{r}-\mathbf{y})$ into Equation (6.46) and rearranging the orders of the integration, the expression for $B$ becomes (see Problem 6.2)

$$B = \frac{1}{(\pi\hbar)^d} \int_{-\infty}^{\infty} d\mathbf{w}\, V_W(\mathbf{r}, \mathbf{w}) f_W(\mathbf{r}, \mathbf{p}-\mathbf{w}). \tag{6.51}$$

Hence in the case in which the effective mass $m_e$ is independent of position, the differential equation for the Wigner function given by Equation (6.44) is

$$\frac{\partial f_W}{\partial t} + \frac{\mathbf{p}}{m_e} \cdot \nabla_{\mathbf{r}} f_W - \frac{1}{(\pi\hbar)^d} \int_{-\infty}^{\infty} d\mathbf{w}\, V_W(\mathbf{r}, \mathbf{w}) f_W(\mathbf{r}, \mathbf{p}-\mathbf{w}) = 0 \tag{6.52}$$

where the function $V_W(\mathbf{r}, \mathbf{w})$ is given by

$$
\begin{aligned}
V_W(\mathbf{r}, \mathbf{w}) &\equiv -\frac{i}{\hbar} \int_{-\infty}^{\infty} d\mathbf{q}\, e^{2i\mathbf{w}\cdot\mathbf{q}/\hbar}\{V(\mathbf{r}+\mathbf{q}) - V(\mathbf{r}-\mathbf{q})\} \\
&= \frac{1}{\hbar} \int_{-\infty}^{\infty} d\mathbf{q}\, \sin(\mathbf{w}\cdot\mathbf{q}/\hbar)\{V(\mathbf{r}+\mathbf{q}) - V(\mathbf{r}-\mathbf{q})\}.
\end{aligned}
\tag{6.53}
$$

Hence the Wigner equation Equation (6.52) is an integro-differential equation, whose kernel is the Wigner potential $V_W$ which is constructed using the physical potential function $V(\mathbf{r})$.

## 6.5.2 Special cases of the Wigner equation

1. If the potential energy $V$ possesses a Taylor expansion, then

$$V(\mathbf{r}+\mathbf{y}) - V(\mathbf{r}-\mathbf{y}) = \sum_{n_1,\dots,n_d=0} \frac{y_1^{n_1}\cdots y_d^{n_d}}{n_1!\cdots n_d!} \frac{\partial^{n_1+\cdots+n_d} V}{\partial x_1^{n_1}\cdots \partial x_d^{n_d}}$$

$$= \sum_{n_1,\dots,n_d=0} \frac{y_1^{n_1}\cdots y_d^{n_d}}{n_1!\cdots n_d!} \partial_{\mathbf{r}1}^{n_1}\cdots \partial_{\mathbf{r}d}^{n_d} V \qquad (6.54)$$

where the summation is taken over the combinations of the $n_1,\dots,n_d$ for which $n_1+\cdots+n_d$ is odd. This expression is then substituted into Equation (6.46), and the integral is evaluated to give (see Problem 6.3)

$$B = \sum_{n_1,\dots,n_d=0} \frac{(\hbar/2i)^{n_1+\cdots+n_d-1}}{n_1!\cdots n_d!} (\partial_{\mathbf{r}1}^{n_1}\cdots \partial_{\mathbf{r}d}^{n_d} V)(\partial_{\mathbf{p}1}^{n_1}\cdots \partial_{\mathbf{p}d}^{n_d} fw).$$

The Wigner equation (6.52) then becomes

$$\frac{\partial fw}{\partial t} + \frac{\mathbf{p}}{m_e}\cdot \nabla_{\mathbf{r}} fw - \sum_{n_1,\dots,n_d=0} \frac{(\hbar/2i)^{n_1+\cdots+n_d-1}}{n_1!\cdots n_d!}(\partial_{\mathbf{r}1}^{n_1}\cdots \partial_{\mathbf{r}d}^{n_d} V)(\partial_{\mathbf{p}1}^{n_1}\cdots \partial_{\mathbf{p}d}^{n_d} fw)$$

$$= 0. \qquad (6.55)$$

2. The quantum nature of the Wigner equation is apparent through the inclusion of terms containing $\hbar$. However, if these terms in the summation in Equation (6.55) are ignored, then the resulting equation should be equivalent to the classical BTE. In order to see this, note that the only terms in this summation which do not contain $\hbar$ are given by the combination $n_1 + \cdots + n_d - 1 = 0$. Since the $n_i$ are not negative, the only combination in the summation which does not contain $\hbar$ is

$$-(\partial_{\mathbf{r}1} V)(\partial_{\mathbf{p}1} fw) - (\partial_{\mathbf{r}2} V)(\partial_{\mathbf{p}2} fw) - (\partial_{\mathbf{r}3} V)(\partial_{\mathbf{p}3} fw)$$
$$= -(\nabla_{\mathbf{r}} V)\cdot (\nabla_{\mathbf{p}} fw)$$
$$= \mathbf{F}\cdot \nabla_{\mathbf{p}} fw$$

where the force is $\mathbf{F} = -\nabla_{\mathbf{r}} V$. Hence the Wigner equation will then reduce to

$$\frac{\partial fw}{\partial t} + \frac{\mathbf{p}}{m_e}\cdot \nabla_{\mathbf{r}} fw + \mathbf{F}\cdot \nabla_{\mathbf{p}} fw = 0 \qquad (6.56)$$

which is the collisionless form of the BTE Equation (6.5).

3. For the case in which the effective mass $m_e$ is independent of position but dependent on direction: $m_e \equiv m_{ei}$ $(i = 1,\dots,d)$, then the kinetic energy operator in the Schrödinger equation (6.39) must be replaced with

$$-\frac{\hbar^2}{2}\sum_{i=1}^{d}\frac{1}{m_{ei}}\partial_{\mathbf{r}i}^2.$$

The second term in the Wigner equation (6.55) must then be replaced with

$$\sum_{i=1}^{d}\frac{p_i}{m_{ei}}\frac{\partial f_W}{\partial x_i}.$$

### 6.5.3 Moments of the Wigner equation

As in the case of the BTE, moments of the Wigner equation may be taken by multiplying it by any function $A(\mathbf{p})$ of the momentum $\mathbf{p}$ and integrating. This process gives

$$\frac{\partial(n\langle A\rangle)}{\partial t}+\nabla_{\mathbf{r}}\cdot\left(n\left\langle\frac{\mathbf{p}}{m_e}A\right\rangle\right)$$

$$-\sum_{n_1,\dots,n_d=0}\frac{(\hbar/2i)^{n_1+\cdots+n_d-1}}{n_1!\cdots n_d!}(\partial_{\mathbf{r}1}^{n_1}\cdots\partial_{\mathbf{r}d}^{n_d}V)n\left\langle(\partial_{\mathbf{p}1}^{n_1}\cdots\partial_{\mathbf{p}d}^{n_d}A)\right\rangle=0$$

$$(6.57)$$

where $n$ is the particle density. The first three moments are found by taking $A(\mathbf{p})$ in turn as 1, $\mathbf{p}$, and $p^2/(2m_e)$, and then integrating (Gardner 1994). The lowest terms in the summation are those given when $n_1+\cdots+n_d=1$, and these will produce terms which are independent of $\hbar$. However, all terms which contain $\hbar$ in the summation occur when $n_1+\cdots+n_d\geq 3$, and the corresponding derivatives of $A(\mathbf{p})$ will vanish for the three particular forms for the function $A(\mathbf{p})$ which were listed above. Hence the three resulting moment equations will not contain $\hbar$, and the equations will therefore be identical to those obtained from the BTE.

## 6.6 Description of a typical device

It is not possible to describe all of the possible devices for which the BTE approach is appropriate. Instead, we will describe here a simple model of a High Electron Mobility Transistor (HEMT), because this model provides a very typical example of the application of the device equations derived from the BTE. The mathematical and computational methods derived in Part II will be appropriate to the solution of the modelling equations for many types of device, and their application to the solution of the modelling equations of the HEMT will illustrate the general methods and techniques involved.

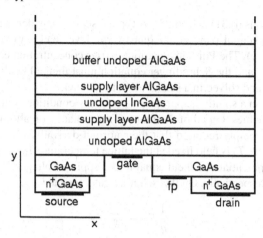

**Fig. 6.1** A simplified cross section in the $x$–$y$ plane of a HEMT. Shown are the first seven layers, two recesses, the source, gate, fieldplate (fp), and drain.

We begin by describing a simplified model of a HEMT. The two dimensional rectangular cross section of such a device is shown in Fig. 6.1. The device illustrated (Hussain et al. 2003) consists of parallel layers of material—the first seven layers are shown in Fig. 6.1—with two recesses along one edge. Four contacts—the source, gate, fieldplate, and drain—are arranged as shown along this edge. Starting from the contact edge, the materials making up the layers of our example are n$^+$ GaAs, GaAs, undoped $Al_uGa_{1-u}As$, a thin supply layer of AlGaAs, an undoped channel of $In_vGa_{1-v}As$, etc., where $u$ is the mole fraction of Al and $v$ is the mole fraction of In. The device has an overall width of 1400 nm along the contact edge, and an overall thickness of 250 nm. Axes $x$ and $y$ are respectively taken along and perpendicular to the contact edge. The length along the third $z$-axis is considered to be large, in the sense that physical quantities vary much more slowly in this third direction than in the other two. Consequently, the solution of the modelling equations on the two dimensional $x$–$y$ cross section will be sufficient for many purposes.

Voltages $V_d$ and $V_g$, measured relative to the voltage on the source, are applied to the drain and gate respectively. With $V_g$ fixed, the current-voltage curve can be found by varying $V_d$. A set of such curves can be found by varying $V_g$, with one curve for each value of $V_g$. The value of $V_g$ is able to control the current through the device by affecting the size of the carrier depletion region around the gate. Very often, a fieldplate contact is added, and this allows the operating voltages to be raised by smoothing out hotspots in the electric field.

Such a device is very effective in producing small transit times of the carriers between the source and drain. It works because quantum wells are formed at the AlGaAs interfaces, and carriers which fall into the wells then travel virtually unhindered by collisions in sheets parallel to the interfaces. The equations based on the moments of the BTE must then be supplemented by having to solve the Schrödinger equation, so that its eigensolutions can be used to calculate the electron densities $n_{3d}$

and $n_{2d}$ of Equations (6.31) and (6.33). A typical set of dependent variables will be the electrostatic potential $\psi(x, y)$, the quasi-fermi level $E_F(x, y)$, and the electron temperature $T_e(x, y)$. The Poisson equation, carrier concentration equation, energy transport equation, and the Schrödinger equation must then all be discretised on an appropriate mesh and solved in a self-consistent manner.

In order to obtain a solution, appropriate boundary conditions must be applied to the dependent variables. On all of the free surfaces, that is, on all non-contact parts of the edges, it is simply required that their normal derivatives be zero there. The electron temperature $T_e$ is held fixed at the lattice temperature $T_0$ on all the contacts. The electrostatic potential $\psi$ is held equal to the applied voltages $V_s$ and $V_d$ on the source and drain respectively. On the Schottky gate, we take

$$\psi = \frac{1}{q} E_h - V_g - \psi_b$$

where $E_h$ is the conduction band offset factor to be described in Sect. 6.7, and $\psi_b$ is the built-in potential which is taken as $-0.8$ V. The electron concentration $n_{3d}$ is nominally taken to be $2.5 N_d$ on the source and drain, and nominally zero on the gate. Equation (6.31) is then solved in order to determine the values of $E_F$ on these contacts.

## 6.7 Material properties

The equations which are used to model a HEMT must be solved for the principal dependent variables, in this case the electrostatic potential $\psi$, the quasi Fermi level $E_F$, and the electron temperature $T_e$. These equations also contain parameters which are specific to the materials being modelled. These parameters include the effective electron mass $m_e$, mobility $\mu$, and the relative permittivity $\epsilon_r$ which appears in the Poisson equation.

Fig. 6.2 illustrates the situation which occurs at a GaAs-$Al_u Ga_{1-u}$As interface, where $u$ is the mole fraction of the Al in the AlGaAs. The AlGaAs of region 1 has a bandgap $E_{g_1}$ which will be a function of $u$. This will differ from the bandgap $E_{g_2}$ of the GaAs in region 2. There will then be a conduction band offset factor $E_h$ between the two, and this value will enter into the potential of the quantum well formed at the interface. The following data is taken from the work of Adachi (1985).

### 6.7.1 GaAs

Clearly, this case corresponds to the value $u = 0$. Then

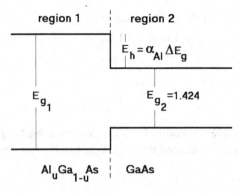

**Fig. 6.2** Bandgap offset factors between $Al_uGa_{1-u}As$ and GaAs, where $u$ is the Al mole fraction. The bandgaps of the $Al_uGa_{1-u}As$ and GaAs are $E_{g_1}$ and $E_{g_2}$ respectively, and $E_h$ is the band offset factor.

$$E_{g_1} = 1.424 \text{ eV},$$
$$E_h = 0.0,$$
$$m_e = 0.067 \, m_{electron},$$
$$\epsilon_r = 13.18,$$

where $m_{electron} = 9.11 \times 10^{-31}$ kg is the electron rest mass

## 6.7.2 AlGaAs

For $Al_uGa_{1-u}As$ we take

$$E_{g_1} = E_{g_2} + 1.155u + 0.37u^2$$
$$= 1.424 + 1.155u + 0.37u^2.$$

Writing $\Delta E_g \equiv E_{g_1} - E_{g_2}$ as the difference in the energy gaps, it follows that

$$\Delta E_g = 1.155u + 0.37u^2.$$

The band offset factor $E_h$ is then taken as some multiple $\alpha_{Al}$ of $\Delta E_g$:

$$E_h = \alpha_{Al}\Delta E_g = \alpha_{Al}(1.155u + 0.37u^2),$$

and we take

$$\alpha_{Al} = 0.65.$$

The effective mass and relative permittivity are taken as

$$m_e = (0.067 + 0.083u)m_{electron},$$
$$\epsilon_r = 13.18 - 3.12u.$$

### 6.7.3 InGaAs

Let $v$ denote the In mole fraction in $In_vGa_{1-v}As$. Let a third region 3 contain the InGaAs (just as region 1 contained the $Al_uGa_{1-u}As$). Region 2 remains the GaAs. Then we take

$$E_{g3} = 1.2$$
$$E_h = \alpha_{In}(E_{g3} - E_{g2})$$
$$\alpha_{In} = 0.7$$
$$m_{e(InGaAs)} = 0.0556$$
$$m_e = \left(0.067 + \frac{v}{0.15}(m_{e(InGaAs)} - 0.067)\right)m_{electron}$$
$$\epsilon_{r(InGaAs)} = 12.94$$
$$\epsilon_r = 13.18 + \frac{v}{0.15}(\epsilon_{r(InGaAs)} - 13.18).$$

Note that the expressions for $m_e$ and $\epsilon_r$ are taken so that they reduce to the values $m_{e(InGaAs)}$ and $\epsilon_{r(InGaAs)}$ when $v = 0.15$. These quantities do not incorporate hydrodynamic strains (Ando and Itoh 1988).

### 6.7.4 The electron mobility

It has already been stated that the Monte Carlo method is computationally expensive when applied to device modelling. However, this method is extremely useful in providing data regarding lifetimes and effective masses. This data can then be pre-programmed for use when solving the differential equations which are based on the moments of the BTE.

Using this method, combined with experimental data on steady state transport characteristics, it is possible to obtain curves for $m_e$, $\tau_e$, and the average single carrier energy $\xi$ plotted against the static electric field $E_{ss}$. Then by using $E_{ss}$ as an intermediate variable, numerical data may be found for $\tau_e$ and $m_e$ in terms of $\xi$, and hence in terms of the electron temperature $T_e$.

The electron mobility $\mu$ is given by

$$\mu(E_{ss}) = \frac{300\mu_0}{T_0}\left(\frac{1 + \frac{v_s E_{ss}^3}{\mu_0 E_0^4(1-5.3\times10^{-4}T_0)}}{1 + (E_{ss}/E_0)^4}\right) \qquad (6.58)$$

**Table 6.1** Mobility parameters

| Material | $E_0$ (V m$^{-1}$) | $v_s$ (m s$^{-1}$) | $\mu_0$ (m$^2$ V$^{-1}$ s$^{-1}$) |
|---|---|---|---|
| GaAs | $4.0 \times 10^5$ | $8.5 \times 10^4$ | 0.8 |
| AlGaAs | $7.0 \times 10^5$ | $5.0 \times 10^4$ | 0.2 |
| InGaAs | $3.0 \times 10^5$ | $1.0 \times 10^5$ | 1.0 |

with the various values of $E_0$, $v_s$ and $\mu_0$ given in Table 6.1.

## 6.8 The Schrödinger equation applied to the HEMT

Quantum wells are produced at the abrupt junctions formed by the mismatch in the bandgaps of AlGaAs and GaAs. The value $E_c$ of the conduction band edge is then a function of the offset factor $E_h$:

$$E_c = E_h - q\psi \qquad (6.59)$$

where the electrostatic potential $\psi$ is measured in volts and is a solution of the Poisson equation (6.30). It can be seen from Equations (6.31) and (6.33) that the total electron density in the quantum wells is partly a function of the eigensolutions of the Schrödinger equation, whose potential is formed by the quantum wells at the material interfaces. Hence the Schrödinger equation must also be solved self-consistently together with the equations which are produced by taking the first three moments of the BTE.

Since we are looking at the solutions of the equations on a two dimensional cross section of the device, the potential of the Schrödinger equation will be a function of both $x$ and $y$, and hence the eigenfunctions will be functions of these two coordinates. The TISE

$$-\frac{\hbar^2}{2}\nabla_{\mathbf{r}} \cdot \left(\frac{1}{m_e}\nabla_{\mathbf{r}}\xi_k\right) + V(x, y)\xi_k = \lambda_k\xi_k \qquad (6.60)$$

will then become

$$-\frac{\hbar^2}{2}\frac{\partial}{\partial y}\left(\frac{1}{m_e}\frac{\partial\xi_k}{\partial y}\right) - \frac{\hbar^2}{2}\frac{\partial}{\partial x}\left(\frac{1}{m_e}\frac{\partial\xi_k}{\partial x}\right) + V(x, y)\xi_k = \lambda_k\xi_k. \qquad (6.61)$$

The potential will be composed of two parts:

$$V(x, y) = E_c + V_{xc} = E_h - q\psi + V_{xc} \qquad (6.62)$$

where $V_{xc}$ is the *exchange correlation energy*, measured in eV, which arises when we change from a single-particle description to a many-particle description. A simple parametrisation suggested by Hedin and Lundqvist (1971) has been used by

Stern and Das Sarma (1984) in the form

$$V_{xc} = -\left(1 + 0.03683\, r_s\, \ln\left[1 + \frac{21}{r_s}\right]\right)\left(\frac{2}{\pi \alpha r_s}\right) R_y^* \tag{6.63}$$

where

$$\alpha = \left(\frac{4}{9\pi}\right)^{1/3}, \qquad a^* = \frac{16\pi^3 \epsilon_0 \epsilon_r \hbar^2}{q^2 m_e}, \qquad r_s = \left(\frac{4}{3}\pi a^{*3} n\right)^{-1/3}.$$

The quantity $R_y^* = q^2/(8\pi \epsilon_0 \epsilon_r a^*)$ is an effective Rydberg constant, and $n$ is again the total electron density.

With the potential given by Equations (6.62) and (6.63), then Equation (6.61) must be solved on the two dimensional region. Boundary conditions are specified by setting $\xi_k(x, y) = 0$ on all edges, and the solution is normalised using the condition

$$\int dx \int dy |\xi_k(x, y)|^2 = 1.$$

In this case, $\xi_k(x, y)$ has dimension m$^{-1}$.

The full two dimensional solution of the Schrödinger equation is computationally very expensive, and many eigensolutions must be calculated in order to cover the full range given by the variation in the value of $E_c$. However, since the material interfaces are all perpendicular to the $y$-direction in our device, it is clear that the variation in $E_c$ is much more rapid along the $y$-direction compared to the variation along the $x$-direction. Consequently, a good approximation is produced by solving the Schrödinger equation in one dimensional strips along the $y$-direction:

$$-\frac{\hbar^2}{2}\frac{\partial}{\partial y}\left(\frac{1}{m_e}\frac{\partial \xi_k}{\partial y}\right) + V(x, y)\xi_k = \lambda_k \xi_k. \tag{6.64}$$

However, we must remember that the potential $V = E_c + V_{xc}$ remains a function of both $x$ and $y$, and hence there will be a separate set of eigensolutions for each value of $x$. In this case, each solution is normalised along its own strip using the condition

$$\int dy |\xi_k(y)|^2 = 1,$$

and now $\xi_k(x, y)$ will have dimension m$^{-1/2}$.

## 6.9 The overall nature of the modelling equations

The modelling equations described in this Chapter are highly nonlinear differential equations which have been derived from the moments of the BTE. They consist of the carrier concentration equation (6.27), the momentum conservation equation (6.28), and the energy transport equation (6.29), together with the Poisson equation (6.30) and the TISE equation (6.64), The parameters such as mobility and lifetime which occur in the first two equations are not constants, but are non-linear functions of the average particle energy $\xi$, which itself is a function of the electron temperature $T_e$. The electron densities $n_{2d}$ and $n_{3d}$, which are given by Equations (6.31) and (6.33), are functions of $\psi$, $E_F$, $T_e$ and the eigensolutions of the Schrödinger equation. These electron densities appear on the right hand side of the Poisson equation which must be solved in order to provide the potential of the Schrödinger equation.

There is no way in which analytic solutions of these equations can be found for realistic devices, and solutions must be obtained numerically. The chapters in Part II detail a variety of the main methods which can be used in the solution of these equations.

## Problems

**Problem 6.1.** Obtain the expression for the quantity $A$ in Equation (6.47) from that in Equation (6.45) by first expanding the gradient operator, and then evaluating some of the integrals by making the replacement $\mathbf{r} \to \mathbf{y}$. Take care to ensure that the correct signs are taken in this replacement—for example, $\nabla_{\mathbf{r}} \Psi(\mathbf{r} - \mathbf{y}) = -\nabla_{\mathbf{y}} \Psi(\mathbf{r} - \mathbf{y})$.

**Problem 6.2.** Derive the expression for the quantity $B$ in Equation (6.51) in terms of the Wigner potential.

**Problem 6.3.** When the potential energy $V$ possesses a Taylor expansion, show that the Wigner equation can be written in the form

$$\frac{\partial f_W}{\partial t} + \frac{\mathbf{p}}{m_e} \cdot \nabla_{\mathbf{r}} f_W - \sum_{n_1,\dots,n_d=0} \frac{(\hbar/2i)^{n_1+\cdots+n_d-1}}{n_1!\cdots n_d!} (\partial_{\mathbf{r}1}^{n_1} \cdots \partial_{\mathbf{r}d}^{n_d} V)(\partial_{\mathbf{p}1}^{n_1} \cdots \partial_{\mathbf{p}d}^{n_d} f_W) = 0.$$

# Part II
# Mathematical and numerical methods

# Chapter 7
# Basic approximation and numerical methods

There exists such a vast literature on the subjects of approximation and numerical methods that it is only possible to include a very small portion in this book. The contents of this chapter will be concerned with only some of the techniques which will be found most useful in applications to device modelling. Later, some sections of programme code written in the C programming language (with a small subset of C++) are given to illustrate the theory, and Sect. 7.1 contains an explanation of some aspects of this language for readers whose first programming language is not C. Later sections cover finite differences, solution of simultaneous equations, time discretisation, phase plane methods, and other techniques which will be useful specifically for device modelling.

## 7.1 Reading the C programmes

The C programming language has often been described as a write-only language. This is because it is possible to compress many instructions into one very small line of code. Although the code writer may derive some smug satisfaction from this compaction, the reader could spend much effort in deciphering this code (and so could the code writer if enough time has elapsed between the writing and attempt at reading). Luckily, it is possible to write C code in a perfectly transparent manner, with multiple statements put on separate lines. This should make the coding perfectly readable to the reader who has experience of other programming languages. In the sections of code presented in this book, an attempt has been made to present the code in as simple a manner as possible; it may not be elegant or in its most compact form, but it should be readable. Many excellent expositions of the C/C++ programming language can be found in the literature (Capper 2001; Deitel and Deitel 2001).

C++ is an extension of C which covers object oriented programming. Such a capability will not be needed in the sections of code presented in these chapters, but

E.A.B. Cole, *Mathematical and Numerical Modelling of*
*Heterostructure Semiconductor Devices: From Theory to Programming,*
DOI 10.1007/978-1-84882-937-4_7, © Springer-Verlag London Limited 2009

some of its simpler attributes are included for extra clarity; for example, the ability to nest comments inside other comments.

The following description is for programmers whose main programming language is not C, and could be omitted by C programmers. Each item in the description explains what the command stands for, so that you can transpose the commands into the programming language of your choice. The list is obviously not exhaustive, but covers most of the commands in the sections of code.

- Comments come in two types. Any text between the entries /* and */ is treated as a comment, and this text can extend over several lines. However, it is not possible to nest these comments. A second type of comment lasts for one line of text only, and the line is started with the entry //. These comments can be nested inside the first type of comment. Hence the section of code

```
/*
  All of this is a comment.
  // This is a nested comment,
  // and this is another nested comment.
  Must remember to feed the cat ...
*/
// ... and the dog.
```

is a comment. This nesting is useful if we wish to comment out whole sections of code while retaining the comments within it.
- Many programmes usually start with a section of code in the following form:

```
#define programme_name "Myprog.cxx"
#define programme_date "24 September 2007"
// The following lines describe the nature of the programme
// ...
// ... end of description.
//
#include<cstdio>
#include<cstddef>
#include<cstdlib>
#include<cmath>
#include<ctime>
#include<iostream>
#include<iomanip>
#include<fstream>
#include<cstring>
//
inline double min(double a, double b){return (a<b)? a : b;}
//
// dictionary
// ----------
  const int yes=1;
  const int YES=1;
  const int no=0;
  const int NO=0;
  const int zero=0;
  const int minimise=1;
  const int maximise=3;
  const int male=0;
  const int female=1;
```

The strings programme_name and programme_date are defined in the first two lines; these are not essential, but are useful if they need to be included in any output. The next few lines are comments detailing what the programme is all

about. Although not essential to the running of the programme, they should give
a complete description of the programme, both to inform the reader and to remind
the writer at a later date. The following lines

```
#include<cstdio>
    ....
#include<cstring>
```

are essential preprocessor directives. These lines are processed by the preprocessor before the programme is compiled. However, if you are a C programmer you
will know this, and if you are not a C programmer you will not need to know.
A C programme consists of a set of functions, called at various times, which may
sometimes return variables using the command `return`. A C programme must
always have a function called `main()` from which other functions are called.
There is always an overhead in making a function call and, if the function is
very small, this overhead is a significant element in the process of evaluating
and returning the function value. A way around this is to declare a very short
function as an `inline` call at the start of the programme. This suggests that
the compiler should put the function directly into the code to avoid the calling
overhead. As we all know, however, compilers have minds of their own, and they
don't always do this if they don't want to. In general, inline functions increase
the size of the generated code but decrease the execution time; like all sweeping
statements, however, this is not always true. The line beginning `inline dou-
ble..` is an example of such a function; it returns the function `min` as a double
value to find the minimum of two double quantities $a$ and $b$. The lines beginning
`//dictionary` are not strictly necessary, but I like to include them to make
writing and reading the code more comfortable, since they assign friendly names
to integers. Note that the C programming language is case sensitive, so that entries such as "YES" and "yes" should not cause trouble if the Caps Lock key is
left on.

- Declarations and commands must always end with a semicolon (;), apart from
  the `#include` and `#define` commands listed above.
- Constant values can be set at the start of the programme as, for example,

```
const int my_integer_value = 63;
const double PI = 3.14159;
```

On the other hand, variables can be declared locally in functions, and their values
can be lost on exit from the function—for example

```
int i,j,k;
double x,y;
```

- Output to the screen can typically be written in the form

```
cout << "\nMy output is " << variable_value << endl;
```

In this example, the element \n starts a new line, the string "My output is" is
then written, followed by the value of `variable_value`. The element `endl`
causes the output to be printed immediately without storing it up until other output comes along, and also puts the cursor on to a new line.

- Variables are incremented in the following way: the command `i++;` denotes the instruction `i=i+1;`, and `j--;` denotes the instruction `j=j-1;`. More generally, `i+=a;` denotes the instruction `i=i+a;`, with a corresponding interpretation of `j-=b;`. However, all of these longer forms can be used instead of the slightly shorter forms.
- There is a concise way of declaring variable values using conditional statements. Consider the case in which we wish to assign the value of a double variable $x$ which depends on the value of an integer $a$. For example, let $x = 1.1$ when $a = 1$, $x = 2.2$ when $a = 2$, and $x = 3.3$ when $a = 3$. The section of code

```
if(a==1) { x=1.1;}
else if(a==2) { x=2.2;}
else { x=3.3;}
```

will work in C (note the "==" entry when comparing quantities in C). However, it is more normal, in the case of simple assignments, to use the form

```
x=(a==1) ? 1.1:
  (a==2) ? 2.2: 3.3;
```

- For-Next loops are typically written in the form

```
for(int i=0;i<=I;i++) {
  do something with the integer i;
}
```

Do-While loops are typically written in the form

```
int i=1;
do {
  do something with the integer i;
  i++;
} while(i<=I);
```

and are always implemented at least once. The equivalent While-Do loop

```
int i=1;
while(i<=I) {
  do something with the integer i;
  i++;
}
```

may not be implemented at all.
- Linked lists can be defined recursively. For example, a class `person` can be defined in the form

```
class person {
            public:
            person *left;
            person *right;
            int gender;
            int age;
            }
```

The declaration of a person in a function is then made as `person Einstein;`, where `Einstein->left` and `Einstein->right` are the persons to the left and right of Einstein in the linked list. This person must be created using the command `Einstein = new person;`, and eventually deleted using the command `delete Einstein;`.

## 7.2 Finite differences

Many semiconductor devices can be described on a rectangular cross-section in the $x$–$y$ plane. In these cases, the most natural coordinate system to use is a rectangular mesh. The use of a mesh which is uniform allows for easier programme coding, but if extra coordinate lines are needed for regions in which the physical variables are rapidly varying (for example, the electric field), it is wasteful to introduce them into regions where the variables are slowly varying. Hence the use of a nonuniform mesh is necessary in the modelling of realistic devices.

### 7.2.1 Description of the mesh

The rectangular mesh to be considered will extend over the regions $x_{min} \le x \le x_{max}$ along the $x$-direction, and $y_{min} \le y \le y_{max}$ along the $y$-direction. The discrete points along the $x$-axis will be labelled $0, 1, \ldots, N_x$, the set of $x$-values will be $\{x_i : i = 0, 1, \ldots, N_x\}$ with $x_0 = x_{min}$, $x_{N_x} = x_{max}$, and the set of mesh spacings will be $\{h_i \equiv x_{i+1} - x_i : i = 0, 1, \ldots, N_x - 1\}$. Similarly, the discrete points along the $y$ axis will be labelled $0, 1, \ldots, N_y$, the set of $y$-values will be $\{y_j : j = 0, 1, \ldots, N_y\}$ with $y_0 = y_{min}$, $y_{N_y} = y_{max}$, and the set of mesh spacings will be $\{k_j \equiv y_{j+1} - y_j : j = 0, 1, \ldots, N_y - 1\}$. The special case of a uniform mesh is given by $h_i \equiv h$ ($i = 0, 1, \ldots, N_x - 1$) and $k_j \equiv k$ ($j = 0, 1, \ldots, N_y - 1$). The value of any function $f(x, y)$ at the meshpoint $(i, j)$ will be denoted by $f_{i,j} \equiv f(x_i, y_j)$. The values of $f$ at the half points $(x_i + \frac{1}{2}h_i, y_j)$, $(x_i, y_j + \frac{1}{2}k_j)$ and $(x_i + \frac{1}{2}h_i, y_j + \frac{1}{2}k_j)$ will be denoted by $f_{i+\frac{1}{2},j}$, $f_{i,j+\frac{1}{2}}$ and $f_{i+\frac{1}{2},j+\frac{1}{2}}$ respectively.

### 7.2.2 Numerical differentiation

In order to derive formulae for numerical derivatives, the notation will be eased by looking at derivatives of a function $f(x)$ of a single variable $x$. The generalisation to functions of more than one variable is straightforward.

The starting point for generating these formulae is the Taylor series expansion

$$f(x + H) = f(x) + Hf'(x) + \frac{1}{2!}H^2 f''(x) + \frac{1}{3!}H^3 f'''(x) + \ldots, \qquad (7.1)$$

from which it follows that

$$\frac{f(x + H) - f(x)}{H} = f'(x) + \frac{1}{2!}Hf''(x) + \frac{1}{3!}H^2 f'''(x) + \ldots. \qquad (7.2)$$

This result shows that the use of the left hand side as an approximation to $f'(x)$ leads to a leading error term $Hf''(x)/2!$. This error term can be made sufficiently small by decreasing the value of $H$, but this introduces inefficiency by introducing too many mesh points. The approximation

$$f'(x) = \frac{f(x+H) - f(x)}{H}$$

is called the *forward difference* formula; it is easily shown that the *backward difference* formula

$$f'(x) = \frac{f(x) - f(x-H)}{H}$$

has the same magnitude of error. A formula which gives a smaller error can be found by combining the pair of equations

$$f(x+H) = f(x) + Hf'(x) + \frac{1}{2!}H^2 f''(x) + \frac{1}{3!}H^3 f'''(x) + \ldots$$

$$f(x-H) = f(x) - Hf'(x) + \frac{1}{2!}H^2 f''(x) - \frac{1}{3!}H^3 f'''(x) + \ldots$$

to give

$$\frac{f(x+H) - f(x-H)}{2H} = f'(x) + \frac{1}{3!}H^2 f'''(x) + \ldots.$$

Hence the error in this *central difference* formula

$$f'(x) = \frac{f(x+H) - f(x-H)}{2H} \tag{7.3}$$

is of the order of $H^2$, which is an improvement on the errors involved in the forward and backward formulae.

In the discretisation on the non-uniform mesh introduced above, it is useful to be able to evaluate derivatives at the half-points. The central difference formula

$$f'_{i+\frac{1}{2}} \equiv \frac{f_{i+1} - f_i}{h_i} \tag{7.4}$$

will have an error of the order of $h_i^2$. Alternatively, it is easily shown that the formula for the derivative $f'_i$ at the full point, given by

$$f'_i = \frac{f_{i+\frac{1}{2}} - f_{i-\frac{1}{2}}}{\frac{1}{2}(h_i + h_{i-1})}, \tag{7.5}$$

has only a first order error proportional to $(h_i - h_{i-1})$ when $h_i \neq h_{i-1}$.

An alternative full-point formula can be derived using the forward and backward Taylor expansions

$$f_{i+1} = f_i + h_i f'_i + \frac{1}{2!}h_i^2 f''_i + \frac{1}{3!}h_i^3 f'''_i + \cdots,$$

$$f_{i-1} = f_i - h_{i-1}f'_i + \frac{1}{2!}h_{i-1}^2 f''_i + \frac{1}{3!}h_{i-1}^3 f'''_i + \cdots.$$

These equations can be combined to give

$$f'_i = \frac{f_{i+1} + (r_i^2 - 1)f_i - r_i^2 f_{i-1}}{(1+r_i)h_i} \tag{7.6}$$

where $r_i \equiv h_i/h_{i-1}$, and the term $-h_i h_{i-1} f'''_i/6$ has been neglected. This reduces to the formula in Equation (7.3) in the case of a uniform mesh.

Very often, the first derivative of a function is specified on a boundary point. Repeated use of the Taylor series of Equation (7.1) at the boundary $x = x_0$ gives

$$f_1 = f_0 + h_0 f'_0 + \frac{1}{2!}h_0^2 f''_0 + \frac{1}{3!}h_0^3 f'''_0 + \cdots$$

$$f_2 = f_0 + (h_0 + h_1)f'_0 + \frac{1}{2!}(h_0 + h_1)^2 f''_0 + \frac{1}{3!}(h_0 + h_1)^3 f'''_0 + \cdots,$$

from which it follows that

$$f'_0 = -\frac{h_1(2h_0 + h_1)f_0 - (h_0 + h_1)^2 f_1 + h_0^2 f_2}{h_0 h_1(h_0 + h_1)} \tag{7.7}$$

in which the quantity proportional to $h_0(h_0 + h_1)f'''$ has been neglected. This formula gives the value of $f'_0$ in terms of the first three adjacent values $f_0$, $f_1$ and $f_2$. If the value of the derivative of $f$ is specified at the point $x = x_0$, say $f'_0 = C'$, then Equation (7.7) can be inverted to give

$$f_0 = \frac{-h_0 h_1(h_0 + h_1)C' + (h_0 + h_1)^2 f_1 - h_0^2 f_2}{h_1(2h_0 + h_1)}. \tag{7.8}$$

Similarly, the corresponding formula at the opposite end $x = x_{N_x}$ can be found as

$$f'_{N_x} = \frac{h_{N_x-2}(2h_{N_x-1} + h_{N_x-2})f_{N_x} - (h_{N_x-2} + h_{N_x-1})^2 f_{N_x-1} + h_{N_x-1}^2 f_{N_x-2}}{h_{N_x-1}h_{N_x-2}(h_{N_x-1} + h_{N_x-2})},$$
$$\tag{7.9}$$

with an expression for $f_{N_x}$ similar to that in Equation (7.8) applying there (see Problem 7.1).

Second derivatives can be constructed from the above formulae for the first derivatives. Applying the half-point formula in Equation (7.5) to the first derivative, it can be shown that the neglected term in the 3-point formula

$$f''_i = \frac{f'_{i+\frac{1}{2}} - f'_{i-\frac{1}{2}}}{\frac{1}{2}(h_i + h_{i-1})} = \frac{\frac{1}{h_i}f_{i+1} - (\frac{1}{h_i} + \frac{1}{h_{i-1}})f_i + \frac{1}{h_{i-1}}f_{i-1}}{\frac{1}{2}(h_i + h_{i-1})} \tag{7.10}$$

is $(h_i - h_{i-1})f'''/3$. Hence this approximation can be of an order of magnitude worse than in the case of a uniform mesh.

One of the important modelling equations which has been encountered is the Poisson equation which, in two dimensions, takes the general form

$$\nabla^2 f \equiv \frac{\partial^2 f}{\partial x^2} + \frac{\partial^2 f}{\partial y^2} = g$$

where $f$ is some function of both $x$ and $y$: $f \equiv f(x, y)$. Using the Equation (7.10) for the second derivatives on a uniform mesh, the Poisson equation can be discretised using the 5-point formula

$$p^2 f_{i-1,j} + p^2 f_{i+1,j} + f_{i,j-1} + f_{i,j+1} - 2(1 + p^2)f_{i,j} = p^2 h^2 g_{i,j} \qquad (7.11)$$

where $p \equiv k/h$. This discretisation has an error of the order of $\max(h^4, k^4)$. A more accurate discretisation which has an error of the order of $\max(h^6, k^6)$ is given by the 9-point formula (Scratton 1987)

$$(10p^2 - 2)(f_{i-1,j} + f_{i+1,j}) + (10 - 2p^2)(f_{i,j-1} + f_{i,j+1})$$
$$+ (1 + p^2)(f_{i-1,j-1} + f_{i+1,j-1} + f_{i-1,j+1} + f_{i+1,j+1} - 20f_{i,j})$$
$$= p^2 h^2 (g_{i-1,j} + g_{i+1,j} + g_{i,j-1} + g_{i,j+1} + 8g_{i,j}) \qquad (7.12)$$

### 7.2.3 Numerical integration

Realistic devices are very rarely modelled using simple equations which can be solved analytically, and any integration which is done must be done numerically. For example, the inclusion of the Schrödinger equation and its eigensolutions into the modelling requires that integrations involving the evaluation of normalising constants and other inner products must be done numerically. Again, the following discussion of numerical integration will be restricted to one dimension—the extension to other dimensions is straightforward.

We will be concerned with the numerical evaluation of a typical integral

$$I \equiv \int_{x_{min}}^{x_{max}} f(x)dx,$$

recognising that the integral can be considered as the area bounded by the $x$-axis between the values $x = x_{min}$ and $x = x_{max}$, and the curve $y = f(x)$. The simplest numerical formula for the integral is given by the *trapezium rule*, in which the area bounded by $x = x_i$ and $x = x_{i+1}$ is approximated by the area

$$\frac{1}{2}(f_i + f_{i+1})h_i$$

of each trapezium. The integral $I$ is then found by adding all such areas:

$$I \approx \frac{1}{2}(f_0 + f_1)h_0 + \frac{1}{2}(f_1 + f_2)h_1 + \ldots + \frac{1}{2}(f_{N_x-1} + f_{N_x})h_{N_x-1}$$

$$= \frac{1}{2}h_0 f_0 + \frac{1}{2}h_{N_x-1}f_{N_x} + \frac{1}{2}\sum_{i=1}^{N_x-1}(h_{i-1} + h_i)f_i. \qquad (7.13)$$

The error involved in this approximation can be found as follows. The exact contribution to the integral from the trapezium bounded by the values $x = x_i$ and $x = x_{i+1}$ is

$$\int_{x_i}^{x_{i+1}} f(x)dx = \int_0^{h_i} f(x_i + t)dt$$

$$= \int_0^{h_i}\left(f_i + tf'_i + \frac{1}{2!}t^2 f''_i + \frac{1}{3!}t^3 f'''_i + \ldots\right)dt$$

$$= f_i h_i + \frac{1}{2}h_i^2 f'_i + \frac{1}{6}h_i^3 f''_i + \frac{1}{24}h_i^4 f'''_i + \ldots, \quad (7.14)$$

while the trapezium rule would give

$$\frac{1}{2}(f_i + f_{i+1})h_i$$

$$= \frac{1}{2}h_i\left(f_i + f_i + h_i f'_i + \frac{1}{2!}h_i^2 f''_i + \frac{1}{3!}h_i^3 f'''_i + \frac{1}{4!}h_i^4 f^{iv}_i + \ldots\right)$$

$$= f_i h_i + \frac{1}{2}h_i^2 f'_i + \frac{1}{4}h_i^3 f''_i + \frac{1}{12}h_i^4 f'''_i + \ldots. \qquad (7.15)$$

Hence the lowest order difference in the expressions of Equation (7.14) and Equation (7.15) is $h_i^3 f''_i/12$, and the total error in the total integral is then

$$\sum_{i=1}^{N_x-1}\frac{1}{12}h_i^3 f''_i = \frac{1}{2}\frac{(x_{max} - x_{min})}{N_x}\sum_{i=1}^{N_x-1}f''_i\frac{h_i^3}{\langle h_i\rangle}$$

where the average mesh spacing is given by $\langle h_i\rangle \equiv (x_{max} - x_{min})/N_x$. Hence the trapezium rule has an error of the order of $\langle h_i^2\rangle$.

More accurate numerical integration should be carried out using *Simpson's rule*, which is

$$\int_{x_{min}}^{x_{max}} f(x)dx = \sum_{i=0}^{N_x} A_i f_i \qquad (7.16)$$

where, on writing $r_i \equiv h_i/h_{i-1}$, the coefficients $A_i$ are given by

$$A_0 = \frac{1}{6}(h_0 + h_1)(2 - r_1)$$

$$A_{N_x} = \frac{1}{6}(h_{N_x-2} + h_{N_x-1})(2 - r_{N_x-1}^{-1})$$

$$A_i = \begin{cases} \frac{1}{6}(h_i + h_{i-1})(2 + r_i + r_i^{-1}) & \text{for } i \text{ odd} \\ \frac{1}{6}(h_i + h_{i-1})(2 - r_i^{-1}) + \frac{1}{6}(h_{i+1} + h_{i+2})(2 - r_{i+2}) & \text{for } i \text{ even.} \end{cases}$$

Using an argument similar to that above, it can be shown that the error in this case is of the order of $\langle h_i^4 \rangle$.

### 7.2.4 Discretisation of the Poisson and Schrödinger equations

The Poisson equation for the electrostatic potential $\psi$ has the form

$$\nabla \cdot (\epsilon_0 \epsilon_r \nabla \psi) = -\rho$$

where $\epsilon_0$ is the permittivity of free space, $\epsilon_r$ is the relative permittivity of the material, and $\rho$ is the charge density. The TISE has the form

$$-\frac{\hbar^2}{2} \nabla \cdot \left( \frac{1}{m_e} \nabla u_n \right) + V(\mathbf{r}) u_n = E_n u_n$$

for a particle with variable effective mass $m_e$. The common feature of these equations is that the terms containing the spatial derivatives have the form $\nabla.(a\nabla\phi)$, where $a$ is some function of position. Hence it is necessary to obtain, at least using two dimensions, a numerical expression for

$$\nabla \cdot (a\nabla\phi) = \frac{\partial}{\partial x}\left(a\frac{\partial\phi}{\partial x}\right) + \frac{\partial}{\partial y}\left(a\frac{\partial\phi}{\partial y}\right). \tag{7.17}$$

Consider the derivatives with respect to $x$ in this equation. Using Equation (7.5), this can be approximated at the mesh point $(x_i, y_j)$ by

$$\frac{\partial}{\partial x}\left(a\frac{\partial\phi}{\partial x}\right) \approx \frac{(a\frac{\partial\phi}{\partial x})_{i+\frac{1}{2},j} - (a\frac{\partial\phi}{\partial x})_{i-\frac{1}{2},j}}{\frac{1}{2}(h_i + h_{i-1})}$$

$$\approx \frac{a_{i+\frac{1}{2},j}(\frac{\phi_{i+1,j}-\phi_{i,j}}{h_i}) - a_{i-\frac{1}{2},j}(\frac{\phi_{i,j}-\phi_{i-1,j}}{h_{i-1}})}{\frac{1}{2}(h_i + h_{i-1})}.$$

On writing $a_{i+\frac{1}{2},j} = (a_{i+1,j} + a_{i,j})/2$ and $a_{i-\frac{1}{2},j} = (a_{i,j} + a_{i-1,j})/2$, this expansion becomes

$$\frac{\partial}{\partial x}\left(a\frac{\partial\phi}{\partial x}\right) \approx \frac{a_{i-1,j}+a_{i,j}}{h_{i-1}(h_{i-1}+h_i)}\phi_{i-1,j}$$
$$-\frac{1}{h_{i-1}+h_i}\left(\frac{a_{i-1,j}+a_{i,j}}{h_{i-1}}+\frac{a_{i,j}+a_{i+1,j}}{h_i}\right)\phi_{i,j}$$
$$+\frac{a_{i,j}+a_{i+1,j}}{h_i(h_{i-1}+h_i)}\phi_{i+1,j},$$

with a corresponding expression for $\partial(a\partial\phi/\partial y)/\partial y$. Hence the complete discretised form of $\nabla\cdot(a\nabla\phi)$ in two dimensions is

$$(\nabla\cdot(a\nabla\phi))_{i,j} = \frac{(a_{i-1,j}+a_{i,j})}{h_{i-1}(h_{i-1}+h_i)}\phi_{i-1,j} + \frac{(a_{i,j}+a_{i+1,j})}{h_i(h_{i-1}+h_i)}\phi_{i+1,j}$$
$$-\frac{1}{h_{i-1}+h_i}\left(\frac{a_{i-1,j}+a_{i,j}}{h_{i-1}}+\frac{a_{i,j}+a_{i+1,j}}{h_i}\right)\phi_{i,j}$$
$$+\frac{(a_{i,j-1}+a_{i,j})}{k_{j-1}(k_{i,j-1}+k_j)}\phi_{i,j-1} + \frac{(a_{i,j}+a_{i,j+1})}{k_j(k_{j-1}+k_j)}\phi_{i,j+1}$$
$$-\frac{1}{k_{j-1}+k_j}\left(\frac{a_{i,j-1}+a_{i,j}}{k_{j-1}}+\frac{a_{i,j}+a_{i,j+1}}{k_j}\right)\phi_{i,j}. \quad (7.18)$$

## 7.3 Solution of simultaneous equations

The discretisation of the modelling equations on a mesh will give rise to a set of simultaneous equations of the form

$$F_i(X_0, X_1, \ldots, X_M) = 0, \quad (i = 0, 1, \ldots, M). \quad (7.19)$$

In these equations, the values of the dependent variables $X$ will be some ordered set of the physical variables at each mesh point, and the total number of these quantities on the mesh will be $(M+1)$. The $(M+1)$ functions $F_i$ will either all be linear, or some of them will be nonlinear. For example, suppose the equations are discretised on a two dimensional mesh consisting of the set of points

$$\{(x_i, y_j) : i = 0, 1, \ldots, N_x; j = 0, 1, \ldots, N_y\},$$

and the defining equations are the Poisson equation, current continuity equation, and the energy transport equation. If the dependent variables are taken to be the electrostatic potential $\psi$, electron density $n$, and the electron energy density $W$, then the numbers of equations and unknowns will be

$$M+1 = 3(N_x+1)(N_y+1).$$

The set of unknowns $X_i$ in Equation (7.19) will be the values of $\psi_{i,j}$, $n_{i,j}$ and $W_{i,j}$ taken in some order. This ordering of the values is important, and it will be shown in Sects. 7.3.1 and 10.5 how a suitable ordering can lead to an efficient programme of solution.

Most of this Section will be concerned with the solution of linear equations, which are of the form

$$\mathbf{AX} = \mathbf{d} \tag{7.20}$$

where $\mathbf{A}$ is an $(M + 1) \times (M + 1)$ matrix with constant coefficients, and $\mathbf{d}$ is a constant $(M+1)$-component vector. A brief introduction is also given to the Newton method for solving nonlinear equations; however, this topic is covered more fully in Chapter 10.

### 7.3.1 Linear equations: direct method

Very often, the matrix $\mathbf{A}$ will be tridiagonal as in, for example, the 3-point discretisation of the Poisson equation in the $x$-direction only. Then each line of the equation will be of the form

$$a_i X_{i-1} + b_i X_i + c_i X_{i+1} = d_i, \quad (i = 0, 1, \ldots, M) \tag{7.21}$$

in which the quantities $a_0$ and $c_M$ do not appear, and $M = 3(N_x + 1) - 1 = 3N_x + 2$. The standard method of solving such a set of equations is by eliminating the variables until we are left with an equation which can be solved for $X_M$, and then back-substituting for the remaining variables. The algorithm for achieving this is

$$X_M = \frac{a_M b'_{M-1}}{c_{M-1} a_M - b_M b'_{M-1}} \left( \frac{d'_{M-1}}{b'_{M-1}} - \frac{d_M}{a_M} \right)$$

$$X_i = \frac{1}{b'_i} (d'_i - c_i X_{i+1}), \quad (i = M - 1, \ldots, 0)$$

$$\tag{7.22}$$

where the quantities $b'_i$ and $d'_i$ are generated by

$$b'_0 = b_0$$
$$d'_0 = d_0$$
$$b'_i = b_i - \frac{a_i}{b'_{i-1}} c_{i-1} \tag{7.23}$$
$$d'_i = d_i - \frac{a_i}{b'_{i-1}} d'_{i-1}.$$

When the problem is such that there is a natural grouping of physical variables at each mesh point, there is a more efficient way of writing and solving the equations. For example, in the above case of the one dimensional solution of the equations of

Poisson, current continuity, and energy transport, it is rather artificial to stretch out the variables $(\psi_0, n_0, W_0, \psi_1, n_1, W_1, \ldots \psi_{N_x}, n_{N_x}, W_{N_x})$ into a vector of length $3(N_x + 1)$ components. It is more sensible to collect the variables at a mesh point into a 3-component column vector $\mathbf{X}_i = (\psi_i, n_i, T_i)^T$ (here, the superscript $T$ denotes the transpose operator), and then Equation (7.21) will itself become a vector equation

$$\mathbf{a}\mathbf{X}_{i-1} + \mathbf{b}_i\mathbf{X}_i + \mathbf{c}_i\mathbf{X}_{i+1} = \mathbf{d}_i, \quad (i = 0, 1, \ldots, N_x)$$

where $\mathbf{a}_i, \mathbf{b}_i, \mathbf{c}_i$ and $\mathbf{d}_i$ are all $3 \times 3$ matrices with constant coefficients. The processes of elimination and back substitution will also apply in this case, except that care must be taken to write the matrices in the correct order. The algorithm in this case is (Press et al. 2002; Varga 1962)

$$\mathbf{X}_{N_x} = \left((\mathbf{b}'_{N_x-1})^{-1}\mathbf{c}_{N_x-1} - (\mathbf{a}_{N_x})^{-1}\mathbf{b}_{N_x}\right)^{-1}\left((\mathbf{b}'_{N_x-1})^{-1}\mathbf{d}'_{N_x-1} - (\mathbf{a}_{N_x})^{-1}\mathbf{d}_{N_x}\right)$$
$$\mathbf{X}_i = (\mathbf{b}'_i)^{-1}(\mathbf{d}'_i - \mathbf{c}_i\mathbf{X}_{i+1}), \quad (i = N_x - 1, \ldots, 0) \tag{7.24}$$

where the quantities $\mathbf{b}'_i$ and $\mathbf{d}'_i$ are generated by

$$\begin{aligned}
\mathbf{b}'_0 &= \mathbf{b}_0 \\
\mathbf{d}'_0 &= \mathbf{d}_0 \\
\mathbf{b}'_i &= \mathbf{b}_i - \mathbf{a}_i(\mathbf{b}'_{i-1})^{-1}\mathbf{c}_{i-1} \\
\mathbf{d}'_i &= \mathbf{d}_i - \mathbf{a}_i(\mathbf{b}'_{i-1})^{-1}\mathbf{d}_{i-1}.
\end{aligned} \tag{7.25}$$

When the matrix $\mathbf{A}$ is not tridiagonal, it is necessary to use *lower-upper decomposition* (LU). In this method, the matrix $\mathbf{A}$ is decomposed into the form $\mathbf{A} = \mathbf{LU}$ where $\mathbf{L}$ and $\mathbf{U}$ are upper and lower triangular matrices. Then

$$\mathbf{d} = \mathbf{AX} = (\mathbf{LU})\mathbf{X} = \mathbf{L}(\mathbf{UX}).$$

Hence on defining $\mathbf{Y} \equiv \mathbf{UX}$, we can first solve for $\mathbf{Y}$ using the equation $\mathbf{LY} = \mathbf{d}$ (this is easily solved by forward substitution due to the triangular nature of $\mathbf{L}$) and then solve for $\mathbf{X}$ using the equation $\mathbf{UX} = \mathbf{Y}$ by simple back substitution. Rounding errors can be a problem in this process, and it is necessary to use partial pivoting to control these (Press et al. 2002; Smith 1985; Wilkinson and Reinsch 1971). These rounding errors can become serious when the matrix $\mathbf{A}$ is large, but it is a straightforward process to control them. For, suppose that the error in the computed value $\mathbf{X}_c$ of $\mathbf{X}$ is $\mathbf{e} = \mathbf{X} - \mathbf{X}_c$. Then $\mathbf{A}(\mathbf{X}_c + \mathbf{e}) = \mathbf{d}$, or $\mathbf{Ae} = \mathbf{d} - \mathbf{AX}_c$. Hence in solving this equation for $\mathbf{e}$, we are solving an equation which is identical to the initial problem which was posed in Equation (7.20), but with a different right hand side. Since the hard work has already been done in producing the LU decomposition of $\mathbf{A}$, it is a simple matter to use this to produce the error $\mathbf{e}$. In fact, the process may be iterated any number of times to produce a sufficiently small value of $\mathbf{e}$.

## 7.3.2 Linear equations: relaxation method

The direct method of Sect. 7.3.1 is useful when the matrix $\mathbf{A}$ is tridiagonal, or when it has a special form which lends itself to a special direct method. Inversion of Equation (7.20) to give $\mathbf{X} = \mathbf{A}^{-1}\mathbf{d}$ can be a costly process, unless the matrix $\mathbf{A}$ is sparse, in which case special sparse-matrix methods apply. A relatively foolproof method of solving Equation (7.20) involves an iteration process. This process can take longer to run than the direct method, but it has the advantage that the modelling equations can be easily modified without having to do too much extra work in the solution process.

The quantity $\mathbf{X}^{(k)}$ will denote the approximation to the solution at the $k$th iteration. In the iterative process, we start with a guess $\mathbf{X}^{(0)}$ at the solution of Equation (7.20). This is then used to obtain a better approximation $\mathbf{X}^{(1)}$, and so on, until suitable convergence has been reached. Consider the iterative process

$$\mathbf{X}^{(k+1)} = (\mathbf{I} - \mathbf{A})\mathbf{X}^{(k)} + \mathbf{d} \tag{7.26}$$

where $\mathbf{I}$ is the $(M + 1) \times (M + 1)$ identity matrix. If this process converges to the solution $\mathbf{X}^{(k)} \to \mathbf{X}$ as $k \to \infty$, then Equation (7.26) gives

$$\mathbf{X} = (\mathbf{I} - \mathbf{A})\mathbf{X} + \mathbf{d} = \mathbf{X} - \mathbf{A}\mathbf{X} + \mathbf{d}$$

which leads to the original equation (7.20). Hence the iterative process should, if it converges, lead to the required solution. It remains to be seen under what circumstances the process converges, and how the process can be speeded up if it does converge.

The basic form of the iterative process given in Equation (7.26) is invariably used in a modified form so that the rate of convergence is maximised. In order to see how to do this, define the matrix $\mathbf{B}$ by

$$\mathbf{B} \equiv \mathbf{I} - \mathbf{A}.$$

Then Equation (7.26) becomes

$$\mathbf{X}^{(k+1)} = \mathbf{B}\mathbf{X}^{(k)} + \mathbf{d}. \tag{7.27}$$

Suppose that the matrix $\mathbf{B}$ has $(M + 1)$ distinct eigenvalues $\lambda_0, \lambda_1, \dots, \lambda_M$ with corresponding eigenfunctions $\mathbf{y}_i$, so that

$$\mathbf{B}\mathbf{y}_i = \lambda_i \mathbf{y}_i, \quad (i = 0, 1, \dots, M).$$

The functions $\mathbf{X}^{(0)}$ and $\mathbf{d}$ can be expanded in terms of these eigenfunctions in the form

$$\mathbf{X}^{(0)} = \sum_{i=0}^{M} \alpha_i \mathbf{y}_i, \qquad \mathbf{d} = \sum_{i=0}^{M} \delta_i \mathbf{y}_i.$$

When these are substituted into Equation (7.27), it is easily verified that

$$\mathbf{X}^{(k)} = \sum_{i=0}^{M} \left( \alpha_i \lambda_i{}^k + \delta_i (1 + \lambda_i + \lambda_i{}^2 + \ldots + \lambda_i{}^{k-1}) \right) \mathbf{y}_i. \qquad (7.28)$$

This result shows that $\mathbf{X}^{(k)}$ is linear in the quantity $\lambda_i{}^k$ for all of the eigenvalues $\lambda_i$. Hence the iterative process will not converge if at least one of these eigenvalues has a magnitude greater than unity. Otherwise, if the process does converge, then Equation (7.28) shows that

$$\mathbf{X}^{(k)} \rightarrow \sum_{i=0}^{M} \frac{\delta_i}{1 - \lambda_i} \mathbf{y}_i. \qquad (7.29)$$

If the *spectral radius* is defined as $\rho_s \equiv |\lambda_{max}| \equiv \max_i |\lambda_i|$, then the convergence will be very slow if $\rho_s$ is close to unity. Therefore the process should be modified by choosing an alternative to the matrix $\mathbf{B}$ such that the alternative matrix has as small a spectral radius as possible. One way of achieving this is to modify the iteration process of Equation (7.27) to

$$\mathbf{X}^{(k+1)} = \mathbf{X}^{(k)} + w \left( \mathbf{d} - (\mathbf{I} - \mathbf{B})\mathbf{X}^{(k)} \right) \qquad (7.30)$$

$$= (1 - w)\mathbf{X}^{(k)} + w(\mathbf{d} + \mathbf{B}\mathbf{X}^{(k)}) \qquad (7.31)$$

$$= \mathbf{B}'\mathbf{X}^{(k)} + w\mathbf{d} \qquad (7.32)$$

where $w$ is some real parameter, and $\mathbf{B}' \equiv \mathbf{I} - w(\mathbf{I} - \mathbf{B})$. The eigenvalues of $\mathbf{B}'$ are $1 - w(1 - \lambda_i)$, and so we attempt to make the iteration process as fast as possible by finding a value of the *relaxation factor* $w$ such that the spectral radius $\rho_s'$ of the new matrix $\mathbf{B}'$ is as small as possible. The value of $\rho_s'$ can be found by performing the iteration in Equation (7.27) with the vector $\mathbf{d}$ taken to be the zero vector. Then Equation (7.28) will give

$$\mathbf{X}^{(k)} \rightarrow (\lambda_{max})^k \sum_{i=0}^{M} \alpha_i \left( \frac{\lambda_i}{\lambda_{max}} \right)^k \mathbf{y}_i \rightarrow (\lambda_{max})^k \alpha_{max} \mathbf{y}_{max} \rightarrow \lambda_{max} \mathbf{X}^{(k-1)}.$$

Therefore by taking a sufficiently large number $k$ of iterations, the spectral radius can be calculated from $\rho_s = |\lambda_{max}| = |\mathbf{X}^{(k)}|/|\mathbf{X}^{(k-1)}|$. This process can suffer from the problem of overflows and underflows, and these can be avoided by rescaling the equations at each iteration. Suppose that the error in the computed solution at the $k$th iteration is $\mathbf{e}^{(k)} = \mathbf{X} - \mathbf{X}^{(k)}$. Then it follows that $\mathbf{e}^{(k+1)} = \mathbf{B}\mathbf{e}^{(k)} \rightarrow \lambda_{max}\mathbf{e}^{(k)}$, or $|\mathbf{e}^{(k+1)}| = \rho_s|\mathbf{e}^{(k)}|$. Hence the number of decimal digits by which the error is asymptotically reduced at each iteration is $|\log_{10} \rho_s|$.

There are strict bounds on the values which the relaxation parameter $w$ can take. It has been shown (Stoer and Bulirsch 1980; Varga 1962; Young 1971) that the relaxation scheme in Equation (7.30) is convergent only if $0 < w < 2$. If $0 < w < 1$ then we have *successive under-relaxation* (SUR), and if $1 < w < 2$ then we have *successive over-relaxation* (SOR). When applied to the equations of device modelling, it is generally found that SUR produces better convergence for the current continuity equation and the energy transport equation, while SOR produces more satisfactory convergence for the Poisson equation. For example, the two dimensional Poisson equation, written on a uniform mesh, is

$$p^2\psi_{i-1,j} + p^2\psi_{i+1,j} + \psi_{i,j-1} + \psi_{i,j+1} - 2(1 + p^2)\psi_{i,j} = p^2h^2g_{i,j} \quad (7.33)$$

where $p \equiv k/h$, and the charge density has been written as $g$ in order to avoid confusion with the notation for the spectral radius $\rho$. It can be shown that the spectral radius for this equation is $\rho_P$ where

$$\rho_P = \frac{p^2 \cos\frac{\pi}{N_x+1} + \cos\frac{\pi}{N_y+1}}{1 + p^2}, \quad (7.34)$$

and the optimal choice for $w$ is

$$w = \frac{2}{1 + \sqrt{1 - \rho_P}}. \quad (7.35)$$

However, although this value is satisfactory in the later stages of the iterative process, it is not satisfactory in the earlier stages during which the iteration process is settling down from a (possibly) wild initial guess. Therefore, it is common to vary these parameters as the iterations progress. For example, it can be seen from Equation (7.33) that the mesh can, in fact, be divided into even and odd sub-meshes. Then a half iteration can be performed on one sub-mesh, to be followed by a second half iteration on the other sub-mesh. The value of $w^{(k)}$ of $w$ can be varied using *Chebyshev acceleration* using the scheme

$$w^{(0)} = 1,$$

$$w^{(\frac{1}{2})} = \frac{1}{1 - \rho_P^2/2},$$

$$w^{(k+\frac{1}{2})} = \frac{1}{1 - \rho_P^2 w^{(k)}/4} \quad \left(k = \frac{1}{2}, 1, \frac{3}{2}, \ldots\right).$$

The value of $w$ given in Equation (7.35) is reached in the limit $k \to \infty$. Another scheme is used by Stern (1970) in the form

$$w^{(k+1)} = \frac{w^{(k)}}{1 - q^{(k)}/q^{(k-1)}} \quad \text{where } q^{(k)} \equiv \max|(\mathbf{B} - \mathbf{I})\mathbf{X}^{(k)} + \mathbf{d}|.$$

When applying the iterative scheme to Equation (7.31), the variables are not updated simultaneously, but each is updated in sequence. The question then arises: is it better to use these updated values as soon as they become available for updating the remaining variables, or should all of the variables at the $(k + 1)$th iteration be found from those at the previous iteration? To illustrate the difference between these two approaches, consider the simple case of the discretisation along the $x$-axis only of the Poisson equation on a uniform mesh with Dirichlet end conditions

$$\psi_{i-1} - 2\psi_i + \psi_{i+1} = h^2 g_i. \tag{7.36}$$

Then

$$\psi_i = \frac{1}{2}(\psi_{i-1} + \psi_{i+1} - h^2 g_i). \tag{7.37}$$

The iteration scheme of Equation (7.31) would then be applied in the form

$$\psi_0^{(k+1)} = C = \text{const.}$$
$$\psi_{N_x}^{(k+1)} = D = \text{const.}$$
$$\psi_i^{(k+1)} = (1-w)\psi_i^{(k)} + w \cdot \frac{1}{2}(\psi_{i-1}^{(k)} + \psi_{i+1}^{(k)} - h^2 g_i). \tag{7.38}$$

In this implementation, all of the $\psi_i^{(k+1)}$ ($i = 1, \ldots, N_x - 1$) are generated from the values at the previous iteration step $k$. This would be programmed as follows:

```
// ------------------------------------
double fi[Nx+1];  // temporary storage array
for(int i=1;i<Nx;i++) {
   fi[i] = (1.0-w)*psi[i] + 0.5*w*(psi[i-1]+psi[i+1]-h*h*g[i]);
}
for(int i=1;i<Nx;i++) {
   psi[i] = fi[i];  // update all of the psi[i]
}
// ------------------------------------
```

On the other hand, if the mesh is swept from left to right, then for a given mesh number $i$, the values to the left will already have been updated. So it is possible to use these updated values as the remainder of the mesh is swept. In this case, Equation (7.38) is modified to

$$\psi_0^{(k+1)} = C = \text{const.} \tag{7.39}$$
$$\psi_{N_x}^{(k+1)} = D = \text{const.} \tag{7.40}$$
$$\psi_i^{(k+1)} = (1-w)\psi_i^{(k)} + w \cdot \frac{1}{2}(\psi_{i-1}^{(k+1)} + \psi_{i+1}^{(k)} - h^2 g_i). \tag{7.41}$$

The programming code would then be simplified to

```
// ------------------------------------
// no need for temporary storage.
// iteration on internal points:
for(int i=1;i<Nx;i++) {
   psi[i] = (1.0-w)*psi[i] + 0.5*w*(psi[i-1]+psi[i+1]-h*h*g[i]);
}
// ------------------------------------
```

On some parts of the device boundary, Neumann boundary conditions are specified instead of the Dirichlet conditions given in Equations (7.39) and (7.40). Suppose that $\psi'_0$ and $\psi'_{N_x}$ are specified as $C'$ and $D'$ respectively—very often these values are taken as zero on the free surfaces. Then the three-point boundary condition given in Equation (7.8), written with $f \equiv \psi$, becomes

$$\psi_0 = \frac{-h_0 h_1 (h_0 + h_1) C' + (h_0 + h_1)^2 \psi_1 - h_0{}^2 \psi_2}{h_1 (2h_0 + h_1)}.$$

The relaxation scheme to replace Equation (7.39) would then be

$$\psi_0{}^{(k+1)} = (1 - w)\psi_0{}^{(k)}$$
$$+ w \left( \frac{-h_0 h_1 (h_0 + h_1) C' + (h_0 + h_1)^2 \psi_1{}^{(k+1)} - h_0{}^2 \psi_2{}^{(k+1)}}{h_1 (2h_0 + h_1)} \right), \quad (7.42)$$

but will have to be performed after the sweep on the internal points. A similar expression will hold at the end $i = N_x$.

These two approaches can be formalised as follows. Generally, if we write

$$\mathbf{A} = \mathbf{L}_0 + \mathbf{D} + \mathbf{U}_0$$

where $\mathbf{D}$ is a diagonal matrix, and $\mathbf{L}_0$ and $\mathbf{U}_0$ are upper and lower triangular matrices with zeros on the diagonals, then the equation

$$\mathbf{A}\mathbf{X} = \mathbf{d}$$

becomes

$$(\mathbf{L}_0 + \mathbf{D} + \mathbf{U}_0)\mathbf{X} = \mathbf{d}$$

which suggests the iteration scheme

$$\mathbf{D}\mathbf{X}^{(k+1)} = \mathbf{d} - (\mathbf{L}_0 + \mathbf{U}_0)\mathbf{X}^{(k)}. \quad (7.43)$$

This is called the *Jacobi* iteration scheme, and corresponds to that in Equation (7.38). On the other hand, we can use updated values as soon as they become available, and this suggests the iteration scheme

$$(\mathbf{L}_0 + \mathbf{D})\mathbf{X}^{(k+1)} = \mathbf{d} - \mathbf{U}_0 \mathbf{X}^{(k)}. \quad (7.44)$$

This is called the *Gauss-Seidel* iteration scheme, and corresponds to that in Equation (7.41).

### 7.3.3 The Newton method: a brief introduction

One of the main methods of solving nonlinear equations is the Newton method. This method will be dealt with fully in Chapter 10, and so only a brief introduction will be given in this section.

This method is concerned with the solution of the set of equations

$$F_i(\mathbf{X}) = 0, \quad (i = 0, 1, \ldots, M)$$

for the $(M + 1)$-component vector $\mathbf{X}$. The functions $F_i$ will generally be nonlinear functions. Again, the Newton method is an iterative process, yielding the approximate solution $\mathbf{X}^{(k)}$ at the $k$th iteration. Let $\delta\mathbf{X}^{(k)}$ be the error in the computed solution at this stage. Then

$$
\begin{aligned}
0 &= F_i(\mathbf{X}^{(k)} + \delta\mathbf{X}^{(k)}) \\
&= F_i(\mathbf{X}^{(k)}) + \sum_{j=0}^{M} \left(\frac{\partial F_i}{\partial X_j^{(k)}}\right) \delta X_j^{(k)} + O(\delta\mathbf{X}^2).
\end{aligned}
$$

Hence neglecting the higher order terms in this equation, the errors $\delta X_j^{(k)}$ will satisfy the set of equations

$$\sum_{j=0}^{M} \left(\frac{\partial F_i}{\partial X_j^{(k)}}\right) \delta X_j^{(k)} = -F_i(\mathbf{X}^{(k)}), \quad (i = 0, \ldots, M),$$

or

$$\mathbf{J}^{(k)} \delta\mathbf{X}^{(k)} = -\mathbf{F}(\mathbf{X}^{(k)}) \tag{7.45}$$

where $\mathbf{J}$ is the $(M+1) \times (M+1)$ *Jacobian matrix* whose elements must be calculated at each iteration. This is a set of *linear* equations for the elements of the vector $\delta\mathbf{X}^{(k)}$ which can be solved using the methods of Sects. 7.3.1 and 7.3.2, either directly or by relaxation. Once $\delta\mathbf{X}^{(k)}$ has been found, the updated estimate

$$\mathbf{X}^{(k+1)} = \mathbf{X}^{(k)} + \delta\mathbf{X}^{(k)}$$

is used in the next round of the iteration. There are several ways in which the Newton method may be modified in order to solve Equation (7.45) in the specific context of device modelling, and these are detailed fully in Chapter 10.

## 7.4 Time discretisation

The main equations of current continuity and energy transport have been presented in Chapter 6. The current continuity equation is

$$\frac{\partial n}{\partial t} = \frac{1}{q}\nabla_\mathbf{r} \cdot \mathbf{J}, \tag{7.46}$$

$$\mathbf{J} = -qn\mu\nabla_\mathbf{r}\psi + k_B\mu T_e\nabla_\mathbf{r}n + ak_B\mu n\nabla_\mathbf{r}T_e + bk_B T_e n\nabla_\mathbf{r}\mu \tag{7.47}$$

where the values of the constants $a$ and $b$ are taken differently by different authors, and the energy transport equation is

$$\frac{\partial W}{\partial t} = \mathbf{J} \cdot \mathbf{E} - \frac{(W - W_0)}{\tau_e} - \nabla_{\mathbf{r}} \cdot \mathbf{s}, \qquad (7.48)$$

$$\mathbf{s} = -\mu W \mathbf{E} - \frac{k_B}{q} \nabla_{\mathbf{r}} (\mu W T_e). \qquad (7.49)$$

In order to discuss time discretisation using the current continuity equation as an example, the values ($a = 0$, $b = 0$) will be used. Up to this point we have concentrated on finding the steady state solutions of these equations by taking the time derivatives to be zero. However, it is often necessary to obtain the transient solutions to these equations when, for example, the applied potentials vary in time. Further, it is possible to obtain the steady state solutions—if they exist—by iterating the transient solution to large times. In fact, there is a close relationship between the application of the iterative Newton method for the steady state solutions and the iteration of the transient solution to the steady state; this relationship will be examined more closely in Chapter 10.

Obtaining the transient solution presents a much larger problem than that of obtaining the steady state solution. This is mainly due to the fact that any sensible transient solution must be based on the solution of complicated implicit equations. As before, suppose that there are $M+1$ dependent variables which are used to model the process, and that these $M + 1$ variables are grouped into an $(M + 1)$-component vector $\mathbf{Y} = (Y_0, Y_1, \ldots, Y_M)$. Then there will be a total of $M + 1$ primary vector equations at each point in space and time, with extra subsidiary equations as appropriate. For example, $M = 2$ in the modelling equations described above, with $Y_0 \equiv \psi$, $Y_1 \equiv n$ and $Y_2 \equiv W$, with the three primary equations being the Poisson equation, current continuity equation, and the energy transport equation. In the transient case, the vector $\mathbf{Y}$ will be a function of space and time: $\mathbf{Y} \equiv \mathbf{Y}(\mathbf{r}, t)$. In general, the modelling equations will have the form

$$\frac{\partial \mathbf{Y}}{\partial t} = \mathbf{F}(\mathbf{Y}) \qquad (7.50)$$

where, for example, the components $F_1$ and $F_2$ of the function $\mathbf{F}$ represent the right hand sides of Equations (7.46) and (7.48) respectively, and the component $F_0$ can be suitably constructed from the Poisson equation. The time coordinate will be discretised into the steps

$$t = t_0 + r \Delta t, \quad (r = 0, 1, \ldots)$$

where $t_0$ is some initial start time and $\Delta t$ is taken to be a constant time step. The size of $\Delta t$ is crucial to the success of any time iteration scheme.

### 7.4.1 Explicit and implicit schemes

Let $\mathbf{Y}^{(r)}$ denote the value of the vector $\mathbf{Y}$ at a given space point and at the $r$'th time step, and let primes denote differentiation with respect to $t$: $\mathbf{Y}' \equiv \partial \mathbf{Y} / \partial t$. Then the

Taylor expansion in $t$ gives

$$\mathbf{Y}(t + \Delta t) = \mathbf{Y}(t) + \mathbf{Y}'(t)\Delta t + \frac{1}{2!}\mathbf{Y}''(t)\Delta t^2 + \frac{1}{3!}\mathbf{Y}'''(t)\Delta t^3 + \dots \quad (7.51)$$

giving the forward difference equation

$$\mathbf{Y}'(t) = \frac{\mathbf{Y}(t + \Delta t) - \mathbf{Y}(t)}{\Delta t} + O(\Delta t).$$

Hence Equation (7.50) can be discretised to the first order in $\Delta t$ in the form

$$\mathbf{Y}^{(r+1)} = \mathbf{Y}^{(r)} + \Delta t \mathbf{F}(\mathbf{Y}^{(r)}). \quad (7.52)$$

For example, using this prescription, the continuity equation will be discretised in the form

$$n^{(r+1)} = n^{(r)} + \Delta t \cdot \frac{1}{q}\nabla_{\mathbf{r}} \cdot \mathbf{J}(\psi^{(r)}, n^{(r)}, W^{(r)}). \quad (7.53)$$

This explicit *Euler* discretisation is extremely easy to use in practice: having found $\mathbf{Y}^{(r)}$, then the right hand side of Equation (7.52) can be calculated and then the updated quantity $\mathbf{Y}^{(r+1)}$ can be found. Unfortunately, this method is not satisfactory and should not be used; unless the time step is excessively small, errors quickly accumulate and the solution becomes inaccurate. In order to see why this is so, and in order to learn lessons to apply to more satisfactory methods, let us carry out a simple *von Neumann stability analysis* on a simplified version of the current continuity equation. This equation will be written in one space dimension only as

$$\frac{\partial n}{\partial t} = \frac{1}{q}\frac{d}{dx}\left(-q\mu n\frac{d\psi}{dx} + k_B\mu T_e\frac{dn}{dx}\right)$$

$$= An + C\frac{dn}{dx} + Q\frac{d^2n}{dx^2} \quad (7.54)$$

where

$$A \equiv -\frac{d}{dx}\left(\mu\frac{d\psi}{dx}\right) = \frac{d}{dx}(\mu E),$$

$$C \equiv -\mu\frac{d\psi}{dx} + \frac{k_B\mu}{q}\frac{d}{dx}(\mu T_e), \quad \text{and} \quad Q \equiv \frac{k_B\mu T_e}{q}.$$

Central differencing can be performed on the space derivatives at the $j$th mesh point, and Equations (7.53) and (7.54) can be combined to give

$$n_j^{(r+1)} = n_j^{(r)} + \Delta t \left(An_j^{(r)} + C\frac{(n_{j+1}^{(r)} - n_{j-1}^{(r)})}{2h}\right.$$

$$\left. + Q\frac{(n_{j+1}^{(r)} + n_{j-1}^{(r)} - 2n_j^{(r)})}{h^2}\right) \quad (7.55)$$

in which it has been assumed that the quantities $A$, $C$ and $Q$ are sufficiently constant with respect to the space and time coordinates. In the von Neumann stability analysis, we look for independent solutions of the form

$$n_j^{(r)} = \xi^r e^{iZjh} \tag{7.56}$$

where $Z$ is a real spatial wave number, and $\xi$ is a complex amplitude. Substitution of this into Equation (7.55) gives

$$\xi = 1 + \Delta t \left( A + iC \frac{\sin(Zh)}{h} - 4Q \frac{\sin^2(\frac{1}{2}Zh)}{h^2} \right),$$

from which it follows that

$$|\xi|^2 = \left[ 1 + \Delta t \left( A - 4Q \frac{\sin^2(\frac{1}{2}Zh)}{h^2} \right) \right]^2 + \Delta t^2 C^2 \frac{\sin^2(Zh)}{h^2}.$$

Looking at Equation (7.56), we require the condition $|\xi| < 1$ for stability, which leads to the conditions that $\Delta t$ must be sufficiently small, and that

$$A - 4Q \frac{\sin^2(\frac{1}{2}Zh)}{h^2} < 0,$$

or

$$\frac{d}{dx}(\mu E) < \frac{4}{h^2} \left( \frac{k_B \mu T_e}{q} \right) \sin^2 \left( \frac{1}{2} Zh \right). \tag{7.57}$$

An alternative method is to replace the simple explicit Euler method by an implicit *Crank-Nicholson* method. A Taylor series expansion on $\mathbf{Y}'(t)$ gives

$$\mathbf{Y}'(t + \Delta t) = \mathbf{Y}'(t) + \mathbf{Y}''(t)\Delta t + \frac{1}{2!}\mathbf{Y}'''(t)\Delta t^2 + \frac{1}{3!}\mathbf{Y}^{(iv)}(t)\Delta t^3 + \dots . \tag{7.58}$$

Elimination of $\mathbf{Y}''$ between this equation and Equation (7.51) gives

$$\frac{1}{2} \left( \mathbf{Y}'(t + \Delta t) + \mathbf{Y}'(t) \right) = \frac{\mathbf{Y}(t + \Delta t) - \mathbf{Y}(t)}{\delta t} + O(\Delta t^2).$$

Hence Equation (7.50) can now be discretised to the second order in $\Delta t$ by

$$\mathbf{Y}^{(r+1)} = \mathbf{Y}^{(r)} + \Delta t \cdot \frac{1}{2} \left( \mathbf{F}(\mathbf{Y}^{(r+1)}) + \mathbf{F}(\mathbf{Y}^{(r)}) \right). \tag{7.59}$$

A von Neumann stability analysis on this scheme, using the above example for the current continuity equation, gives (see Problem 7.2)

$$|\xi|^2 = \frac{[1 + \frac{1}{2}\Delta t(A - 4Q \frac{\sin^2(\frac{1}{2}Zh)}{h^2})]^2 + \frac{1}{4}\Delta t^2 C^2 \frac{\sin^2(Zh)}{h^2}}{[1 - \frac{1}{2}\Delta t(A - 4Q \frac{\sin^2(\frac{1}{2}Zh)}{h^2})]^2 + \frac{1}{4}\Delta t^2 C^2 \frac{\sin^2(Zh)}{h^2}} \tag{7.60}$$

which is less than unity if, again, Equation (7.57) still holds, but this time with no restriction on the size of $\Delta t$.

Hence the method based on Equation (7.59) will converge whatever the size of $\Delta t$, but with the restriction imposed by Equation (7.57). The drawback of the method based on Equation (7.59) is that it is totally implicit, with the quantities $\mathbf{Y}^{(r+1)}$ being contained on the right hand side of the equation. This makes the scheme extremely difficult to use. An alternative *semi-implicit* scheme (Snowden 1988) is formed by replacing the fully implicit scheme of Equation (7.59) with

$$Y_j{}^{(r+1)} = Y_j{}^{(r)} + \frac{1}{2}\left(F_j(Y_0{}^{(r)}, \ldots, Y_j{}^{(r+1)}, \ldots, Y_M{}^{(r)})\right.$$
$$\left. +F_j(Y_0{}^{(r)}, \ldots, Y_j{}^{(r)}, \ldots, Y_M{}^{(r)})\right),$$
$$(j = 0, 1, \ldots, M). \tag{7.61}$$

For example, the current continuity equation would become

$$n^{(r+1)} = n^{(r)} + \Delta t \cdot \frac{1}{2q}\left(\nabla_{\mathbf{r}} \cdot \mathbf{J}(\psi^{(r)}, n^{(r+1)}, W^{(r)}) + \nabla_{\mathbf{r}} \cdot \mathbf{J}(\psi^{(r)}, n^{(r)}, W^{(r)})\right), \tag{7.62}$$

with the index $(r+1)$ appearing only in the term $n^{(r+1)}$ on the right hand side. This equation would then be iterated to find $n^{(r+1)}$ using the relaxation methods of the previous section. This semi-implicit scheme is much more stable than the explicit scheme, but care must still be taken to keep the mesh spacing $h$ and the time step $\Delta t$ sufficiently small. Unfortunately, for realistic semiconductor devices, there is no way of determining in advance what these upper bounds should be. A further complication can be seen from the condition of Equation (7.57). This condition depends on the value of the quantity $dE/dx$, that is, on the magnitude and direction of the electric field. A much more satisfactory approach involves the method of *upwinding*, in which the direction of the electric field is automatically accounted for. This upwinding approach is discussed fully in Chapter 9.

### 7.4.2 The ADI method

We have seen that the explicit scheme

$$\mathbf{Y}^{(r+1)} = \mathbf{Y}^{(r)} + \Delta t \mathbf{F}(\mathbf{Y}^{(r)})$$

requires an unfeasably small value of the time step $\Delta t$ in order to produce meaningful results. The Crank-Nicholson method given by

$$\mathbf{Y}^{(r+1)} = \mathbf{Y}^{(r)} + \Delta t \cdot \frac{1}{2}\left(\mathbf{F}(\mathbf{Y}^{(r+1)}) + \mathbf{F}(\mathbf{Y}^{(r)})\right)$$

produces stability for all values of $\Delta t$. Further, it can be shown that the backward time difference formula

$$\mathbf{Y}^{(r+1)} = \mathbf{Y}^{(r)} + \Delta t \mathbf{F}(\mathbf{Y}^{(r+1)})$$

is stable for all values of $\Delta t$. These last two schemes are implicit schemes, and require either relaxation methods for their solution, or methods based on the inversion of large non-triangular matrices. An alternative method is the *alternating direction implicit* (ADI) method, in which it may be possible to reduce the problem to a series of solutions involving only triangular matrices. This is accomplished by taking half time steps $\frac{1}{2}\Delta t$.

For example, consider an equation whose type is similar to that of the heat conduction equation

$$\frac{\partial T}{\partial t} = \frac{\partial^2 T}{\partial x^2} + \frac{\partial^2 T}{\partial y^2} + \sigma(x, y) \tag{7.63}$$

in two spatial dimensions, in which the function $\sigma(x, y)$ is a source term. The 5-point discretisation on a uniform mesh gives

$$\left(\frac{\partial T}{\partial t}\right)_{i,j} = \frac{T_{i-1,j}(t) + 2T_{i+1,j}(t) - 2T_{i,j}(t)}{h^2}$$

$$+ \frac{T_{i,j-1}(t) + 2T_{i,j+1}(t) - 2T_{i,j}(t)}{k^2}$$

$$+ \sigma_{i,j}. \tag{7.64}$$

Notice that the first three terms on the right hand side are all evaluated at a constant value of $j$, and the next three terms are all evaluated at a constant value of $i$. Let $T^{(r+\frac{1}{2})}$ denote the value of $T$ at the intermediate time $t_0 + (r + \frac{1}{2})\Delta t$. The ADI method is performed in two stages.

- Perform a half time step $r \rightarrow r + \frac{1}{2}$:

$$T_{i,j}^{(r+\frac{1}{2})} = T_{i,j}^{(r)} + \frac{\Delta t}{2}\left(\frac{T_{i-1,j}^{(r+\frac{1}{2})} + 2T_{i+1,j}^{(r+\frac{1}{2})} - 2T_{i,j}^{(r+\frac{1}{2})}}{h^2}\right.$$

$$\left. + \frac{T_{i,j-1}^{(r)} + 2T_{i,j+1}^{(r)} - 2T_{i,j}^{(r)}}{k^2} + \sigma_{i,j}\right)$$

and solve the quantities $T_{1,j}^{(r+\frac{1}{2})}, T_{2,j}^{(r+\frac{1}{2})}, \ldots, T_{N_x-1,j}^{(r+\frac{1}{2})}$ for each value of $j$ using the tridiagonal matrix method, by treating those values evaluated at time step $r$ as constant.

- Perform a further half time step $r + \frac{1}{2} \rightarrow r + 1$:

$$T_{i,j}^{(r+1)} = T_{i,j}^{(r+\frac{1}{2})} + \frac{\Delta t}{2} \left( \frac{T_{i-1,j}^{(r+\frac{1}{2})} + 2T_{i+1,j}^{(r+\frac{1}{2})} - 2T_{i,j}^{(r+\frac{1}{2})}}{h^2} \right.$$

$$\left. + \frac{T_{i,j-1}^{(r+1)} + 2T_{i,j+1}^{(r+1)} - 2T_{i,j}^{(r+1)}}{k^2} + \sigma_{i,j} \right)$$

and solve the quantities $T_{i,1}^{(r+1)}, T_{i,2}^{(r+1)}, \ldots, T_{i,N_y-1}^{(r+1)}$ for each value of $i$ using the tridiagonal matrix method, by treating those values evaluated at time step $r + \frac{1}{2}$ as constant.

This method therefore involves two passes at each time step, each pass involving a solution using the tridiagonal matrix method.

The ADI method has been illustrated using a simple heat flow equation, for which the splitting into two half-passes is simple to perform. However, in its application to the more complicated equations which are used to model semiconductor devices, the splitting of the equations in this way is not so obviously straightforward. In these cases, the modeller's intuition must play a large part in obtaining a splitting which produces convergence.

## 7.5 Function updating and fitting

In device modelling, there is a constant two-way interplay between the use of differential equations involving continuous functions, and the discretisation of these functions on a mesh. For example, the electrostatic potential $\psi(x, y)$ is a continuous function of $x$ and $y$, yet discretised values $\psi_{i,j}$ are obtained computationally. If the boundary conditions are changed—say, a contact potential is changed—then a new initial guess for the function $\psi(x, y)$ must be found in terms of these values $\psi_{i,j}$, and then this new functional form must itself be discretised. Again, at the interface between two materials with different doping values, the junction may be considered to be abrupt for some purposes, but it is necessary to produce a smoothly graded function for computational purposes.

### 7.5.1 Updating due to altered boundary conditions

When obtaining current-voltage characteristics of devices, it is common to take small increments in the applied voltages, and to use the solution of the dependent variables as a first guess at the next step. However, it is necessary to carry out some preparatory work on the previous solution before the next step is taken. For example, suppose a function $f(y)$ has been found which satisfies the conditions

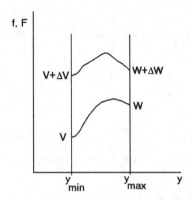

**Fig. 7.1** The function $f(y)$ with fixed values $V$ and $W$ at the ends $y = y_{min}$ and $y = y_{max}$, and the modified function $F(y)$ with new fixed values $V + \Delta V$ and $W + \Delta W$.

$$f(y_{min}) = V \quad \text{and} \quad f(y_{max}) = W, \tag{7.65}$$

where $V$ and $W$ are given values. Now suppose that it is required to find a new function $F(y)$ which satisfies the new conditions

$$F(y_{min}) = V + \Delta V \quad \text{and} \quad F(y_{max}) = W + \Delta W \tag{7.66}$$

and which follows the shape of the original function $f(y)$ as closely as possible. This new function is to be used as an initial guess in the next stage of the iteration procedure. There is no unique way of doing this—for example, we may wish to take the function difference $F(y) - f(y)$ in as simple a way as possible, or we may wish to preserve the derivatives of the functions at $y = y_{min}$ and $y = y_{max}$. This situation is shown in Fig. 7.1.

Look for a relation in the form

$$F(y) = f(y) + P + Qg(y)$$

where $P$ and $Q$ are constants, and $g(y)$ is any function which satisfies the conditions $g(y_{min}) = 0$ and $g(y_{max}) = 1$. It will also be required that $g(y)$ and its derivative $g'(y)$ are continuous in order to preserve any continuity in the function $f(y)$. Then applying the conditions in Equations (7.65) and (7.66), we find

$$V + \Delta V = V + P \quad \text{and} \quad W + \Delta W = W + P + Q,$$

giving

$$F(y) = f(y) + \Delta V + (\Delta W - \Delta V)g(y), \tag{7.67}$$

where the function $g(y)$ must satisfy the conditions

$$g(y_{min}) = 0 \quad \text{and} \quad g(y_{max}) = 1. \tag{7.68}$$

The discretised version of this result is

$$F_j = f_j + \Delta V + (\Delta W - \Delta V)g(y_j). \tag{7.69}$$

The function $g(y)$ is arbitrary, but must satisfy the conditions of Equation (7.68). In particular, two simple cases can be considered.

- The simplest case occurs when the function $g(y)$ is linear:

$$g(y) = \frac{y - y_{min}}{y_{max} - y_{min}}. \tag{7.70}$$

However, since $g'(y) \neq 0$, this form of $g$ does not preserve the derivatives at the ends $y = y_{min}$ and $y = y_{max}$, so that if $f'$ were zero at these ends, then the new function $F$ would not have this property.
- It is easily verified that the form

$$g(y) = \sin^2\left(\frac{\pi(y - y_{min})}{2(y_{max} - y_{min})}\right) \tag{7.71}$$

does have zero derivative at the ends $y = y_{min}$ and $y = y_{max}$, thus transferring the values of the end derivatives of the function $f$ to the new function $F$.

The following section of code applies to the above case in which $g(y)$ has the linear form of Equation (7.70). In this case, Equation (7.69) can be written

$$F_j = f_j + \left(\frac{\Delta W - \Delta V}{y_{max} - y_{min}}\right)y_j + \frac{\Delta V y_{max} - \Delta W y_{min}}{y_{max} - y_{min}}. \tag{7.72}$$

```
// -----------------------------------------
double y[Ny+1];
// mesh points are y[j] for j=0,1,Ny.
// mesh values y[j] are already specified, with y[0]=ymin and y[Ny]=ymax.
double f[Ny+1];  // already found
double F[Ny+1];
double DeltaW,DeltaV;  // specified increments
double A = (DeltaW-DeltaV)/(y[Ny]-y[0]);
double B = (DeltaV*y[Ny]-DeltaW*y[0])/(y[Ny]-y[0]);
for(int j=1;j<Ny;j++) {
  F[j] = f[j] + A*y[j] + B;
}
// -----------------------------------------
```

A separate case arises when the value of $V$ is changed by an amount $\Delta V$ at $y = y_{min}$, but the derivative has to be preserved at the end $y = y_{max}$ without the value of $W$ being prescribed there. An example of this situation occurs when the boundary opposite a contact is a free surface at which the normal derivative of $f$ is zero. In this case, we may take the function $g(y)$ to be a constant. Since this constant is evaluated at the end $y = y_{min}$ as $g(y_{min}) = \Delta V$, then the updated function is simply

$$F_j = f_j + \Delta V. \tag{7.73}$$

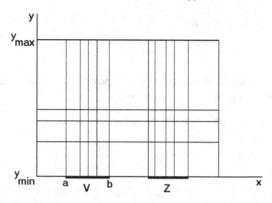

**Fig. 7.2** A simplified rectangular two dimensional cross section with two contacts, on which the potentials are $V$ and $Z$, showing several mesh lines. Values on the lines parallel to the $y$-direction which emanate from the contacts are updated first, followed by the lines parallel to the $x$-direction. Finally, values are updated on lines parallel to the $y$-direction which do not emanate from the contacts.

The use of this process in the modelling of a device can be illustrated by the following example. Consider the simplified cross section of such a device as shown in Fig. 7.2, in which two contacts are placed on the surface $y = y_{min}$. More realistic devices will have more contacts on this face, and some of them will be recessed. Suppose that a function $f(x, y)$ has been found such that $f$ takes the values $V$ and $Z$ on the contacts, and has zero normal derivatives on the remaining free surfaces. The values of $V$ and $Z$ are now changed to $V + \Delta V$ and $Z + \Delta Z$, and it is required to find the function $F(x, y)$ which fits in with these updated potentials, while following the original function $f$ as closely as possible; this new function will perhaps be used as an initial guess in any subsequent iterations for these new potentials. The new function $F$ could be found using the following three steps.

1. At the contact lying in the range $a \leq x \leq b$ on which the potential is $V$, use the result in Equation (7.73) to upgrade the function values along the mesh lines in the $y$-direction which emanate from the contact. Do the same for the second contact.
2. Having found the updated values along these mesh lines, use the result in Equation (7.69), in either its linear or nonlinear form, to update along lines parallel to the $x$-direction to fill in the intermediate points.
3. There are remaining lines running along the $y$-direction which terminate at both ends on free surfaces. Update along these remaining lines by interpolating the updated mesh values found in the previous two steps.

### 7.5.2 Discretising mixed boundary conditions

It is possible to have mixed boundary conditions on a single boundary. Care must be taken, when applying these conditions on a discrete mesh, to apply them in such a way that there are smooth transitions in the computed solution at the points at which these conditions occur. For example, in the simplified cross section shown in Fig. 7.2, the value of $f$ may be specified as $f = V$ on the contact $a < x < b$, but the normal derivative of $f$ may be specified as some constant $C'$ on the adjacent free surfaces. It should then be ensured that there is a smooth change in these conditions at the end points of the contact; this is done by taking a weighted average of the two conditions at that point.

Suppose that the $i$th meshpoint along the $x$-direction is situated at the contact end at $a$: $x_i = a$. The normal derivative of $f$ is specified as the value $C'$ at the point $x_{i-1}$, and the value $f = V$ is specified at the point $x_{i+1}$. Then using Equation (7.8), the conditions on either side of $x = x_i$ are

$$f_{i-1,0} = \frac{-k_0 k_1 (k_0 + k_1) C' + (k_0 + k_1)^2 f_{i-1,1} - k_0{}^2 f_{i-1,2}}{k_1 (2k_0 + k_1)} \tag{7.74}$$

$$f_{i+1,0} = V. \tag{7.75}$$

Hence the weighted average of these two conditions can be taken at $x = x_i$:

$$f_{i,0} = \left( \frac{h_i}{h_{i-1} + h_i} \right) f_{i-1,0} + \left( \frac{h_{i-1}}{h_{i-1} + h_i} \right) f_{i+1,0}. \tag{7.76}$$

In the case of a uniform mesh, this becomes

$$f_{i-1,0} = \frac{-2kC' + 4f_{i-1,1} - f_{i-1,2}}{3}$$

$$f_{i+1,0} = V \tag{7.77}$$

$$f_{i,0} = \frac{1}{2}(f_{i-1,0} + f_{i+1,0}).$$

### 7.5.3 Modelling abrupt junctions

Suppose that an interface between two different materials is situated at the position $y = Y$. For example, if the doping density $N(y)$ is fixed at the value $N_-$ for $y < Y$ and $N_+$ for $y > Y$, then it is required to produce a smooth functional transition of $N(y)$ at $y = Y$. The alternative option—to place a mesh point at $y = Y$ and to use an abrupt change there—can produce problems in any subsequent numerical scheme. We would like to produce a function $N(y)$ which is continuous and has a continuous derivative $N'(y)$ at $y = Y$. Further, we would like to have some control over the width and shape of the transition region. There is obviously no unique way of achieving this, but the method based on that of Stern and Das Sarma (1984) produces a simple form with the above properties. In this scheme, a length $b$ is chosen such that

$$N(y) = \begin{cases} N_- & \text{for } y \leq Y - b \\ N_+ & \text{for } y \geq Y + b, \end{cases}$$

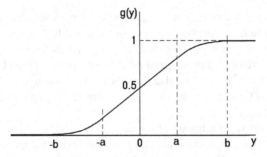

**Fig. 7.3** The function $g(y)$, which is continuous and has a continuous first derivative, defined in the five separate regions $y < -b$, $-b \leq y < -a$, $-a \leq y \leq a$, $a < y \leq b$, and $b < y$.

and the junction is smoothly graded over the intermediate region $Y - b \leq y \leq Y + b$.

Consider the continuous function $g(y)$ shown in Fig. 7.3. It consists of five separate sections: $g(y) = 0$ for $y \leq -b$, $g(y) = 1$ for $y \geq b$, a straight line section for $-a \leq y \leq a$ for some suitably chosen positive value $a < b$, and smooth curves in the remaining regions $-b \leq y \leq -a$ and $a \leq y \leq b$. Now define a function $h(y)$ in the regions $-b \leq y \leq -a$ and $a \leq y \leq b$ as

$$h(y) \equiv \frac{|y| + b + 2a + (\frac{b-a}{\pi})\cos(\frac{\pi(2|y|-a-b)}{2(b-a)})}{2(a+b)}.$$

(7.78)

Then the function $g(y)$ defined by

$$g(y) \equiv \begin{cases} 0 & \text{for } y < -b \\ h(y) & \text{for } -b \leq y < -a \\ \frac{1}{2} + \frac{y}{a+b} & \text{for } -a \leq y \leq a \\ 1 - h(y) & \text{for } a < y \leq b \\ 1 & \text{for } b < y \end{cases}$$

(7.79)

is continuous and has a continuous first derivative for all values of $y$. Hence the functional form

$$N(y) = N_+ g(y - Y) + N_- (1 - g(y - Y))$$

(7.80)

satisfies the requirement of a continuously graded junction whose first derivative is also continuous. The width and shape of the junction is controlled by the choices of the values of the parameters $b$ and $a$.

## Problems

**Problem 7.1.** Show that the three-point formula for the derivative $f'$ of the function $f$ at the right-hand end $i = N_x$ is

$$f'_{N_x} = \frac{h_{N_x-2}(2h_{N_x-1}+h_{N_x-2})f_{N_x} - (h_{N_x-2}+h_{N_x-1})^2 f_{N_x-1} + h_{N_x-1}^2 f_{N_x-2}}{h_{N_x-1}h_{N_x-2}(h_{N_x-1}+h_{N_x-2})}.$$

Invert this formula to obtain an expression for $f_{N_x}$.

**Problem 7.2.** Perform a von Neumann stability analysis for the Crank-Nicholson method.

**Problem 7.3.** The function $h(y)$ is defined as

$$h(y) \equiv \frac{|y| + b + 2a + (\frac{b-a}{\pi})\cos(\frac{\pi(2|y|-a-b)}{2(b-a)})}{2(a+b)},$$

where $0 < a < b$. The function $g(y)$ is defined as

$$g(y) \equiv \begin{cases} 0 & \text{for } y < -b \\ h(y) & \text{for } -b \le y < -a \\ \frac{1}{2} + \frac{y}{a+b} & \text{for } -a \le y \le a \\ 1 - h(y) & \text{for } a < y \le b \\ 1 & \text{for } b < y. \end{cases}$$

Show that $g(y)$ is continuous and has a continuous first derivative for all values of $y$.

# Chapter 8
# Fermi and associated integrals

It has been shown in Chapters 4 and 5 how the Fermi integrals play an important role in the evaluation of electron number and energy densities. In particular, the associated Fermi integrals $I_r(a, b)$ are used in the evaluation of these quantities in quantum wells. For example, it can be seen from Equations (5.46) and (5.47) that the expressions for the electron density $n$ and the energy density $W$ inside a quantum well contain the integrals

$$I_r\left(\frac{1}{k_B T}(E_F - \lambda_L), \frac{1}{k_B T}(\lambda_L - E_c)\right)$$

where $r = 1/2$ and $r = 3/2$ respectively. These terms are of the form $I_r(a, b)$ where

$$a \equiv \frac{1}{k_B T}(E_F - \lambda_L), \qquad b \equiv \frac{1}{k_B T}(\lambda_L - E_c). \tag{8.1}$$

Apart from the quantity $k_B$, all of the terms which go to make up the quantities $a$ and $b$ will depend on position. Furthermore, the values of these terms will change as the equations iterate to solution. Hence the integrals $I_r(a, b)$ must be evaluated continually at each iteration and at each space point.

The numerical evaluation of these integrals comprises only a small part of the overall numerical simulation which, in the case of the HEMT, consists typically of the consistent solutions of the equations of Poisson, current continuity, energy transport, and Schrödinger. It is therefore necessary to obtain simple approximations to these integrals, consistent with maintaining accuracy and speed of evaluation.

## 8.1 Definition of the Fermi integrals

The standard Fermi integrals $F_r(a)$ has been introduced in Sect. 4.1, and the associated Fermi integral $I_r(a, b)$ has been introduced in Sect. 5.4.2. These definitions are repeated here for convenience.

E.A.B. Cole, *Mathematical and Numerical Modelling of*
*Heterostructure Semiconductor Devices: From Theory to Programming,*
DOI 10.1007/978-1-84882-937-4_8, © Springer-Verlag London Limited 2009

### 8.1.1 The standard Fermi integrals

The standard Fermi integral is defined by

$$F_r(a) \equiv \frac{1}{\Gamma(r+1)} \int_0^\infty \frac{x^r}{1+e^{x-a}} dx \quad \text{where} \tag{8.2}$$

$$\Gamma(r+1) \equiv \int_0^\infty x^r e^{-x} dx. \tag{8.3}$$

It is a straightforward matter to show that the following properties hold:

$$\lim_{a \to -\infty} F_r(a) = e^a \quad \text{for all } r,$$

$$\lim_{a \to +\infty} F_r(a) = \frac{1}{(r+1)\Gamma(r+1)} a^r,$$

$$F_0(a) = \ln(1 + e^a),$$

$$\frac{d}{da} F_r(a) = F_{r-1}(a) \quad \text{for all } r > 0,$$

$$\Gamma(r+1) = r\Gamma(r),$$

and

$$\Gamma(1) = 1, \qquad \Gamma(2) = 1, \qquad \Gamma\left(\frac{1}{2}\right) = \sqrt{\pi}, \qquad \Gamma\left(\frac{3}{2}\right) = \frac{1}{2}\sqrt{\pi},$$

$$\Gamma\left(\frac{5}{2}\right) = \frac{3}{4}\sqrt{\pi}.$$

### 8.1.2 The associated Fermi integrals

The associated integrals are defined by

$$I_r(a, b) \equiv \frac{1}{\Gamma(r+1)} \int_b^\infty \frac{x^r}{1+e^{x-(a+b)}} dx \tag{8.4}$$

$$= \frac{1}{\Gamma(r+1)} \int_0^\infty \frac{(x+b)^r}{1+e^{x-a}} dx. \tag{8.5}$$

These integrals can be shown to have the properties

$$I_r(a, 0) = F_r(a), \tag{8.6}$$

$$I_0(a, b) = F_0(a) = \ln(1 + e^a), \tag{8.7}$$

$$I_1(a, b) = F_1(a) + b\ln(1 + e^a), \tag{8.8}$$

$$I_n(a, b) = F_n(a) + b \sum_{r=1}^{n} \frac{1}{r!} b^{r-1} F_{n-r}(a), \quad n \text{ integer } \geq 1, \quad (8.9)$$

$$\frac{\partial}{\partial b} I_r(a, b) = I_{r-1}(a, b), \quad (8.10)$$

$$\frac{\partial}{\partial a} I_r(a, b) = \frac{1}{\Gamma(r + 1)} \frac{b^r}{1 + e^{-a}} + I_{r-1}(a, b). \quad (8.11)$$

This last result is obtained by differentiating with respect to $a$ and then integrating by parts. Very often the integral in Equation (8.5) occurs with a different lower limit $c$. By making the substitution $y = x - c$, such an integral can be expressed as

$$\int_c^\infty \frac{(x + b)^r}{1 + e^{x-a}} dx = \int_0^\infty \frac{(y + c + b)^r}{1 + e^{y+c-a}} dy$$

$$= \int_0^\infty \frac{(y + c + b)^r}{1 + e^{y-(a-c)}} dy$$

$$= \Gamma(r + 1) I_r(a - c, c + b). \quad (8.12)$$

## 8.2 Approximation of the associated integrals

The evaluation of the $I$-integrals is computationally very expensive, and clearly it would be useful to have a technique whereby these integrals may be evaluated rapidly. To achieve this goal, we will seek a way of writing them in terms of simple polynomials of the arguments $a$ and $b$. Define the function $L_r(a, b)$ by

$$L_r(a, b) \equiv I_r(a, b) - F_r(a). \quad (8.13)$$

Then $L_r(a, b)$ has the properties (see Problem 8.1)

$$L_r(a, 0) = 0, \quad (8.14)$$

$$L_0(a, b) = 0, \quad (8.15)$$

$$L_1(a, b) = b \ln(1 + e^a). \quad (8.16)$$

Again,

$$F_r(a) + L_r(a, b) = I_r(a, b) = \frac{1}{\Gamma(r + 1)} \int_b^\infty \frac{z^r}{1 + e^{z-(a+b)}} dz$$

$$= \frac{1}{\Gamma(r + 1)} \left( \int_0^\infty - \int_0^b \right) \frac{z^r}{1 + e^{z-(a+b)}} dz$$

$$= F_r(a + b) - \frac{1}{\Gamma(r + 1)} \int_0^b \frac{z^r}{1 + e^{z-(a+b)}} dz.$$

Hence

$$L_r(a, b) = F_r(a + b) - F_r(a) - \frac{1}{\Gamma(r + 1)} \int_0^b \frac{z^r}{1 + e^{z-(a+b)}} dz. \tag{8.17}$$

In view of the properties in Equations (8.14)–(8.16), now define the intermediate function $c_r(a, b)$ by (Cole 2001)

$$I_r(a, b) = F_r(a) + L_r(a, b)$$
$$= F_r(a) + c_r(a, b) \ln(1 + e^a). \tag{8.18}$$

It is easily shown that $c_r(a, b)$ satisfies the properties

$$c_r(a, b) = \frac{F_r(a + b) - F_r(a) - \frac{1}{\Gamma(r+1)} \int_0^b \frac{z^r}{1 + e^{z-(a+b)}} dz}{\ln(1 + e^a)}, \tag{8.19}$$

and

$$c_r(a, 0) = 0, \qquad c_0(a, b) = 0, \qquad c_1(a, b) = b.$$

If the functions $c_r(a, b)$ can be evaluated rapidly, then the integral $I_r(a, b)$ may be evaluated rapidly using Equation (8.18) when the standard Fermi integrals $F_r(a)$ have been evaluated.

In order to achieve this rapid evaluation, the functions $c_r(a, b)$ will be approximated by new simpler functions $c_r'(a, b)$ which will be taken in the form

$$c_r'(a, b) = b^{p_r}(\alpha_r + \beta_r a), \tag{8.20}$$

where the numbers $p_r$, $\alpha_r$ and $\beta_r$ are to be determined numerically. Clearly, such a form will not be unique; other forms will be considered later.

Let $a_{min}$ and $a_{max}$ be the minimum and maximum values of the argument $a$ which will be encountered in the numerical simulation. Choose a number $(A + 1)$ of representative values $\{a_i : i = 0, 1, \ldots, A\}$ of $a$ in this range. Similarly, let $b_{min}$ and $b_{max}$ be the minimum and maximum values of the argument $b$ which will be encountered in the numerical simulation. Choose a number $(B+1)$ of representative values $\{b_j : j = 0, 1, \ldots, B\}$ of $b$ in this range. Let the chosen representative values be ordered in their respective ranges, so that

$$a_{min} = a_0 < a_1 < a_2 < \ldots < a_A = a_{max}$$
$$b_{min} = b_0 < b_1 < b_2 < \ldots < b_B = b_{max}. \tag{8.21}$$

Then for a given value of the index $r$, evaluate $c_r(a_i, b_j)$ directly using Equation (8.19) for each pair of values $a_i$ and $b_j$. If we define the quantities

$$c_{rij,\alpha_r \beta_r p_r}' \equiv b_j^{p_r}(\alpha_r + \beta_r a_i), \tag{8.22}$$
$$c_{rij} \equiv c_r(a_i, b_j), \tag{8.23}$$

then we will look for those values of $\alpha_r$, $\beta_r$ and $p_r$ which minimise the sum of squares

$$S_r \equiv \sum_{i=0}^{A} \sum_{j=0}^{B} [c'_{rij,\alpha_r \beta_r p_r} - c_{rij}]^2$$

$$= \sum_{i=0}^{A} \sum_{j=0}^{B} \left[ b_j^{2p_r} (\alpha_r + \beta_r a_i)^2 + c_{rij}^2 - 2c_{rij} b_j^{p_r} (\alpha_r + \beta_r a_i) \right]$$

$$= \alpha_r^2 \sum_{i=0}^{A} \sum_{j=0}^{B} b_j^{2p_r} + \beta_r^2 \sum_{i=0}^{A} \sum_{j=0}^{B} b_j^{2p_r} a_i^2 + 2\alpha_r \beta_r \sum_{i=0}^{A} \sum_{j=0}^{B} b_j^{2p_r} a_i$$

$$- 2\alpha_r \sum_{i=0}^{A} \sum_{j=0}^{B} b_j^{p_r} c_{rij} - 2\beta_r \sum_{i=0}^{A} \sum_{j=0}^{B} c_{rij} b_j^{p_r} a_i + \sum_{i=0}^{A} \sum_{j=0}^{B} c_{rij}^2. \quad (8.24)$$

This process will produce an optimal set of values $\alpha_r$, $\beta_r$ and $p_r$ for any Fermi index $r$.

One measure of the accuracy of the method will be given by the *average relative error* $E_r$ defined by

$$E_r \equiv \frac{\sum_{i=0}^{A} \sum_{j=0}^{B} | I_r(a_i, b_j) - F_r(a_i) - b_j^{p_r} (\alpha_r + \beta_r a_i) \ln(1 + e^{a_i}) |}{\sum_{i=0}^{A} \sum_{j=0}^{B} | I_r(a_i, b_j) |}. \quad (8.25)$$

Although there is no unique way of defining such an error, the expression in Equation (8.25) is taken to give greater weight to larger values of the integral, since these larger values contribute most to the values of the electron number and energy densities.

## 8.3 Implementation of the approximation scheme

Before a full iteration scheme is carried out for a particular device, the values of the quantities $p_r$, $\alpha_r$ and $\beta_r$ must be calculated for each possible value of the index $r$ which will be used in the model. These values need be calculated only once, and not at every iteration. This calculation will follow the scheme based on Equations (8.20)–(8.24).

### 8.3.1 Method of implementation

The following steps should be carried out at the beginning of any full device simulation:

1. The values $a_{min}$, $a_{max}$, $b_{min}$ and $b_{max}$ must be found in order to provide the search ranges for $a$ and $b$. This task is accomplished by running a small number—

say, 100—of iterations of the full set of device equations without using the correction terms; that is, by using a lower limit of zero in all of the $I_r$-integrals. These iterations will not provide an accurate start to the full iteration procedure, and will involve the illegal double-counting of states in the quantum wells. However, by noting the minimum and maximum values of the quantities $a$ and $b$ which are calculated using such expressions as in Equation (8.1), estimates of $a_{min}$, $a_{max}$, $b_{min}$ and $b_{max}$ can be made.

2. Split up the ranges $(a_{min}, a_{max})$ and $(b_{min}, b_{max})$ into $A$ and $B$ intervals respectively, and calculate the intermediate values $a_i$ and $b_j$ using Equations (8.21).

3. For a given value of the index $r$,

   - Decide on search ranges for each of the quantities $p_r$, $\alpha_r$ and $\beta_r$, and take intermediate values in these ranges.
   - Calculate the value of $c'$ using Equation (8.20) for each representative triple of the quantities $p_r$, $\alpha_r$ and $\beta_r$.
   - Calculate the exact set of values of the quantities $c_{rij} \equiv c_r(a_i, b_j)$ using Equation (8.19).
   - Calculate the sum of squares $S_r$ using Equation (8.24).
   - Sweep the representative values of $p_r$, $\alpha_r$ and $\beta_r$ to find the minimum value of $S_r$.

4. Repeat step 3 for new value of the index $r$.

## 8.3.2 Results of the implementation

A simple HEMT device was modelled. Using the method of step 1 above, it was found that the values of $a$ and $b$ were bounded by the values

$$a_{min} = -8.0, \qquad a_{max} = 8.0, \qquad b_{min} = 0.0001, \qquad b_{max} = 8.0.$$

Note that the value of $b_{min}$ is taken as a very small positive value, rather than zero, to avoid underflow in the computation. The search ranges $a_{min} \leq a \leq a_{max}$ and $b_{min} \leq b \leq b_{max}$ were split into $A = 10$ and $B = 10$ representative values respectively. The search ranges for $p_r$, $\alpha_r$ and $\beta_r$ were found experimentally; taking the number of steps in each range to be 300, it was found that the values

$$-2.0 \leq \alpha_r \leq 4.0, \qquad -1.0 \leq \beta_r \leq 1.0,$$

and

$$0.0 \leq p_r \leq 2r \quad (r > 0) \quad \text{or}$$
$$-1.0 \leq p_r \leq 1.0 \quad (r \leq 0)$$

provided acceptable results. Results are given in Table 8.1. This Table lists the optimal triples $(p_r, \alpha_r, \beta_r)$ for all of the values of the index $r$ which appear in the

**Table 8.1** Calculated optimal values of $p_r$, $\alpha_r$, $\beta_r$ and the average (percentage) relative error $E_r$ for various values of $r$.

| $r$ | $p_r$ | $\alpha_r$ | $\beta_r$ | $E_r$ (%) |
|------|--------|---------|---------|---------|
| $-1/2$ | 0.2154 | $-0.3860$ | 0.0294 | 13.45 |
| $1/2$ | 0.6968 | 0.5138 | $-0.0112$ | 1.93 |
| $3/2$ | 1.2937 | 1.4202 | 0.0279 | 3.48 |
| 2 | 1.5564 | 1.8126 | 0.0681 | 7.35 |
| $5/2$ | 1.8109 | 2.1184 | 0.1156 | 11.41 |

modelling equations, while the final column shows the percentage errors relating to each optimal triple.

### 8.3.3 Improvements to the scheme

The approximation method described above is somewhat crude in the sense that arbitrary numbers $(A+1)$ and $(B+1)$ of intermediate values are taken of the quantities $a$ and $b$ in their respective ranges. Arbitrary numbers of intermediate values are also taken for the quantities $p_r$, $\alpha_r$ and $\beta_r$ in their respective ranges. Again, the form of the approximation given in Equation (8.20) is not the only one that could be taken.

It will be shown in Chapter 14 on Genetic Algorithms how the optimisation process can be performed in a much more satisfactory manner using the method of simulated annealing. This method will provide a much more efficient optimisation process, and will allow guesses other than that in Equation (8.20) to be considered and compared.

## 8.4 Calculation of the standard Fermi integrals

The approximations to the associated Fermi integrals have been calculated by minimising the sum of squares $S_r$ which was defined in Equation (8.24). This calculation itself requires the calculation of the exact quantities $c_{rij} \equiv c_r(a_i, b_j)$ from the result in Equation (8.19), and so the standard Fermi integrals $F_r(a)$ must be provided. It has been shown (Blakemore 1962; Blakemore 1982) that good analytic approximations for these integrals take the form

$$F_r(a) \approx (\exp(-a) + \xi_r(a))^{-1}, \qquad (8.26)$$

for which Bednarczyk and Bednarczyk (1978) take

$$\xi_{\frac{1}{2}}(a) \equiv 0.75\pi^{1/2}(a^4 + 50 + 33.6a\{1 - 0.68\exp[-0.17(a+1)^2]\})^{-3/8} \quad (8.27)$$

and Aymerich-Humet et al. (1981) take

$$\xi_{\frac{3}{2}}(a) \equiv 15(\pi/2)^{1/2}(a + 2.64 + (|a - 2.64|^{9/4} + 14.9)^{-5/2}. \qquad (8.28)$$

Rational Chebyshev approximations of the integrals have been studied by Jones (1966) and Cody and Thacher (1967). Polynomial approximations have been given by Battocletti (1965) and Arpigny (1963).

## Problems

**Problem 8.1.** Show that the function $L_r(a, b)$, which is defined as

$$L_r(a, b) \equiv \frac{1}{\Gamma(r + 1)} \int_0^\infty \frac{(x + b)^r}{1 + e^{x - a}} dx - F_r(a),$$

satisfies the properties

$$L_r(a, 0) = 0,$$
$$L_0(a, b) = 0,$$
$$L_1(a, b) = b \ln(1 + e^a).$$

**Problem 8.2.** In the HEMT simulation described in Sect. 8.3.2, it was found that the minimum and maximum values of the arguments $a$ and $b$ were

$$a_{min} = -8.0, \qquad a_{max} = 8.0, \qquad b_{min} = 0.0001, \qquad b_{max} = 8.0.$$

Values $A = 10$ and $B = 10$ were chosen, and the search ranges for $p_r$, $\alpha_r$ and $\beta_r$ were taken as

$$-2.0 \le \alpha_r \le 4.0, \qquad -1.0 \le \beta_r \le 1.0,$$

and

$$0.0 \le p_r \le 2r \quad (r > 0) \quad \text{or}$$
$$-1.0 \le p_r \le 1.0 \quad (r \le 0).$$

1. Write a programme to repeat this calculation.
2. Vary the search range values $A$ and $B$.
3. Vary the search ranges for the values of $p_r$, $\alpha_r$ and $\beta_r$. What happens if the ranges are made too short?
4. Investigate different forms for the function $c_r'(a, b)$.
5. Investigate the possibility that a choice of a new function $S_r$, to replace that in Equation (8.24), will make any appreciable difference to the computed optimal values of $p_r$, $\alpha_r$ and $\beta_r$.
6. Investigate different forms for the measure of the error given in Equation (8.25). Will the results of taking different forms affect your choice for the function $c_r'(a, b)$?

# Chapter 9
# The upwinding method

It has already been shown in Sect. 7.4.1 that, when using the von Neumann stability analysis on the time-discretisation scheme, the magnitude and direction of the electric field $\mathbf{E}$ has an important bearing on the stability of the scheme. The upwinding method is a method of discretisation which utilises the "flow" of an influence from neighbouring spatial points. This method has previously been applied to the electron continuity equation only by making use of Bernoulli functions, but it will be seen how the method can be generalised to use with the energy transport equation using the C-functions which will be introduced in this Chapter.

## 9.1 Description of the upwinding approach

It has been seen in Sect. 6.3.2 that current density $\mathbf{J}$ takes the form

$$\mathbf{J} = -qn\mu\nabla_{\mathbf{r}}\psi + k_B\mu T_e\nabla_{\mathbf{r}}n + ak_B\mu n\nabla_{\mathbf{r}}T_e + bk_B T_e n\nabla_{\mathbf{r}}\mu \qquad (9.1)$$

where $q > 0$ is the magnitude of the electron charge, $k_B$ is Boltzmann's constant. $\psi$ is the electrostatic potential, $T_e$ is the electron temperature, $n$ is the electron density, and $\mu$ is the electron mobility. The constant quantities $(a, b)$ have been taken by various authors to have the values $(1,1)$, $(\frac{1}{2},0)$, or $(1,0)$. Due to a surfeit of extra suffices which will be introduced in this chapter, the suffix $e$ on $T_e$ will be dropped for the remainder of the chapter, since there is only one temperature involved, and no confusion should arise.

In order to illustrate the use of the upwinding process, consider a crude one-dimensional steady state calculation, using a uniform mesh, in which the quantities $\mathbf{J}$, the electron temperature $T$, $\mu$, and the electric field $\mathbf{E} = -\nabla_{\mathbf{r}}\psi$ are constants. That is, take

$$J = qn\mu E + k_B\mu T\frac{dn}{dx} = \text{constant}, \qquad (9.2)$$

or

E.A.B. Cole, *Mathematical and Numerical Modelling of*
*Heterostructure Semiconductor Devices: From Theory to Programming*,
DOI 10.1007/978-1-84882-937-4_9, © Springer-Verlag London Limited 2009

$$\frac{dn}{dx} + \frac{qE}{k_BT}n = \frac{J}{\mu k_BT}.$$

In view of the crude approximation being used in this example, this equation has the solution

$$n = Ae^{-\frac{qE}{k_BT}x} + \frac{J}{q\mu E}.$$

Hence the values of $n$ at the discretised mesh points $x = x_i$ and $x = x_i + h$ are

$$n_i = Ae^{-\frac{qE}{k_BT}x_i} + \frac{J}{q\mu E} \quad \text{and} \quad n_{i+1} = Ae^{-\frac{qE}{k_BT}(x_i+h)} + \frac{J}{q\mu E}$$

Then for $x$ in the range $x_i \leq x \leq x_{i+1}$, it follows that

$$\frac{n - n_i}{n_{i+1} - n_i} = \frac{1 - e^{\frac{-P(x-x_i)}{h}}}{1 - e^{-P}}$$

where $P \equiv qEh/(k_BT)$ is called the *Peclet Number*. The quantities $P$ and $E$ have the same sign, and therefore

- if $P$ (or $E$) is large and positive then $n \approx n_{i+1}$;
- if $P$ (or $E$) is large and negative then $n \approx n_i$;
- if $P$ (or $E$) = 0 then $n = \frac{1}{2}(n_i + n_{i+1})$.

The drift part of the current can then be replaced with

$$J_{drift} \equiv q\mu En = \begin{cases} q\mu En_i & \text{if } E < 0 \text{ (influence from left)} \\ q\mu En_{i+1} & \text{if } E > 0 \text{ (influence from right)} \\ q\mu E(n_i + n_{i+1})/2 & \text{if } E = 0. \end{cases} \quad (9.3)$$

Thus a crude either-or situation can be used to give the value of $J_{drift}$ in the range $x_i \leq x \leq x_{i+1}$, with the value of $n$ at $x_i$ selected if the field sweeps from left to right, or the value of $n$ at $x_{i+1}$ is selected if the field sweeps from right to left.

The following sections will provide a more comprehensive description of the upwinding method applied to the full current continuity equation, and also to the energy transport equation. This description will involve a continuous distribution, rather than an either-or distribution, so that derivatives exist when calculating the Jacobian matrices for the direct method.

## 9.2  Upwinding applied to device equations

The current density **J** and energy flux **s** are given in the general forms

$$\mathbf{J} = \alpha_C(\nabla_\mathbf{r}\psi)n + \beta_C T(\nabla_\mathbf{r}n) + \gamma_C n\nabla_\mathbf{r}T$$
$$\mathbf{s} = \alpha_E(\nabla_\mathbf{r}\psi)W + \beta_E T(\nabla_\mathbf{r}W) + \gamma_E W\nabla_\mathbf{r}T \quad (9.4)$$

where the quantities $\alpha_C, \ldots, \gamma_E$ are not generally constant, but will be functions of the mobility, Fermi integrals etc. These two vectors are special cases of the general vector $\mathbf{v}$ which is given by

$$\mathbf{v} = \alpha(\nabla_\mathbf{r}\psi)\theta + \beta T(\nabla_\mathbf{r}\theta) + \gamma(\nabla_\mathbf{r}T)\theta + bT\theta(\nabla_\mathbf{r}\beta) \qquad (9.5)$$

where $b = 0$ or $1$, and $(\mathbf{v}, \theta) \equiv (\mathbf{J}, n)$ in the case of the current density, or $(\mathbf{v}, \theta) \equiv (\mathbf{s}, W)$ in the case of the energy flux.

A non-uniform two dimensional mesh is taken with the variable steplengths $h_i \equiv x_{i+1} - x_i$ in the $x$-direction and $k_j \equiv y_{j+1} - y_j$ in the $y$-direction. The divergence $\nabla_\mathbf{r} \cdot \mathbf{v}$ at the meshpoint $(i, j)$ will have the form

$$\nabla_\mathbf{r} \cdot \mathbf{v} = \frac{(v_x)_{i+\frac{1}{2},j} - (v_x)_{i-\frac{1}{2},j}}{(h_i + h_{i-1})/2}$$

$$+ \frac{(v_y)_{i,j+\frac{1}{2}} - (v_y)_{i,j-\frac{1}{2}}}{(k_j + k_{j-1})/2}. \qquad (9.6)$$

The following analysis will be carried out on the $x$-component only; exactly the same considerations will apply to the $y$-component. On multiplying Equation (9.5) by $\beta^{b-1}$, the $x$-component can be written in the form

$$\frac{\partial}{\partial x}(\beta^b\theta) + \left( \frac{\gamma}{\beta}\frac{1}{T}\frac{\partial T}{\partial x} + \frac{\alpha}{\beta}\frac{1}{T}\frac{\partial \psi}{\partial x} \right)(\beta^b\theta) = \frac{1}{T}(\beta^{b-1}v_x). \qquad (9.7)$$

In order to proceed, it will now be assumed that the quantities $\partial\psi/\partial x$, $\beta^{b-1}$, $\partial T/\partial x$, $\alpha/\beta$ and $\gamma/\beta$ are all constants in the interval $(i, j) \rightarrow (i+1, j)$. These assumptions are justified as follows:

- the first two quantities are those which are held constant in the original Scharfetter-Gummel scheme for the case of constant temperature (Scharfetter and Gummel 1968). These authors showed that substantial errors are introduced for quite small values of the mesh spacings when it is assumed that $\psi$ itself is held constant in the interval. The assumption that only the derivative is constant in the interval leads to much smaller errors in the solution, and allows the use of much larger mesh spacings;
- Tang (1984) applied the same argument to the electron temperature $T$;
- for the remaining quantities $\alpha/\beta$ and $\gamma/\beta$, these are strictly constant in the non-degenerate case in which the effective mass is constant, while in the degenerate case they depend on the ratios of certain Fermi integrals.

Under these assumptions, Equation (9.7) possesses an integrating factor

$$\exp\left[\int\left(\frac{\gamma}{\beta}\frac{1}{T}\frac{\partial T}{\partial x}+\frac{\alpha}{\beta}\frac{1}{T}\frac{\partial\psi}{\partial x}\right)dx\right]$$

$$=\exp\left[\left(\frac{\gamma}{\beta}\right)\ln|T|+\frac{\alpha}{\beta}\frac{\partial\psi}{\partial x}\int\frac{1}{T}\left(\frac{\partial T}{\partial x}\right)^{-1}dT\right]$$

$$=\exp\left[\left(\frac{\gamma}{\beta}\right)\ln|T|+\frac{\alpha}{\beta}\frac{\partial\psi}{\partial x}\left(\frac{\partial T}{\partial x}\right)^{-1}\ln|T|\right]$$

$$=|T|^{r}$$

where the power $r$ is given by

$$r\equiv\frac{\gamma}{\beta}+\frac{\alpha}{\beta}\frac{\partial\psi}{\partial x}\left(\frac{\partial T}{\partial x}\right)^{-1}.$$

Hence Equation (9.7) can be written

$$\frac{\partial}{\partial x}\left(\beta^{b}\theta|T|^{r}\right)=\left(\beta^{b-1}v_{x}\right)_{i+\frac{1}{2},j}|T|^{r-1}\text{sign}(T). \tag{9.8}$$

This is integrated from $x=x_i$ to $x=x_{i+1}=x_i+h_i$ to give

$$\left[\beta^{b}\theta|T|^{r}\right]_{x_i}^{x_{i+1}}=\left(\beta^{b-1}v_{x}\right)_{i+\frac{1}{2},j}\int_{x_i}^{x_{i+1}}|T|^{r-1}\text{sgn}(T)dx$$

$$=\left(\beta^{b-1}v_{x}\right)_{i+\frac{1}{2},j}\left(\frac{\partial T}{\partial x}\right)^{-1}\int|T|^{r-1}\text{sgn}(T)dT$$

$$=\left(\beta^{b-1}v_{x}\right)_{i+\frac{1}{2},j}\left(r\frac{\partial T}{\partial x}\right)^{-1}\left[|T|^{r}\right]_{x_i}^{x_{i+1}}, \tag{9.9}$$

where the result

$$\int|T|^{r-1}\text{sgn}(T)dT=\frac{1}{r}|T|^{r}$$

has been used. Now, at the midpoint of the interval $x_i\leq x\leq x_{i+1}$, we have

$$r\frac{\partial T}{\partial x}=\left(\frac{\gamma}{\beta}\right)\frac{\partial T}{\partial x}+\left(\frac{\alpha}{\beta}\right)\frac{\partial\psi}{\partial x}$$

$$=\left(\frac{\gamma}{\beta}\right)_{i+\frac{1}{2},j}\frac{T_{i+1,j}-T_{i,j}}{h_i}+\left(\frac{\alpha}{\beta}\right)_{i+\frac{1}{2},j}\frac{\psi_{i+1,j}-\psi_{i,j}}{h_i},$$

$$r=\left(\frac{\gamma}{\beta}\right)_{i+\frac{1}{2},j}+\left(\frac{\alpha}{\beta}\right)_{i+\frac{1}{2},j}\frac{\psi_{i+1,j}-\psi_{i,j}}{T_{i+1,j}-T_{i,j}}.$$

Remembering that

$$|T|^{r}=\exp(r\ln|T|),$$

it follows that

$$
\left(\beta^{b-1}v_x\right) = \frac{[(\frac{\gamma}{\beta})_{i+\frac{1}{2},j}(T_{i+1,j} - T_{i,j}) + (\frac{\alpha}{\beta})_{i+\frac{1}{2},j}(\psi_{i+1,j} - \psi_{i,j})]}{h_i(e^{r\ln|T_{i+1,j}|} - e^{r\ln|T_{i,j}|})}
$$
$$
\times [(\beta^b\theta)_{i+1,j}e^{r\ln|T_{i+1,j}|} - (\beta^b\theta)_{i,j}e^{r\ln|T_{i,j}|}]
$$
$$
= \frac{[(\frac{\gamma}{\beta})_{i+\frac{1}{2},j}(T_{i+1,j} - T_{i,j}) + (\frac{\alpha}{\beta})_{i+\frac{1}{2},j}(\psi_{i+1,j} - \psi_{i,j})]}{h_i(1 - e^{r\ln|T_{i,j}/T_{i+1,j}|})}\left(\beta^b\theta\right)_{i+1,j}
$$
$$
- \frac{[(\frac{\gamma}{\beta})_{i+\frac{1}{2},j}(T_{i+1,j} - T_{i,j}) + (\frac{\alpha}{\beta})_{i+\frac{1}{2},j}(\psi_{i+1,j} - \psi_{i,j})]}{h_i(e^{r\ln|T_{i+1,j}/T_{i,j}|} - 1)}\left(\beta^b\theta\right)_{i,j}
$$
$$
= \frac{1}{h_i}\frac{[-(\frac{\gamma}{\beta})_{i+\frac{1}{2},j}(T_{i+1,j} - T_{i,j}) - (\frac{\alpha}{\beta})_{i+\frac{1}{2},j}(\psi_{i+1,j} - \psi_{i,j})]}{e^{r\ln|T_{i,j}/T_{i+1,j}|} - 1}
$$
$$
\times \left(\beta^b\theta\right)_{i+1,j}
$$
$$
- \frac{1}{h_i}\frac{[-(\frac{\gamma}{\beta})_{i+\frac{1}{2},j}(T_{i+1,j} - T_{i,j}) - (\frac{\alpha}{\beta})_{i+\frac{1}{2},j}(\psi_{i+1,j} - \psi_{i,j})]}{e^{r\ln|T_{i+1,j}/T_{i,j}|} - 1}
$$
$$
\times \left(\beta^b\theta\right)_{i,j}. \tag{9.10}
$$

## 9.3 Upwinding in terms of the C-function

The upwinding method will be based on the result in Equation (9.10). In its present form, however, this equation is too unwieldy for coding purposes. By recognising that there is a pattern to the terms which appear in this equation, it is possible to define a function—the C-function—which enables the coding to be performed much more simply.

### 9.3.1 Definition of the C-function

For real values $u$, $x$, $y$ and $z$, the C-function is defined by

**Definition 9.1.**

$$
C_u(x, y, z) \equiv \begin{cases} \dfrac{x - u(y - z)}{\exp(\frac{x - u(y - z)}{z - y}\ln|z/y|) - 1} & (y \neq z) \\[2ex] \dfrac{x}{e^{\frac{x}{y}} - 1} & (y = x) \end{cases} \tag{9.11}
$$

Then it follows (see Problem 9.1) that Equation (9.10) can be written in terms of the C-function as

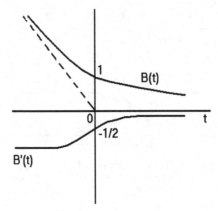

**Fig. 9.1** Plots of the Bernoulli function $B(t)$ and its derivative $B'(t)$. The dashed line has slope $-1$.

$$v_{(x)i+\frac{1}{2},j} = \frac{1}{h_i}(\beta^{1-b})_{i+\frac{1}{2},j}$$

$$\times \left[ C_{(\frac{\gamma}{\beta})_{i+\frac{1}{2},j}} \left( -\left(\frac{\alpha}{\beta}\right)_{i+\frac{1}{2},j} (\psi_{i+1,j} - \psi_{i,j}), T_{i+1,j}, T_{i,j} \right) (\beta^b \theta)_{i+1,j} \right.$$

$$\left. - C_{(\frac{\gamma}{\beta})_{i+\frac{1}{2},j}} \left( \left(\frac{\alpha}{\beta}\right)_{i+\frac{1}{2},j} (\psi_{i+1,j} - \psi_{i,j}), T_{i,j}, T_{i+1,j} \right) (\beta^b \theta)_{i,j} \right].$$

$$(9.12)$$

### 9.3.2 Properties of the C-function and related functions

The C-function has been defined in Equation (9.11), and the upwinding formula in Equation (9.12) has been written in terms of it. However, computer coding of the C-function as defined in Equation (9.11) is more efficiently performed using two related functions—the Bernoulli function and the p-function.

**Definition 9.2.** The Bernoulli function $B(t)$ is defined as

$$B(t) \equiv \frac{t}{e^t - 1}. \tag{9.13}$$

Fig. 9.1 shows plots of $B(t)$ and its derivative $B'(t)$. It is a straightforward matter to prove that the Bernoulli function has the following properties (see Problem 9.2):

$$B(0) = 1,$$

$$B(t) \approx 1 - \frac{1}{2}t \quad \text{for } |t| \ll 1,$$

$$B(t) \to 0 \quad \text{as } t \to \infty,$$

$$B(t) \to -t \quad \text{as } t \to -\infty,$$

$$B(-t) = B(t) + t, \tag{9.14}$$

$$B'(t) < 0,$$

$$B'(t) = \frac{B(t)}{t}(1 - B(t) - t),$$

$$B'(t) \approx \frac{1}{4}t - \frac{1}{2} \quad \text{for } |t| \ll 1.$$

In any numerical implementation using the function $B(t)$, care must be taken to avoid overflows and underflows when evaluating the function. Selberherr (1984) uses a piecewise approximation in the form

$$B(t) = \begin{cases} -t & \text{for } t \le t_1 \\ \frac{t}{e^t - 1} & \text{for } t_1 < t \le t_2 \\ 1 - \frac{1}{2}t & \text{for } t_2 < t \le t_3 \\ \frac{te^{-t}}{1 - e^{-t}} & \text{for } t_3 < t \le t_4 \\ te^{-t} & \text{for } t_4 < t \le t_5 \\ 0 & \text{for } t_5 < t \end{cases}$$

where the values of $t_1, \ldots, t_5$ delimit certain regions of the $t$-axis. A clever aspect of this method is that $t_1, \ldots, t_5$ are determined by the particular machine on which the numerics are performed; for example, the value of $t_1$ may be found by asking the machine to solve the equation $-t = t/(e^t - 1)$ to its own accuracy. I found the values

$$t_1 = -t_4 = -22.874, \qquad t_2 = -t_3 = -2.527 \times 10^{-3}, \qquad t_5 = 89.416$$

when implementing the routine.

A second useful intermediate function is the p-function:

**Definition 9.3.** The p-function $p(y, z)$ is defined for $y > 0$ and $z > 0$ as

$$p(y, z) \equiv \begin{cases} \frac{z - y}{\ln z - \ln y} & \text{for } z \neq y \\ y & \text{for } z = y \end{cases}. \tag{9.15}$$

It is then a straightforward matter to prove that the p-function has the following properties:

$$p(x, y) = p(y, x), \tag{9.16}$$

$$p(ax, ay) = ap(x, y) \quad \text{for } a, x, y > 0, \tag{9.17}$$

$$p(1, y) = \frac{y - 1}{\ln y} = \frac{1}{B(\ln y)}, \tag{9.18}$$

$$p(1, e^y) = \frac{1}{B(y)}, \tag{9.19}$$

$$p(x, y) = xp(1, y/x) = \frac{x}{B(\ln y/x)} = \frac{y}{B(\ln x/y)} \tag{9.20}$$

$$\min(x, y) \le p(x, y) \le \max(x, y). \tag{9.21}$$

Result (9.21) is proved as follows: suppose that $x < y$. Then $y/x > 1$ and $x/y < 1$, so that $B(\ln y/x) < 1$ and $B(\ln x/y) > 1$. Hence

$$p(x, y) = xp(1, y/x) = \frac{x}{B(\ln y/x)} > x$$

$$p(x, y) = yp(1, x/y) = \frac{y}{B(\ln x/y)} < y.$$

A physical interpretation of the function $p(x, y)$ is given by comparing it with the arithmetic and geometric means of $x$ and $y$. Defining the arithmetic mean $m_A(x, y)$ and the geometric mean $m_G(x, y)$ by

$$m_A(x, y) \equiv \frac{1}{2}(x + y),$$

$$m_G(x, y) \equiv \sqrt{xy} \quad (x, y > 0).$$

Then it may be shown that, if $x, y > 0$, then

$$m_G(x, y) \le p(x, y) \le m_A(x, y), \tag{9.22}$$

with equality holding only if $x = y$ (see Problem 9.3). Hence the function $p(x, y)$, which will be called the *logarithmic mean*, lies between the arithmetic and geometric means. Plots of these means are given in Fig. 9.2.

Returning to the C-function, the definitions of the Bernoulli function and the p-function allow it to be written as

$$C_u(x, y, z) = p(y, z)B\left(\frac{x - u(y - z)}{p(y, z)}\right). \tag{9.23}$$

Further, it is a simple matter to prove that

$$C_u(-x, z, y) - C_u(x, y, z) = x - u(y - z). \tag{9.24}$$

The actual coding of the C-function is then performed through the related Bernoulli function and the p-function.

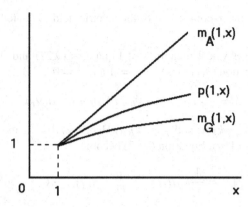

**Fig. 9.2** Plots of the geometric mean $m_G(1, x)$, the arithmetic mean $m_A(1, x)$ and the function $p(1, x)$, plotted for $x \geq 1$.

## 9.4 Upwinding using the C-function

The notation will be simplified by writing

$$C^+_{i+\frac{1}{2},j} \equiv C_{(\frac{\gamma}{\beta})_{i+\frac{1}{2},j}} \left( -\left(\frac{\alpha}{\beta}\right)_{i+\frac{1}{2},j} (\psi_{i+1,j} - \psi_{i,j}), T_{i+1,j}, T_{i,j} \right), \quad (9.25)$$

$$C^-_{i+\frac{1}{2},j} \equiv C_{(\frac{\gamma}{\beta})_{i+\frac{1}{2},j}} \left( \left(\frac{\alpha}{\beta}\right)_{i+\frac{1}{2},j} (\psi_{i+1,j} - \psi_{i,j}), T_{i,j}, T_{i+1,j} \right). \quad (9.26)$$

The particular superscripts $+$ and $-$ are chosen in these definitions because the value of $\alpha/\beta$ is normally negative. Using this notation, Equation (9.12) can be written

$$v_{(x)i+\frac{1}{2},j} = \frac{1}{h_i} (\beta^{1-b})_{i+\frac{1}{2},j} \left[ C^+_{i+\frac{1}{2},j} (\beta^b \theta)_{i+1,j} - C^-_{i+\frac{1}{2},j} (\beta^b \theta)_{i,j} \right]. \quad (9.27)$$

Further, it can be seen from Equation (9.24) that

$$
\begin{aligned}
C^-_{i+\frac{1}{2},j} - C^+_{i+\frac{1}{2},j} &= -\left(\frac{\alpha}{\beta}\right)_{i+\frac{1}{2},j} (\psi_{i+1,j} - \psi_{i,j}) - \left(\frac{\gamma}{\beta}\right)_{i+\frac{1}{2},j} (T_{i+1,j} - T_{i,j}) \\
&= h_i \left[ -\left(\frac{\alpha}{\beta}\right)_{i+\frac{1}{2},j} \left(\frac{\psi_{i+1,j} - \psi_{i,j}}{h_i}\right) \right. \\
&\quad \left. -\left(\frac{\gamma}{\beta}\right)_{i+\frac{1}{2},j} \left(\frac{T_{i+1,j} - T_{i,j}}{h_i}\right) \right] \\
&\approx h_i \left[ \left(\frac{\alpha}{\beta}\right)_{i+\frac{1}{2},j} E_{(x)i+\frac{1}{2},j} - \left(\frac{\gamma}{\beta}\right)_{i+\frac{1}{2},j} \left(\frac{\partial T}{\partial x}\right)_{i+\frac{1}{2},j} \right], \quad (9.28)
\end{aligned}
$$

where $E_{(x)i+\frac{1}{2},j}$ is the $x$-component of the electric field evaluated at the half point $(i + \frac{1}{2}, j)$.

To illustrate the physical meanings of Equations (9.27) and (9.28), consider a special case of Equation (9.1) in which $a = 1$ and $b = 0$:

$$\mathbf{J} = -qn\mu\nabla_r\psi + k_B\mu T\nabla_r n + k_B\mu n\nabla_r T. \tag{9.29}$$

Then $\alpha = -q\mu$, $\beta = k_B\mu$, and $\gamma = k_B\mu$. Hence $\gamma/\beta = 1$ and $\alpha/\beta = -q/k_B$. Then with $(\mathbf{v}, \theta) \equiv (\mathbf{J}, n)$, Equation (9.27) becomes

$$J_{(x)i+\frac{1}{2},j} = \frac{1}{h_i}k_B\mu_{i+\frac{1}{2},j}(C^+_{i+\frac{1}{2},j}n_{i+1,j} - C^-_{i+\frac{1}{2},j}n_{i,j}) \tag{9.30}$$

and Equation (9.28) becomes

$$C^-_{i+\frac{1}{2},j} - C^+_{i+\frac{1}{2},j} = -h_i\left(\frac{q}{k_B}E_{(x)} + \frac{\partial T}{\partial x}\right)_{i+\frac{1}{2},j}. \tag{9.31}$$

It can be seen from Equation (9.30) that the current is dominated by either the contribution from $n_{i+1,j}$ or $n_{i,j}$ depending on whether $C^+_{i+\frac{1}{2},j} > C^-_{i+\frac{1}{2},j}$ or $C^-_{i+\frac{1}{2},j} > C^+_{i+\frac{1}{2},j}$. These inequalities in turn are determined by the sign of the right hand side of Equation (9.31).

For the isothermal case in which $T$ is a constant, then

$$C^-_{i+\frac{1}{2},j} - C^+_{i+\frac{1}{2},j} = -h_i\frac{q}{k_B}E_{(x)i+\frac{1}{2},j}. \tag{9.32}$$

In this case, the term involving $n_{i+1,j}$ will dominate the term $n_{i,j}$ if $C^+_{i+\frac{1}{2},j} > C^-_{i+\frac{1}{2},j}$, that is, if $E_{(x)} > 0$ at the mid-point of the interval. On the other hand, the term $n_{i,j}$ will dominate the term $n_{i+1,j}$ if $C^-_{i+\frac{1}{2},j} > C^+_{i+\frac{1}{2},j}$, that is, if $E_{(x)} < 0$ at this point.

For the non-isothermal case in which Equation (9.31) applies, it is the sign of the combination

$$\frac{q}{k_B}E_{(x)} + \frac{\partial T}{\partial x}$$

evaluated at the mid-point of the interval which determines which term in the expression for $n$ should dominate. More generally still, it is the sign of the combination

$$\left(\frac{\alpha}{\beta}\right)_{i+\frac{1}{2},j}E_{(x)i+\frac{1}{2},j} - \left(\frac{\gamma}{\beta}\right)_{i+\frac{1}{2},j}\left(\frac{\partial T}{\partial x}\right)_{i+\frac{1}{2},j}$$

on the right hand side of Equation (9.28) that determines which term in Equation (9.27) will dominate.

In this way, the original either-or situation of Equation (9.3) has been replaced by the continuous gradation involved in Equation (9.30).

## 9.5 Numerical diffusion

Result (9.22) shows that three different means of the temperature $T$ can be defined at the half point. They are the logarithmic mean $T_{(L)i+\frac{1}{2},j}$, the arithmetic mean $T_{(A)i+\frac{1}{2},j}$, and the geometric mean $T_{(G)i+\frac{1}{2},j}$ which are defined by

$$T_{(L)i+\frac{1}{2},j} \equiv p(T_{i,j}, T_{i+1,j})$$
$$T_{(A)i+\frac{1}{2},j} \equiv \frac{1}{2}(T_{i,j} + T_{i+1,j}) \quad \text{and} \tag{9.33}$$
$$T_{(G)i+\frac{1}{2},j} \equiv \sqrt{T_{i,j}T_{i+1,j}}.$$

It can now be shown that, in certain circumstances, it is more appropriate to use the logarithmic mean at the half-point rather than the more usual arithmetic mean of the adjacent gridpoint values. This certainly applies to the expression for the electron temperature $T$ at the half-point when the C-function method is used (Cole and Snowden 2000). To simplify the analysis, take a uniform mesh with $h_i \equiv h$, and take $b = 0$ in Equation (9.27). Taking $(\mathbf{v}, \theta) \equiv (\mathbf{J}, n)$, this equation gives

$$J_{(x)i+\frac{1}{2},j} = \frac{1}{h}\beta_{i+\frac{1}{2},j}(C^+_{i+\frac{1}{2},j}n_{i+1,j} - C^-_{i+\frac{1}{2},j}n_{i,j}). \tag{9.34}$$

The current continuity equation at gridpoint $(i, j)$ is

$$\left(\frac{dn}{dt}\right)_{i,j} = \frac{1}{q}(\nabla_{\mathbf{r}} \cdot \mathbf{J})_{i,j} = \frac{1}{q}\left(\frac{\partial J_{(x)}}{\partial x} + \frac{\partial J_{(y)}}{\partial y}\right)_{i,j}, \tag{9.35}$$

and the contribution to the term $\nabla_{\mathbf{r}} \cdot \mathbf{J}$ from the $x$-derivative will be

$$\left(\frac{\partial J_{(x)}}{\partial x}\right)_{i,j} = \frac{1}{h}\left(J_{(x)i+\frac{1}{2},j} - J_{(x)i-\frac{1}{2},j}\right)$$
$$= \frac{1}{h^2}\beta_{i+\frac{1}{2},j}C^+_{i+\frac{1}{2},j}n_{i+1,j} + \frac{1}{h^2}\beta_{i-\frac{1}{2},j}C^-_{i-\frac{1}{2},j}n_{i-1,j}$$
$$- \frac{1}{h^2}(\beta_{i+\frac{1}{2},j}C^-_{i+\frac{1}{2},j} + \beta_{i-\frac{1}{2},j}C^+_{i-\frac{1}{2},j})n_{i,j}. \tag{9.36}$$

On substituting the expansion

$$n_{i+1,j} = n_{i,j} + h\left(\frac{\partial n}{\partial x}\right)_{i,j} + \frac{1}{2}h^2\left(\frac{\partial^2 n}{\partial x^2}\right)_{i,j} + \cdots$$

together with a similar expression for $n_{i-1,j}$, Equation (9.36) becomes

$$
\left(\frac{\partial J_{(x)}}{\partial x}\right)_{i,j} = \frac{1}{h^2}\left(\beta_{i+\frac{1}{2},j}C^{+}_{i+\frac{1}{2},j} - \beta_{i+\frac{1}{2},j}C^{-}_{i+\frac{1}{2},j} - \beta_{i-\frac{1}{2},j}C^{+}_{i-\frac{1}{2},j}\right.
$$

$$
+ \beta_{i-\frac{1}{2},j}C^{-}_{i-\frac{1}{2},j}\Big)n_{i,j} + \frac{1}{h}(\beta_{i+\frac{1}{2},j}C^{+}_{i+\frac{1}{2},j}
$$

$$
- \beta_{i-\frac{1}{2},j}C^{-}_{i-\frac{1}{2},j})\left(\frac{\partial n}{\partial x}\right)_{i,j} + \frac{1}{2}(\beta_{i+\frac{1}{2},j}C^{+}_{i+\frac{1}{2},j} + \beta_{i-\frac{1}{2},j}C^{-}_{i-\frac{1}{2},j})
$$

$$
\times \left(\frac{\partial^2 n}{\partial x^2}\right)_{i,j} + \cdots \tag{9.37}
$$

Now introduce the function $A(t)$ defined by

$$
A(t) \equiv \frac{1}{2}(B(t) + B(-t)). \tag{9.38}
$$

This function has the properties

$$
A(t) = A(-t),
$$
$$
A(0) = 1,
$$
$$
A(t) \to \frac{1}{2}|t| \quad \text{as } |t| \to \infty,
$$
$$
A(t) = 1 + O(t^2) \quad \text{for } |t| \ll 1, \quad \text{and}
$$
$$
A'(0) = 0.
$$

The function $A(t)$ is plotted in Fig. 9.3. On defining

$$
P_{(x)i+\frac{1}{2},j} \equiv -\left(\frac{\alpha}{\beta}\right)_{i+\frac{1}{2},j}(\psi_{i+1,j} - \psi_{i,j}) - \left(\frac{\gamma}{\beta}\right)_{i+\frac{1}{2},j}(T_{i+1,j} - T_{i,j}) \tag{9.39}
$$

it can be easily shown that

$$
C^{+}_{i+\frac{1}{2},j} = T_{(L)i+\frac{1}{2},j}B\left(\frac{P_{(x)i+\frac{1}{2},j}}{T_{(L)i+\frac{1}{2},j}}\right)
$$

$$
= T_{(L)i+\frac{1}{2},j}A\left(\frac{P_{(x)i+\frac{1}{2},j}}{T_{(L)i+\frac{1}{2},j}}\right) + \frac{1}{2}P_{(x)i+\frac{1}{2},j}, \tag{9.40}
$$

$$
C^{-}_{i+\frac{1}{2},j} = T_{(L)i+\frac{1}{2},j}B\left(-\frac{P_{(x)i+\frac{1}{2},j}}{T_{(L)i+\frac{1}{2},j}}\right)
$$

$$
= T_{(L)i+\frac{1}{2},j}A\left(\frac{P_{(x)i+\frac{1}{2},j}}{T_{(L)i+\frac{1}{2},j}}\right) - \frac{1}{2}P_{(x)i+\frac{1}{2},j}. \tag{9.41}
$$

The *diffusion coefficient*, denoted by $D_{(x)i,j}$, is defined as the coefficient of the term $(\partial^2 n/\partial x^2)_{i,j}$ in Equation (9.37). Using the logarithmic mean $T_{(L)i+\frac{1}{2},j} = p(T_{i,j}, T_{i+1,j})$ together with Equations (9.40) and (9.41), it follows that

$$
q D_{(x)i,j} = \frac{1}{2}\beta_{i+\frac{1}{2},j} T_{(L)i+\frac{1}{2},j} A\left(\frac{1}{T_{(L)i+\frac{1}{2},j}} P_{(x)i+\frac{1}{2},j}\right)
$$

$$
+ \frac{1}{2}\beta_{i-\frac{1}{2},j} T_{(L)i-\frac{1}{2},j} A\left(\frac{1}{T_{(L)i-\frac{1}{2},j}} P_{(x)i-\frac{1}{2},j}\right)
$$

$$
- \frac{1}{2h}\left(\beta_{i+\frac{1}{2},j}\frac{1}{h} P_{(x)i+\frac{1}{2},j} - \beta_{i-\frac{1}{2},j}\frac{1}{h} P_{(x)i-\frac{1}{2},j}\right) h^2. \tag{9.42}
$$

Neglecting the last term in Equation (9.42) and taking the arithmetic mean of the first two terms, it follows that

$$
D_{(x)i,j} = \frac{\beta_{i,j} T_{i,j}}{q} A\left(\frac{1}{T_{i,j}} P_{(x)i,j}\right) = M_{(x),i,j}\frac{\beta_{i,j} T_{i,j}}{q}, \tag{9.43}
$$

where

$$
M_{(x)i,j} \equiv A\left(\frac{1}{T_{i,j}} P_{(x)i,j}\right) \geq 1. \tag{9.44}
$$

Similar considerations apply to the flux in the $y$-direction, and therefore the total diffusion term in Equation (9.36) is

$$
M_{(x)i,j}\frac{\beta_{i,j} T_{i,j}}{q}\left(\frac{\partial^2 n}{\partial x^2}\right)_{i,j} + M_{(y)i,j}\frac{\beta_{i,j} T_{i,j}}{q}\left(\frac{\partial^2 n}{\partial y^2}\right)_{i,j}
$$

where both $M_{(x)i,j}$ and $M_{(y)i,j}$ are *diffusion magnification factors* which are greater than 1.0. The physical diffusion is $\beta T/q$, and therefore the C-function method provides an artificial numerical diffusion which acts to dampen the instabilities. Kreskovski (1987) has introduced a hybrid central difference scheme for the one-dimensional isothermal case, and produced an approximation to the diffusion magnification factor. This is plotted as the function $K(t)$ in Fig. 9.3.

It can now be shown that the use of the arithmetic mean rather than the logarithmic mean produces magnification factors $M_{(x)}$ and $M_{(y)}$ which can become *less* than 1.0 for certain values of combinations of the differences in $\psi$ and $T$ evaluated at neighbouring points. The retained terms in Equation (9.42) can be written

$$
q D_{(x),i,j} = \frac{1}{2}\beta_{i+\frac{1}{2},j} T_{(A)i+\frac{1}{2},j}\frac{1}{L_{(x)i+\frac{1}{2},j}} A\left(L_{(x)i+\frac{1}{2},j}\frac{1}{T_{(A)i+\frac{1}{2},j}} P_{(x)i+\frac{1}{2},j}\right)
$$

$$
+ \frac{1}{2}\beta_{i-\frac{1}{2},j} T_{(A)i-\frac{1}{2},j}\frac{1}{L_{(x)i-\frac{1}{2},j}} A\left(L_{(x)i-\frac{1}{2},j}\frac{1}{T_{(A)i-\frac{1}{2},j}} P_{(x)i-\frac{1}{2},j}\right)
$$

$$
\tag{9.45}
$$

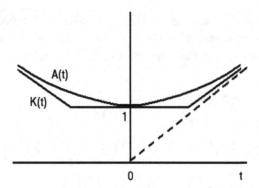

**Fig. 9.3** Plots of the functions $A(t)$ and the Kreskovski function $K(t)$. The functions are symmetric about the vertical axis. The dashed line has slope $1/2$.

where

$$L_{(x)i+\frac{1}{2},j} \equiv \frac{T_{(A)i+\frac{1}{2},J}}{T_{(L)i+\frac{1}{2},J}} \geq 1.$$

The quantity $L_{(x),i+\frac{1}{2},j}$ can be expressed as

$$L_{(x),i+\frac{1}{2},j} = \frac{1}{2\varepsilon_{i+\frac{1}{2},j}} \ln \left( \frac{1+\varepsilon_{i+\frac{1}{2},j}}{1-\varepsilon_{i+\frac{1}{2},j}} \right) \equiv L_{(x)}(\varepsilon_{i+\frac{1}{2},j})$$

where

$$\varepsilon_{i+\frac{1}{2},j} \equiv \frac{(T_{i+1,j} - T_{i,j})}{(T_{i+1,j} + T_{i,j})}.$$

This function $L$ has the properties that $L(\eta) = L(-\eta)$, $L(0) = 1$ (the isothermal case), and $L(\eta) \to \infty$ as $\eta \to \pm 1$. Hence interpreting the temperature $T_{(A)i+\frac{1}{2},j}$ in the numerator of the equation as the value of the temperature at the half point, the effective diffusion magnification factor is now

$$M_{(A)(x)i,j} \equiv \frac{A((L(\varepsilon)\frac{1}{T_{i,j}}\rho_{(x)i,j})}{L(\varepsilon)}$$

where

$$\varepsilon \equiv \frac{(T_{i+\frac{1}{2},j} - T_{i-\frac{1}{2},j})}{(T_{i+\frac{1}{2},j} + T_{i-\frac{1}{2},j})}.$$

Fig. 9.4 shows the factor $M_{(A)} = A(L(\varepsilon)\rho/T)/L(\varepsilon)$ plotted against $\varepsilon$ and $\rho/T$. It demonstrates that $M_{(A)}$ can be reduced below the value 1.0 for certain values of $\varepsilon$ and $\rho/T$. In fact it may be verified that $M_{(A)} < 1$ when $|\rho/T| < 2|\varepsilon|$, but this never happens in the isothermal case in which $\varepsilon = 0$.

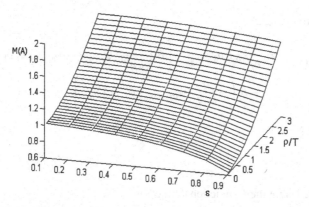

**Fig. 9.4** Plot of the diffusion magnification factor $M_{(A)}$ as a function of $\varepsilon$ and $\rho/T$ for the case in which the electron temperature is interpolated using the arithmetic mean.

## 9.6 The limit of uniform temperature

The preceding formulae simplify when the electron temperature $T$ is uniform throughout the system. In this case, $T_{i,j} \equiv T$, which is independent of the meshpoint $(i, j)$. Equations (9.4) and (9.5) become

$$\mathbf{J} = \alpha_C (\nabla_\mathbf{r} \psi) n + \beta_C T (\nabla_\mathbf{r} n)$$
$$\mathbf{s} = \alpha_E (\nabla_\mathbf{r} \psi) W + \beta_E T (\nabla_\mathbf{r} W) \tag{9.46}$$

and

$$\mathbf{v} = \alpha (\nabla_\mathbf{r} \psi) \theta + \beta T (\nabla_\mathbf{r} \theta) + b T \theta (\nabla_\mathbf{r} \beta). \tag{9.47}$$

The $C$-function which was defined in Equation (9.11) will now become

$$C_u(x, T, T) = p(T, T) B \left( \frac{x}{p(T, T)} \right) = T B \left( \frac{x}{T} \right) \tag{9.48}$$

since $p(T, T) = T$ by the definition (9.15). Equation (9.12) will then become

$$v_{(x)i+\frac{1}{2},j} = \frac{1}{h_i} T (\beta^{1-b})_{i+\frac{1}{2},j}$$

$$\times \left[ B \left( -\frac{1}{T} \left( \frac{\alpha}{\beta} \right)_{i+\frac{1}{2},j} (\psi_{i+1,j} - \psi_{i,j}) \right) (\beta^b \theta)_{i+1,j} \right.$$

$$\left. - B \left( \frac{1}{T} \left( \frac{\alpha}{\beta} \right)_{i+\frac{1}{2},j} (\psi_{i+1,j} - \psi_{i,j}) \right) (\beta^b \theta)_{i,j} \right]. \tag{9.49}$$

The special case of Equation (9.29) for which $a = 1$ and $b = 0$ will then become

$$\mathbf{J} = -q n \mu \nabla_\mathbf{r} \psi + k_B \mu T \nabla_\mathbf{r} n \tag{9.50}$$

which leads to the result

$$
J_{(x)i+\frac{1}{2},j} = \frac{1}{h_i} k_B T \mu_{i+\frac{1}{2},j} \times \left[ B\left( \frac{q}{k_B T}(\psi_{i+1,j} - \psi_{i,j}) \right) n_{i+1,j} \right.
$$
$$
\left. - B\left( -\frac{q}{k_B T}(\psi_{i+1,j} - \psi_{i,j}) \right) n_{i,j} \right]. \tag{9.51}
$$

## Problems

**Problem 9.1.** Using the $C$-function defined as

$$
C_u(x, y, z) \equiv \begin{cases} \dfrac{x - u(y-z)}{\exp(\frac{x - u(y-z)}{z-y} \ln |z/y|) - 1} & (y \neq z) \\[2ex] \dfrac{x}{e^{\frac{x}{y}} - 1} & (y = x) \end{cases} ,
$$

show that Equation (9.10) can be written as

$$
v_{(x)i+\frac{1}{2},j} = \frac{1}{h_i} (\beta^{1-b})_{i+\frac{1}{2},j}
$$
$$
\times \left[ C_{(\frac{\gamma}{\beta})_{i+\frac{1}{2},j}} \left( -\left(\frac{\alpha}{\beta}\right)_{i+\frac{1}{2},j} (\psi_{i+1,j} - \psi_{i,j}), T_{i+1,j}, T_{i,j} \right) (\beta^b \theta)_{i+1,j} \right.
$$
$$
\left. - C_{(\frac{\gamma}{\beta})_{i+\frac{1}{2},j}} \left( \left(\frac{\alpha}{\beta}\right)_{i+\frac{1}{2},j} (\psi_{i+1,j} - \psi_{i,j}), T_{i,j}, T_{i+1,j} \right) (\beta^b \theta)_{i,j} \right].
$$

**Problem 9.2.** Show that the Bernoulli function $B(t)$, which is defined as

$$
B(t) \equiv \frac{t}{e^t - 1},
$$

has the properties

$$
B(0) = 1,
$$
$$
B(t) \approx 1 - \frac{1}{2} t \quad \text{for } |t| \ll 1,
$$
$$
B(t) \to 0 \quad \text{as } t \to \infty,
$$
$$
B(t) \to -t \quad \text{as } t \to -\infty,
$$
$$
B(-t) = B(t) + t,
$$
$$
B'(t) < 0,
$$
$$
B'(t) = \frac{B(t)}{t}(1 - B(t) - t),
$$
$$
B'(t) \approx \frac{1}{4} t - \frac{1}{2} \quad \text{for } |t| \ll 1.
$$

**Problem 9.3.** The p-function $p(y, z)$ is defined for $y > 0$ and $z > 0$ as

$$p(y, z) \equiv \begin{cases} \frac{z-y}{\ln z - \ln y} & \text{for } z \neq y \\ y & \text{for } z = y \end{cases}.$$

If the geometric mean $m_G(y, z)$ and the arithmetic mean $m_A(y, z)$ of two real numbers $y$ and $z$ are defined by

$$m_A(y, z) \equiv \frac{1}{2}(y + z),$$
$$m_G(y, z) \equiv \sqrt{yz} \quad (y, z > 0),$$

show that, if $y, z > 0$, then

$$m_G(y, z) \leq p(y, z) \leq m_A(y, z),$$

with equality holding only if $y = z$.

# Chapter 10
# Solution of equations: the Newton and reduced method

The Newton method is an iterative method for the solution of a set of nonlinear equations

$$F_i(X_0, X_1, \ldots, X_M) = 0 \quad (i = 0, 1, \ldots, M)$$

for the unknowns $X_0, X_1, \ldots, X_M$. In the case $M = 0$ of a single variable, the method involves division by a derivative, but in the case $M \neq 0$ of more than one variable the method requires the inversion of the Jacobian matrix $\mathbf{J}$, and this can be very time consuming. It will be shown that the direct Newton method is equivalent to iterating a set of equations to a time steady state. This correspondence will be used to suggest a *reduced* Newton method in which matrix inversion is kept to a minimum. Superscripts will always denote the iteration number.

## 10.1 The Newton method for one variable

The general problem here is to solve the single equation

$$F(X) = 0 \tag{10.1}$$

for the single variable $X$ using iteration. If $X^{(k)}$ is the $k$th iterated value, write $X^{(k+1)} = X^{(k)} + \delta^{(k)}$. Then, to the first order in the expansion of $F(X^{(k)} + \delta^{(k)})$, the Newton iteration scheme is

$$F(X^{(k+1)}) = 0 \quad \Rightarrow \quad F(X^{(k)}) + \delta^{(k)} F'(X^{(k)}) \approx 0$$

$$\Rightarrow \quad \delta^{(k)} \approx -\frac{F(X^{(k)})}{F'(X^{(k)})}.$$

Hence

$$X^{(k+1)} = X^{(k)} - \frac{F(X^{(k)})}{F'(X^{(k)})}. \tag{10.2}$$

E.A.B. Cole, *Mathematical and Numerical Modelling of Heterostructure Semiconductor Devices: From Theory to Programming*, DOI 10.1007/978-1-84882-937-4_10, © Springer-Verlag London Limited 2009

By calculating both $F(X^{(k)})$ and $F'(X^{(k)})$ at the most recently found iterated value $X^{(k)}$, the value at the next iteration step can be found. This iteration process as it stands can be too harsh in the sense that consecutive values of $X^{(k)}$ can oscillate wildly about the exact solution. In order to avoid this overshoot, the scheme in Equation (10.2) is replaced with

$$X^{(k+1)} = X^{(k)} - \alpha \frac{F(X^{(k)})}{F'(X^{(k)})}, \tag{10.3}$$

where $\alpha$ is some pre-assigned value in the range $0 < \alpha \le 1$. This scheme is iterated until $F(X^{(k)})$ falls below some specified value, or until some other stopping criteria are specified.

Alternatively, consider the differential equation

$$\frac{dF(t)}{dt} = -a(t)F(t), \tag{10.4}$$

where $a(t) > 0$ is some positive function of a real parameter $t$, and $F(t) \equiv F(x(t))$. This has the solution

$$F(t) = F(0) \exp\left(-\int_0^t a(t')dt'\right)$$
$$\to 0 \quad \text{as } t \to \infty.$$

Hence as $t \to \infty$, the solution of Equation (10.4) gives rise to the same equation as the initial problem which was stated in Equation (10.1). Therefore the time iteration of Equation (10.4) (treating $t$ as a "time") should provide the required solution to Equation (10.1). By taking a time step $\Delta t$ and using a simple forward difference expression for $dX/dt$, it follows that

$$\frac{dF}{dt} = -a(t)F \quad \Rightarrow \quad \frac{dF}{dX}\frac{dX}{dt} = -a(t)F$$

$$\Rightarrow \quad \frac{dX}{dt} = -a(t)\frac{F(X)}{F'(X)}$$

$$\Rightarrow \quad \frac{X^{(k+1)} - X^{(k)}}{\Delta t} = -a^{(k)}\frac{F(X^{(k)})}{F'(X^{(k)})}$$

$$\Rightarrow \quad X^{(k+1)} = X^{(k)} - (\Delta t)a^{(k)}\frac{F(X^{(k)})}{F'(X^{(k)})}. \tag{10.5}$$

Comparing Equations (10.5) and (10.3), the overshoot factor $\alpha$ can be identified with the quantity $(\Delta t)a^{(k)}$. Without loss of generality, the value of $\Delta t$ can be taken as $\Delta t = 1$, (that is, it is subsumed into specification of the function $a^{(k)}$), and then

$$X^{(k+1)} = X^{(k)} - a^{(k)}\frac{F(X^{(k)})}{F'(X^{(k)})}, \quad 0 < a^{(k)} \le 1. \tag{10.6}$$

This allows the overshoot factor $a^{(k)}$ to depend on the iteration number $k$. It is common to take $a^{(k)}$ to be small in the fragile initial stages of the iteration procedure, and to increase it gradually as the iterations progress.

## 10.2 Error analysis of the Newton method

If $X$ is the solution of the equation $F(X) = 0$, the error at the $k$th iteration step is $\xi^{(k)} \equiv X^{(k)} - X$. Then the scheme in Equation (10.2) gives

$$X^{(k+1)} - X = X^{(k)} - X - \frac{F(X^{(k)})}{F'(X^{(k)})},$$

or

$$\xi^{(k+1)} = \xi^{(k)} - \frac{F(X^{(k)})}{F'(X^{(k)})}. \tag{10.7}$$

This gives a relation between the errors at consecutive iteration steps $k$ and $k+1$ in terms of the values of the function and its derivative evaluated at $X^{(k)}$. To get the corresponding result in terms of function evaluation at the exact solution $X$, the two expansions

$$F(X^{(k)}) = F(X + \xi^{(k)}) = F(X) + \xi^{(k)} F'(X) + \frac{1}{2}(\xi^{(k)})^2 F''(X) + \cdots$$

$$F'(X^{(k)}) = F'(X + \xi^{(k)}) = F'(X) + \xi^{(k)} F''(X) + \frac{1}{2}(\xi^{(k)})^2 F'''(X) + \cdots$$

can be used. Then since $F(X) = 0$, Equation (10.7) gives

$$\begin{aligned}
\xi^{(k+1)} &= \frac{\xi^{(k)} F'(X^{(k)}) - F(X^{(k)})}{F'(X^{(k)})} \\
&\approx \frac{\xi^{(k)} F'(X) + (\xi^{(k)})^2 F''(X) + \frac{1}{2}(\xi^{(k)})^3 F'''(X) - \xi^{(k)} F'(X) - \frac{1}{2}(\xi^{(k)})^2 F''(X)}{F'(X) + \xi^{(k)} F''(X) + \cdots} \\
&\approx \frac{1}{2}(\xi^{(k)})^2 \frac{F''(X)}{F'(X)}
\end{aligned} \tag{10.8}$$

so that convergence is quadratic in the error $\xi^{(k)}$.

This error analysis may be extended to the case in which $F'(X^{(k)})$ in Equation (10.7) is replaced by a numerical derivative. The exact form will depend on whether central differences or backward differences are used for the derivative (Press et al. 2002).

1. *Central difference analysis*

Let $\Delta$ be a positive increment to the value of $X$. Then, in terms of this increment, the central difference formula for $F'(X)$ is

$$F'(X^{(k)}) = \frac{F(X^{(k)} + \Delta) - F(X^{(k)} - \Delta)}{2\Delta}.$$

In order to ease the notation, we will write $\xi \equiv \xi^{(k)}$, $F \equiv F(X)$ etc. Then

$$F(X^{(k)}) = F(X + \xi) = F + \xi F' + \frac{1}{2}\xi^2 F'' + \dots$$

$$F(X^{(k)} + \Delta) = F(X + (\xi + \Delta)) = F + (\xi + \Delta)F' + \frac{1}{2}(\xi + \Delta)^2 F'' + \dots$$

$$F(X^{(k)} - \Delta) = F(X + (\xi - \Delta)) = F + (\xi - \Delta)F' + \frac{1}{2}(\xi - \Delta)^2 F'' + \dots$$

Hence

$$F(X^{(k)} + \Delta) - F(X^{(k)} - \Delta) = 2\Delta F' + \frac{1}{2}[(\xi + \Delta)^2 - (\xi - \Delta)^2]F''$$

$$= 2\Delta F' + 2\xi \Delta F'',$$

and it follows from Equation (10.7) that

$$\xi^{(k+1)} = \xi^{(k)} - \frac{F(X^{(k)}) \cdot 2\Delta}{F(X^{(k)} + \Delta) - F(X^{(k)} - \Delta)}$$

$$= \frac{2\xi \Delta F' + 2\xi^2 \Delta F'' - 2\xi \Delta F' - \Delta\xi^2 F''}{2\Delta F' + 2\xi \Delta F''}$$

$$\approx (\xi^{(k)})^2 \frac{F''}{2F'}.$$

Hence the convergence is again quadratic.

2. *Forward difference analysis*

In terms of the positive increment $\Delta$, the forward difference formula for $F'(X)$ is

$$F'(X^{(k)}) = \frac{F(X^{(k)} + \Delta) - F(X^{(k)})}{\Delta}.$$

Again to ease the notation we will write $\xi \equiv \xi^{(k)}$, $F \equiv F(X)$ etc. Then

$$F(X^{(k)}) = F(x + \xi) = F + \xi F' + \frac{1}{2}\xi^2 F'' + \dots$$

$$F(X^{(k)} + \Delta) = F(x + (\xi + \Delta)) = F + (\xi + \Delta)F' + \frac{1}{2}(\xi + \Delta)^2 F'' + \dots$$

Hence

$$F(X^{(k)} + \Delta) - F(X^{(k)})$$
$$= \Delta F' + \frac{1}{2}[(\xi + \Delta)^2 - \xi^2]F''$$
$$= \Delta F' + \frac{1}{2}(2\xi + \Delta)\Delta F'',$$

giving

$$\xi^{(k+1)} = \xi^{(k)} - \frac{F(X^{(k)})\Delta}{F(X^{(k)} + \Delta) - F(X^{(k)})}$$
$$= \frac{\xi \Delta F' + \frac{1}{2}\xi(2\xi + \Delta)\Delta F'' - \xi \Delta F' - \frac{1}{2}\Delta \xi^2 F''}{\Delta F' + \frac{1}{2}(2\xi + \Delta)\Delta F''}$$
$$\approx \left(\frac{1}{2}\xi^2 + \frac{1}{2}\xi\Delta\right)\frac{F''}{F'}$$
$$= \frac{1}{2}(\xi^2 + \xi\Delta)\frac{F''}{F'}$$

It follows that if $\Delta$ is too large, the term $\xi\Delta$ will swamp the term $\xi^2$ and so convergence will be *linear*. However, if $\Delta$ is too small, the round-off errors will swamp.

## 10.3 The multi-variable Newton method

The $(M+1)$ values $X_0, X_1, \ldots, X_M$ will be arranged into an $(M+1)$ column vector $\mathbf{X} = (X_0, X_1, \ldots, X_M)^T$. These values will be sought which satisfy the generally nonlinear equations

$$F_i(\mathbf{X}) = 0 \quad (i = 0, 1, \ldots, M), \quad \text{or}$$
$$\mathbf{F}(\mathbf{X}) = \mathbf{0}. \tag{10.9}$$

Writing $\mathbf{X}^{(k+1)} = \mathbf{X}^{(k)} + \boldsymbol{\delta}^{(k)}$, and defining the $N \times N$ Jacobian matrix $\mathbf{J}^{(k)}$ whose elements are given by

$$J_{ij}^{(k)} \equiv \frac{\partial F_i(\mathbf{X}^{(k)})}{\partial X_j^{(k)}},$$

then the Newton iterative scheme is, for $i = 0, \ldots, M$,

$$F_i(\mathbf{X}^{(k+1)}) = 0 \quad \Rightarrow \quad F_i(\mathbf{X}^{(k)}) + \sum_{j=0}^{M} J_{ij}^{(k)} \delta_j^{(k)} \approx 0$$

$$\Rightarrow \quad \sum_{j=0}^{M} J_{ij}^{(k)} \delta_j^{(k)} \approx -F_i(\mathbf{X}^{(k)})$$

$$\Rightarrow \quad \mathbf{J}^{(k)} \delta^{(k)} \approx -\mathbf{F}(\mathbf{X}^{(k)}). \tag{10.10}$$

This matrix equation will have solution

$$\delta^{(k)} = -(\mathbf{J}^{(k)})^{-1} \mathbf{F}(\mathbf{X}^{(k)}),$$

giving

$$\mathbf{X}^{(k+1)} = \mathbf{X}^{(k)} - (\mathbf{J}^{(k)})^{-1} \mathbf{F}(\mathbf{X}^{(k)}). \tag{10.11}$$

To avoid overshoot, a *scalar* overshoot factor $\alpha$ can again be introduced, and then the iteration scheme will be modified to

$$\mathbf{X}^{(k+1)} = \mathbf{X}^{(k)} - \alpha(\mathbf{J}^{(k)})^{-1} \mathbf{F}(\mathbf{X}^{(k)}), \quad (0 < \alpha \le 1). \tag{10.12}$$

It can be seen that this method requires the inversion of the $(M+1) \times (M+1)$ matrix $\mathbf{J}^{(k)}$ at each iteration step $k$, and this process can be computationally very expensive.

Alternatively, consider the differential equation

$$\frac{d\mathbf{F}}{dt} = -\mathbf{a}(t)\mathbf{F} \tag{10.13}$$

where the eigenvalues of the $(M+1) \times (M+1)$ matrix $\mathbf{a}(t)$ are all positive. Then, as in the case for a single variable, it will follow that $\mathbf{F}(\mathbf{X}(t)) \to \mathbf{0}$ as $t \to \infty$. Hence as $t \to \infty$, the solution of Equation (10.13) provides a solution to the original problem involved in solving Equation (10.9). Using a calculation similar to that leading to Equation (10.5), it follows that

$$\frac{d\mathbf{F}}{dt} = -\mathbf{a}(t)\mathbf{F} \quad \Rightarrow \quad \mathbf{J}^{(k)} \frac{d\mathbf{X}}{dt} = -\mathbf{a}(t)\mathbf{F} \tag{10.14}$$

$$\Rightarrow \quad \frac{d\mathbf{X}}{dt} = -(\mathbf{J}^{(k)})^{-1} \mathbf{a}(t)\mathbf{F}(\mathbf{X})$$

$$\Rightarrow \quad \frac{\mathbf{X}^{(k+1)} - \mathbf{X}^{(k)}}{\Delta t} = -(\mathbf{J}^{(k)})^{-1} \mathbf{a}^{(k)} \mathbf{F}(\mathbf{X}^{(k)})$$

$$\Rightarrow \quad \mathbf{X}^{(k+1)} = \mathbf{X}^{(k)} - (\mathbf{J}^{(k)})^{-1} \mathbf{a}^{(k)} \mathbf{F}(\mathbf{X}^{(k)}), \tag{10.15}$$

where again the value of $\Delta t$ has been subsumed into the matrix $\mathbf{a}^{(k)}$. Comparing this result with that of Equation (10.12), it can be seen that the matrix $\mathbf{a}^{(k)}$ acts as a *matrix* overshoot factor. It must have positive eigenvalues, and will generally depend on the iteration number $k$.

It is important to ensure that the overshoot factor is introduced at the correct stage of the process. In general, the inverse $(\mathbf{J}^{(k)})^{-1}$ of the Jacobian matrix will not commute with $\mathbf{a}^{(k)}$, and the order of these two matrices must be preserved in Equation (10.15). This fact indicates that if a matrix overshoot factor is to be introduced into Equation (10.11), it must be introduced at the earlier stage of Equation (10.10):

$$\mathbf{J}^{(k)}\boldsymbol{\delta}^{(k)} = -\mathbf{a}^{(k)}\mathbf{F}(\mathbf{X}^{(k)}). \tag{10.16}$$

However, this complication may not arise if $\mathbf{a}$ is chosen to be diagonal, because the two matrices may then commute.

Instead of using the iterative scheme in Equation (10.15) in which the Jacobian $\mathbf{J}^{(k)}$ has to be inverted, it is possible to go back a step to produce an iterative scheme in which this inversion is not necessary. Essentially, this is an iterative scheme to be applied at each Newton step: on discretising Equation (10.14) with $\Delta t = 1$, it follows that

$$\mathbf{J}^{(k)}\mathbf{X}^{(k+1)} = \mathbf{J}^{(k)}\mathbf{X}^{(k)} - \mathbf{a}^{(k)}\mathbf{F}^{(k)}. \tag{10.17}$$

This is a set of linear equations for the unknown $\mathbf{X}^{(k+1)}$, given in terms of a right hand side which can be calculated from the values found at the previous iteration step. The system given by Equation (10.17) can be iterated to a solution at iteration step $(k + 1)$, using an SUR or SOR relaxation factor $w$, with the scheme

$$u_i^{(k+1)} \equiv \frac{1}{J_{ii}^{(k)}}\left((\mathbf{J}^{(k)}\mathbf{X}^{(k)})_i - (\mathbf{a}^{(k)}\mathbf{F}^{(k)})_i - \sum_{j\neq i} J_{ij}^{(k)} X_j^{(k+1)}\right),$$
$$X_i^{(k+1)} = (1 - w)X_i^{(k+1)} + wu_i^{(k+1)}.$$

## 10.4 The reduced Newton method

The Newton method produces the solution to the set of (generally) nonlinear equations

$$\mathbf{F}(\mathbf{X}) = \mathbf{0}. \tag{10.18}$$

However, the iteration scheme of Equation (10.15) requires that the $(M+1) \times (M+1)$ Jacobian matrix $\mathbf{J}^{(k)}$ be inverted at each step. This is a very time-consuming part of the simulation, especially when it is realised that $M$ could be of the order of $10^5$. It will now be shown how a *reduced* Newton scheme will make this load lighter.

Let $\mathbf{c} = (c_0, c_1, \ldots, c_M)^T$ be a column vector with real constant components. Define a related set of functions $G_i$ by

$$G_i(X_i) \equiv F(c_0, c_1, c_{i-1}, X_i, c_{i+1}, \ldots, c_M), \quad (i = 0, 1, \ldots, M), \tag{10.19}$$

and consider the differential equation

$$\frac{d\mathbf{G}}{dt} = -\mathbf{g}(t)\mathbf{G} \tag{10.20}$$

where the $(M + 1) \times (M + 1)$ matrix $\mathbf{g}(t)$ is such that its eigenvalues are all positive. Then $\mathbf{G}(\mathbf{X}(t)) \to \mathbf{0}$ as $t \to \infty$, and so the solution of Equation (10.20) as $t \to \infty$ provides a solution to the original problem of solving Equation (10.18). Equation (10.20) becomes

$$\mathbf{K}\frac{d\mathbf{X}}{dt} = -\mathbf{g}(t)\mathbf{G}$$

where the $(M + 1) \times (M + 1)$ matrix $\mathbf{K}$ has elements given by

$$(\mathbf{K})_{ij} \equiv \frac{\partial G_i}{\partial X_j} = \frac{dG_i}{dX_i}\delta_{ij}.$$

That is, the matrix $\mathbf{K}$ is diagonal. Then

$$\frac{d\mathbf{X}}{dt} = -(\mathbf{K})^{-1}\mathbf{g}(t)\mathbf{G}$$

and this result, on using forward time discretisation with $\Delta t = 1$, leads to the iteration scheme

$$\mathbf{X}^{(k+1)} = \mathbf{X}^{(k)} - (\mathbf{K}^{(k)})^{-1}\mathbf{g}^{(k)}\mathbf{G}^{(k)}. \tag{10.21}$$

This is essentially the original Newton method in which the Jacobian matrix $\mathbf{J}$ is replaced with the matrix $\mathbf{K}$ which has only diagonal elements. Hence in this method it is only necessary to invert a diagonal reduced Jacobian matrix $\mathbf{K}^{(k)}$ at each iteration step.

## 10.5 The Newton method applied to device modelling

The application of this reduced method will be illustrated using a two dimensional mesh. Special considerations will apply to the method for the one dimensional and three dimensional cases.

### 10.5.1 The reduced method applied to device modelling

The total number $(M + 1)$ of equations and unknowns is calculated as follows. Depending on the simulation scheme which is used, there will be a number $N_e$ of equations satisfied at each mesh point, and there will be $N_e$ primary macroscopic variables. For example for a HEMT model, there will be three equations—Poisson, current continuity and energy transport (excluding the TISE), and there will be three primary macroscopic variables—the electrostatic potential $\psi$, the quasi Fermi level $E_F$ and the electron temperature $T$ (or in the case of the MESFET, the variables will be $\psi$, the particle density $n$ and $T$). Hence $N_e = 3$ in this case. For the drift-diffusion model, $N_e = 2$. On a rectangular grid $(i = 0, \ldots, N_x; j = 0, \ldots, N_y)$ there will be $(N_x - 1)(N_y - 1)$ interior points, and so there will be a total of

$$M + 1 = N_e(N_x - 1)(N_y - 1)$$

equations satisfied internally for that number of unknowns.

It will be convenient to cluster the macroscopic variables at a meshpoint $(i, j)$ into a column vector $\mathbf{X}_{ij} = (X_{ij0}, X_{ij1}, \ldots, X_{ijN_e-1})^T$ so that, in the example of the HEMT,

$$\mathbf{X}_{ij} = (\psi_{ij}, E_{Fij}, T_{ij})^T. \tag{10.22}$$

At each internal meshpoint $(i, j)$ there will be a set of $N_e$ equations

$$\mathbf{F}_{ij}(\{\mathbf{X}_{lm}\}) = \mathbf{0}, \quad (i = 1, \ldots, N_x - 1; j = 1, \ldots, N_y - 1) \tag{10.23}$$

or in component form,

$$F_{ij,\nu}(\{\mathbf{X}_{lm}\}) = 0, \quad (i = 1, \ldots, N_x - 1; j = 1, \ldots, N_y - 1; \nu = 1, \ldots, N_e). \tag{10.24}$$

Now consider the equation

$$\frac{d}{dt}\mathbf{F}_{ij} = -\mathbf{a}_{ij}(t)\mathbf{F}_{ij} \tag{10.25}$$

in which $\mathbf{a}_{ij}(t)$ is an $N_e \times N_e$ matrix with positive eigenvalues. Then $\mathbf{F}_{ij} \to \mathbf{0}$ as $t \to \infty$, and so the solutions of Equation (10.25) as $t \to \infty$ are the same as for the original set in Equation (10.23). Equation (10.25) may be written in component form as

$$\frac{d}{dt}F_{ij,\nu} = -\sum_{\mu=1}^{N_e}(\mathbf{a}_{ij})_{\nu\mu}F_{ij,\mu}$$

which may be expanded as

$$\sum_{l=1}^{N_x-1}\sum_{m=1}^{N_y-1}\sum_{\mu=1}^{N_e}\frac{\partial F_{ij,\nu}}{\partial X_{lm,\mu}}\frac{dX_{lm,\mu}}{dt} = -\sum_{\mu=1}^{N_e}(\mathbf{a}_{i,j})_{\nu\mu}F_{ij,\mu}. \tag{10.26}$$

In a five-point discretisation scheme at each meshpoint $(i, j)$ in the two dimensional case, the vector $\mathbf{F}_{ij}$ will depend only on the vectors $\mathbf{X}_{(i-1)j}, \mathbf{X}_{(i+1)j}, \mathbf{X}_{ij}, \mathbf{X}_{i(j-1)}$ and $\mathbf{X}_{i(j+1)}$. Hence at each meshpoint $(i, j)$, we can define five $N_e \times N_e$ matrices $\mathbf{A}_{ij}, \mathbf{B}_{ij}, \mathbf{C}_{ij}, \mathbf{D}_{ij}$ and $\mathbf{E}_{ij}$ such that

$$(\mathbf{A}_{ij})_{\nu\sigma} \equiv \frac{\partial F_{ij,\nu}}{\partial X_{(i-1)j,\sigma}},$$

$$(\mathbf{B}_{ij})_{\nu\sigma} \equiv \frac{\partial F_{ij,\nu}}{\partial X_{(i+1)j,\sigma}},$$

$$(\mathbf{C}_{ij})_{v\sigma} \equiv \frac{\partial F_{ij,v}}{\partial X_{ij,\sigma}}, \tag{10.27}$$

$$(\mathbf{D}_{ij})_{v\sigma} \equiv \frac{\partial F_{ij,v}}{\partial X_{i(j-1),\sigma}}, \quad \text{and}$$

$$(\mathbf{E}_{ij})_{v\sigma} \equiv \frac{\partial F_{ij,v}}{\partial X_{i(j+1),\sigma}}.$$

By assuming that all other partial derivatives are zero at meshpoint $(i, j)$ in the five-point discretisation, then Equation (10.26) can be written in the vector form

$$\mathbf{A}_{ij}(t)\dot{\mathbf{X}}_{(i-1)j} + \mathbf{B}_{ij}(t)\dot{\mathbf{X}}_{(i+1)j} + \mathbf{C}_{ij}(t)\dot{\mathbf{X}}_{ij} + \mathbf{D}_{ij}(t)\dot{\mathbf{X}}_{i(j-1)} + \mathbf{E}_{ij}(t)\dot{\mathbf{X}}_{i(j+1)}$$
$$= -\mathbf{a}_{ij}(t)\mathbf{F}_{ij}(t), \tag{10.28}$$

where the notation $\dot{\mathbf{X}} \equiv d\mathbf{X}/dt$ has been used. On taking the forward time discretisation with $\Delta t = 1$, this becomes

$$-\mathbf{a}_{ij}^{(k)}\mathbf{F}_{ij}^{(k)} = \mathbf{A}_{ij}^{(k)}\mathbf{X}_{(i-1)j}^{(k+1)} + \mathbf{B}_{ij}^{(k)}\mathbf{X}_{(i+1)j}^{(k+1)} + \mathbf{C}_{ij}^{(k)}\mathbf{X}_{ij}^{(k+1)} + \mathbf{D}_{ij}^{(k)}\mathbf{X}_{i(j-1)}^{(k+1)}$$
$$+ \mathbf{E}_{ij}^{(k)}\mathbf{X}_{i(j+1)}^{(k+1)} - \mathbf{A}_{ij}^{(k)}\mathbf{X}_{(i-1)j}^{(k)} - \mathbf{B}_{ij}^{(k)}\mathbf{X}_{(i+1)j}^{(k)} - \mathbf{C}_{ij}^{(k)}\mathbf{X}_{ij}^{(k)}$$
$$- \mathbf{D}_{ij}^{(k)}\mathbf{X}_{i(j-1)}^{(k)} - \mathbf{E}_{ij}^{(k)}\mathbf{X}_{i(j+1)}^{(k)},$$
$$(i = 1, \ldots, N_x - 1; j = 1, \ldots, N_y - 1). \tag{10.29}$$

This set of equations can then be solved using SOR or SUR relaxation using the scheme

$$\mathbf{u}_{ij}^{(k+1)} \equiv (\mathbf{C}_{ij}^{(k)})^{-1}\Big[-\mathbf{A}_{ij}^{(k)}\mathbf{X}_{(i-1)j}^{(k+1)} - \mathbf{B}_{ij}^{(k)}\mathbf{X}_{(i+1)j}^{(k+1)} - \mathbf{D}_{ij}^{(k)}\mathbf{X}_{i(j-1)}^{(k+1)} - \mathbf{E}_{ij}^{(k)}\mathbf{X}_{i(j+1)}^{(k+1)}$$
$$+ \mathbf{A}_{ij}^{(k)}\mathbf{X}_{(i-1)j}^{(k)} + \mathbf{B}_{ij}^{(k)}\mathbf{X}_{(i+1)j}^{(k)} + \mathbf{C}_{ij}^{(k)}\mathbf{X}_{ij}^{(k)} + \mathbf{D}_{ij}^{(k)}\mathbf{X}_{i(j-1)}^{(k)} + \mathbf{E}_{ij}^{(k)}\mathbf{X}_{i(j+1)}^{(k)}$$
$$- \mathbf{a}_{ij}^{(k)}\mathbf{F}_{ij}^{(k)}\Big], \tag{10.30}$$

$$\mathbf{X}_{ij}^{(k+1)} = (\mathbf{I} - \mathbf{w})\mathbf{X}_{ij}^{(k+1)} + \mathbf{w}\mathbf{u}_{ij}^{(k+1)}. \tag{10.31}$$

The matrix $\mathbf{w}$ is diagonal: $\mathbf{w} = \text{diag}(w_0, w_1, \ldots, w_{N_e-1})$, with each element $w_i$ being an SOR or SUR relaxation factor. Note that the matrix $\mathbf{C}_{ij}^{(k)}$ must be inverted at each meshpoint and at each iteration number $k$. However, since the matrix is only $N_e \times N_e$, (or 3×3 for the HEMT or hydrodynamic MESFET) the matrix inversion will not be too heavy.

The reduced method in this case simply amounts to using the algorithm in Equation (10.29) by ignoring the matrices $\mathbf{A}, \mathbf{B}, \mathbf{D}$ and $\mathbf{E}$:

$$\mathbf{X}_{ij}^{(k+1)} = \mathbf{X}_{ij}^{(k)} - (\mathbf{C}_{ij}^{(k)})^{-1}\mathbf{a}_{ij}^{(k)}\mathbf{F}_{ij}^{(k)}. \tag{10.32}$$

In this case, the matrix $\mathbf{C}$ has a small size, there will be no need to use SOR/SUR iteration. The matrix $\mathbf{a}_{ij}^{(k)}$ contains the Newton overshoot factor, and this matrix can be made independent of the meshpoint number $(i, j)$, although it would be interesting to investigate the consequences of it being dependent. It can also be taken to be diagonal, and it could be made to depend on the iteration number:

$$\mathbf{a}_{ij}^{(k)} = \text{diag}(a_0^{(k)}, a_1^{(k)}, \ldots, a_{N_e-1}^{(k)}). \tag{10.33}$$

For the one dimensional problem, with discretisation in the $x$-direction only, the equations are simplified since the matrices $\mathbf{D}$ and $\mathbf{E}$ do not exist. The algorithms to be used then will consist of Equations (10.30) and (10.31) with $\mathbf{D} = \mathbf{E} = \mathbf{0}$, or the reduced algorithm Equation (10.32). On the other hand, discretisation in three dimensions will require the introduction of two extra matrices $\mathbf{H}$ and $\mathbf{K}$ for the two extra spatial directions.

### 10.5.2 Example

As an example of the application of the reduced method, consider the case of a MESFET in which the independent macroscopic variables are $\psi$, $n$ and $T$. Take $N_e = 3$, $\mathbf{X} = (X_0, X_1, X_2)^T \equiv (\psi, n, T)^T$, and the equations at each meshpoint $(i, j)$ ordered as

$$
\begin{aligned}
F_{ij,0} = 0 \quad &\Leftrightarrow \quad \text{Poisson}_{ij} = 0 \\
F_{ij,1} = 0 \quad &\Leftrightarrow \quad \text{current continuity}_{ij} = 0 \\
F_{ij,2} = 0 \quad &\Leftrightarrow \quad \text{energy transport}_{ij} = 0.
\end{aligned}
$$

At meshpoint $(i, j)$, the matrices $\mathbf{A}$ to $\mathbf{E}$ have the following correspondence:

$$
\begin{aligned}
\mathbf{A} \quad &\Leftrightarrow \quad \text{derivatives at } (i - 1, j) \\
\mathbf{B} \quad &\Leftrightarrow \quad \text{derivatives at } (i + 1, j) \\
\mathbf{C} \quad &\Leftrightarrow \quad \text{derivatives at } (i, j) \\
\mathbf{D} \quad &\Leftrightarrow \quad \text{derivatives at } (i, j - 1) \\
\mathbf{E} \quad &\Leftrightarrow \quad \text{derivatives at } (i, j + 1).
\end{aligned}
$$

The *rows* of the matrices relate to the equation (Poisson, current continuity, and energy transport); the *columns* of the matrices relate to the variable which is doing the differentiation ($\psi$, $n$, and $T$). For example, the Poisson equation on a uniform mesh has the discretised form

$$0 = F_{ij,0} \equiv \epsilon_0 \epsilon_r \left( \frac{1}{h^2}(\psi_{i-1,j} - 2\psi_{ij} + \psi_{i+1,j}) + \frac{1}{k^2}(\psi_{i,j-1} - 2\psi_{i,j} + \psi_{i,j+1}) \right)$$
$$+ q(N_d - n_{i,j}),$$

from which it follows that the $(0, 0)$, $(0, 1)$ and $(0, 2)$ matrix elements will be

$$A_{00} = \frac{\partial F_{ij,0}}{\partial \psi_{i-1,j}} = \frac{\epsilon_0 \epsilon_r}{h^2}$$

$$A_{01} = \frac{\partial F_{ij,0}}{\partial n_{i-1,j}} = 0 \quad A_{02} = \frac{\partial F_{ij,0}}{\partial T_{i-1,j}} = 0$$

$$B_{00} = \frac{\partial F_{ij,0}}{\partial \psi_{i+1,j}} = \frac{\epsilon_0 \epsilon_r}{h^2}$$

$$B_{01} = \frac{\partial F_{ij,0}}{\partial n_{i+1,j}} = 0 \quad B_{02} = \frac{\partial F_{ij,0}}{\partial T_{i+1,j}} = 0$$

$$C_{00} = \frac{\partial F_{ij,0}}{\partial \psi_{i,j}} = -2\epsilon_0 \epsilon_r \left( \frac{1}{h^2} + \frac{1}{k^2} \right)$$

$$C_{01} = \frac{\partial F_{ij,0}}{\partial n_{i,j}} = -q \quad C_{02} = \frac{\partial F_{ij,0}}{\partial T_{i,j}} = 0$$

$$D_{00} = \frac{\partial F_{ij,0}}{\partial \psi_{i,j-1}} = \frac{\epsilon_0 \epsilon_r}{k^2}$$

$$D_{01} = \frac{\partial F_{ij,0}}{\partial n_{i,j-1}} = 0 \quad D_{02} = \frac{\partial F_{ij,0}}{\partial T_{i,j-1}} = 0$$

$$E_{00} = \frac{\partial F_{ij,0}}{\partial \psi_{i,j+1}} = \frac{\epsilon_0 \epsilon_r}{k^2}$$

$$E_{01} = \frac{\partial F_{ij,0}}{\partial n_{i,j+1}} = 0 \quad E_{02} = \frac{\partial F_{ij,0}}{\partial T_{i,j+1}} = 0.$$

# Chapter 11
# Solution of equations: the phaseplane method

The phaseplane method is an iterative method which allows the defining equations to be easily modified without having to do much preparatory work, in the form of re-writing of code, for the resulting numerical solutions (Cole 2004). Extra variables are defined to enable the system of second order differential equations to be written as a system of first order equations, and the enlarged system is iterated in the phase plane which is spanned by this enlarged set of variables.

## 11.1 The basis of the phaseplane method

As in previous chapters,, the notation will be such that lower case $x$ and $y$ will denote spatial coordinates, while upper case $X$ and $Y$ will represent physical variables such as $\psi$, $E_F$, and temperature $T$. If there are $N_x + 1$ grid points along the $x$-direction, and $N_y + 1$ along the $y$-direction, and if there are $N_e$ dependent variables defined at each grid point, then the *internal* simultaneous equations may be written formally as

$$F_i(X_0, \ldots, X_M) = 0, \quad (i = 0, \ldots, M) \tag{11.1}$$

where $M + 1 = 3(N_x - 1)(N_y - 1)$ on a two dimensional cross section. We look for the *equilibrium* solutions $X_i = X_{ei}$ of Equations (11.1). Writing $\mathbf{X} \equiv (X_0, \ldots, X_M)^T$, $\mathbf{X}_e \equiv (X_{e0}, \ldots, X_{eM})^T$ and $\mathbf{F} \equiv (F_0, \ldots, F_M)^T$, one of the most common method of solving the set of equations

$$\mathbf{F}(\mathbf{X}_e) = \mathbf{0} \tag{11.2}$$

is to use the Newton method which has been described in Chapter 10. This method requires that the partial derivatives comprising the elements of the Jacobian matrix $\mathbf{J}$ must be calculated, and then $\mathbf{J}$ inverted, at each iteration step. Since $\mathbf{J}$ is normally very large and sparse, then special sparse matrix techniques can be employed in the manipulation of the Newton iteration equations (Yousef 1996). It has been shown in Chapter 10 that the primary variables at each point may be grouped into an $N_e$-

E.A.B. Cole, *Mathematical and Numerical Modelling of Heterostructure Semiconductor Devices: From Theory to Programming*, DOI 10.1007/978-1-84882-937-4_11, © Springer-Verlag London Limited 2009

component vector, and a modification to the Newton method involves the inversion of only an $N_e \times N_e$ matrix at each point. However, this still requires that the partial derivatives be evaluated.

The phase plane method is again an iterative process. However, it involves only the calculation of functions $\mathbf{F}$, and not of the analytical expressions for their partial derivatives. It will be shown that this method contains the modified Newton method as a special case. The phase plane method will involve the use of parameters which are adjusted to give the most rapid convergence. One of these parameters quantifies the deviation of the method from the Newton method, while another parameter is equivalent to the relaxation parameters of the successive over/under relaxation scheme in the corresponding method for linear equations. Since the equations are highly non-linear, there is no rigorous method of determining the parameter values, but trial and error will produce values that can radically affect the convergence.

## 11.2 The phaseplane method for one variable

For the case of the solution of a single equation for a single unknown, we seek to find the solution $X = X_e$ to the equation

$$F(X_e) = 0. \tag{11.3}$$

In the phaseplane method, an artificial time $t$ is introduced, and we look for the equilibrium value $X = X_e$ of the second order differential equation

$$\ddot{X} + k(X)\dot{X} = \gamma(X)F(X) \tag{11.4}$$

by following the solution from some initial condition to equilibrium. Here, $k(X)$ is some function of $X$ which will provide damping in the system, and the function $\gamma(X)$ must satisfy $\gamma(X_e) \neq 0$. This differential equation may be written equivalently in terms of the two first order equations by defining a new variable $Y \equiv \dot{X}$. This system is

$$\begin{aligned} \dot{X} &= Y, \\ \dot{Y} &= \gamma(X)F(X) - k(X)Y. \end{aligned} \tag{11.5}$$

An equilibrium point $(X, Y) = (X_e, Y_e)$ of this system is given as a solution of the equations $\dot{X} = 0$ and $\dot{Y} = 0$. For this particular system, an equilibrium point will be $(X_e, 0)$. In the phaseplane method, trajectories $(X(t), Y(t))$ are plotted in the $X$–$Y$ phase plane for different initial points in the plane. Each equilibrium point is classified according to a linear approximation of the equations near the point, and classification can be made regarding the stability or otherwise of each point. In particular, all trajectories in the vicinity of an equilibrium point will converge on that point in the case of a *stable node*, or will spiral to the point in the case of a *stable spiral*.

Conditions on the functions $k(X)$ and $\gamma(X)$ which appear in Equation (11.5) must be found which allow the solution to converge in a stable way to the equilibrium point $(X_e, 0)$ in the $X$–$Y$ phase plane. The notation will be that, for any function $g(X)$ of $X$, $g_e$ denotes the quantity $g(X_e)$. If $\xi$ represents a "small" deviation of $X$ from the equilibrium value $X_e$, then linearisation of Equation (11.4) about the equilibrium point, using $X = X_e + \xi$ for $|\xi/X_e| \ll 1$, gives

$$\ddot{\xi} + k_e \dot{\xi} - (\gamma F)'_e \xi = 0$$

to the first nonvanishing order in $\xi$. On looking for a solution $\xi = re^{\lambda t}$ it is found that the value of $\lambda$ must satisfy the quadratic equation

$$\lambda^2 + k_e \lambda - (\gamma F)'_e = 0.$$

This equation has the two possible solutions

$$\lambda_{1,2} \equiv \frac{-k_e \pm \sqrt{k_e^2 + 4(\gamma F)'_e}}{2}.$$

In order for the trajectory near the equilibrium point to represent a stable solution, the two values of $\lambda$ must both be negative if $\lambda$ is real, or must have negative real parts if $\lambda$ is complex. It follows that the conditions

$$k_e > 0 \quad \text{and} \quad (\gamma F)'_e < 0 \tag{11.6}$$

must apply. Hence the full set of conditions for a return to a stable solution are

$$\text{complex roots:} \quad k_e > 0 \text{ and } 4(\gamma F)'_e < -k_e^2$$
$$\text{real roots:} \quad k_e > 0 \text{ and } -k_e^2 \leq 4(\gamma F)'_e < 0. \tag{11.7}$$

## 11.3 Discretisation of the equation

In discretising the equation

$$\ddot{X} + k(X)\dot{X} = \gamma(X)F(X),$$

superscripts $l$ will denote the appropriate value at the time $l(\Delta t)$, where $\Delta t$ is the timestep. The discretisation will take the form

$$\frac{X^{(l+1)} - 2X^{(l)} + X^{(l-1)}}{(\Delta t)^2} + k(X^{(l)})\frac{(X^{(l+1)} - X^{(l-1)})}{2(\Delta t)} = \gamma(X^{(l)})F(X^{(l)}),$$

which can be re-arranged to give

$$X^{(l+1)} - 2X^{(l)} + X^{(l-1)} + \frac{1}{2}(\Delta t)k(X^{(l)})(X^{(l+1)} - X^{(l-1)}) = (\Delta t)^2 \gamma(X^{(l)})F(X^{(l)}).$$

$$(11.8)$$

On defining the functions $a(X)$ and $f(X)$ by

$$a(X) \equiv \frac{1}{2}(\Delta t)k(X) \quad \text{and} \quad f(X) \equiv \frac{1}{2}(\Delta t)^2 \gamma(X)F(X), \qquad (11.9)$$

Equation (11.8) becomes

$$(1 + a(X^{(l)}))X^{(l+1)} - 2X^{(l)} - 2f(X^{(l)}) + (1 - a(X^{(l)}))X^{(l-1)} = 0, \qquad (11.10)$$

which can be re-arranged to give

$$X^{(l+1)} = \frac{2f(X^{(l)}) + 2X^{(l)} - (1 - a(X^{(l)}))X^{(l-1)}}{1 + a(X^{(l)})}. \qquad (11.11)$$

Note that the functions $f(X)$ and $a(X)$ have not yet been specified. But since $(\Delta t) > 0$ and $k_e > 0$, we know that the function $f(X)$ must satisfy the condition

$$f_e \equiv f(X_e) = 0, \qquad a_e > 0. \qquad (11.12)$$

The iteration scheme will be based on Equation (11.11). But first, it is necessary to obtain conditions on the parameters involved in order to obtain a stable solution.

## 11.4 The condition for a stable solution

Assuming that the solution has been iterated to a state close to the equilibrium point $X_e$, define the "small" quantity $\xi^{(l)}$ by

$$\xi^{(l)} \equiv X^{(l)} - X_e. \qquad (11.13)$$

On writing $f'_e \equiv f'(X_e)$, expansion of Equation (11.10) up to the first order in $\xi^{(l)}$ gives

$$(1 + a_e)(X_e + \xi^{(l+1)}) - 2(X_e + \xi^{(l)}) + (1 - a_e)(X_e + \xi^{(l-1)}) = 2f_e + 2f'_e\xi^{(l)},$$

or

$$(1 + a_e)\xi^{(l+1)} - 2(1 + f'_e)\xi^{(l)} + (1 - a_e)\xi^{(l-1)} = 0. \qquad (11.14)$$

If it is assumed that $\xi^{(l)}$ satisfies the power law $\xi^{(l)} = A^l$ for some real or complex number $A$, Equation (11.14) leads to the quadratic equation

$$(1 + a_e)A^2 - 2(1 + f'_e)A + 1 - a_e = 0.$$

This equation has the two possible solutions

$$A = \frac{1 + f'_e \pm \sqrt{(1 + f'_e)^2 - (1 - a_e^2)}}{1 + a_e}$$

$$= \frac{1 + f'_e \pm \sqrt{f'^2_e + 2f'_e + a_e^2}}{1 + a_e}. \tag{11.15}$$

The stability analysis will then depend on whether the two roots of this quadratic equation are complex, real, or identical.

1. *Complex roots*

For complex roots, the quantity under the root sign in Equation (11.15) must be negative:

$$f'^2_e + 2f'_e + a_e^2 < 0 \quad \text{and} \quad f'_e < 0. \tag{11.16}$$

Then Equation (11.15) will give

$$|A|^2 = \frac{(1 + f'_e)^2 - (f'^2_e + 2f'_e + a_e^2)}{(1 + a_e)^2}$$

$$= \frac{1 - a_e^2}{(1 + a_e)^2} = \frac{1 - a_e}{1 + a_e}.$$

Since $a_e > 0$ and $|A|^2 > 0$, it must follow that

$$0 < a_e < 1.$$

Condition (11.16) can be written as

$$(f'_e + 1)^2 + a_e^2 < 1,$$

or alternatively

$$f'_e > -1 - \sqrt{1 - a_e^2} \quad \text{and} \quad f'_e < -1 + \sqrt{1 - a_e^2}.$$

These conditions may be combined in the form

$$f'_e = -1 + (2b - 1)\sqrt{1 - a_e^2}, \quad (0 < b < 1), \tag{11.17}$$

and hence convergence in the case of complex roots is given when

$$0 < a_e < 1, \quad (f'_e + 1)^2 + a_e^2 < 1. \tag{11.18}$$

Under these conditions, the equilibrium point will be a spiral point according to the linear approximation.

## 2. *Distinct real roots*

For real roots, the quantity under the root sign in Equation (11.15) must be non-negative. For distinct real roots, it must follow that

$$f_e'^2 + 2f_e' + a_e^2 > 0,$$

or

$$(f_e' + 1)^2 + a_e^2 > 1$$

with the quantity $A$ being given by Equation (11.15). Now examine the separate cases $A < 1$ and $A > -1$:

- 
$$A < 1 \quad \Rightarrow \quad 1 + f_e' + \sqrt{f_e'^2 + 2f_e' + a_e^2} < 1 + a_e$$
$$\Rightarrow \quad \sqrt{f_e'^2 + 2f_e' + a_e^2} < a_e - f_e'$$
$$\Rightarrow \quad \sqrt{(f_e' - a_e)^2 + 2f_e'(a_e + 1)} < a_e - f_e',$$

  which is satisfied if $f_e' < 0$, or $f_e' + 1 < 1$.

- 
$$A > -1 \quad \Rightarrow \quad 1 + f_e' - \sqrt{f_e'^2 + 2f_e' + a_e^2} > -1 - a_e$$
$$\Rightarrow \quad \sqrt{f_e'^2 + 2f_e' + a_e^2} < 2 + f_e' + a_e \qquad (11.19)$$
$$\Rightarrow \quad 2 + f_e' + a_e > 0$$
$$\Rightarrow \quad f_e' > -2 - a_e, \text{ or } f_e' + 1 > -1 - a_e. \qquad (11.20)$$

By squaring and cancelling both sides of Equation (11.19), it follows that

$$0 < 4 + 4a_e + 2f_e' + 2a_e f_e' = 2(2 + f_e')(1 + a_e).$$

Hence

$$f_e' > -2, \quad \text{or} \quad f_e' + 1 > -1 \qquad (11.21)$$

which is stronger than condition (11.20).

Under these conditions, the equilibrium point will be a stable node according to the linear approximation.

## 3. *Identical roots*

If the two roots of Equation (11.15) are identical, then the quantity under the root sign must be zero:

$$(f_e' + 1)^2 + a_e^2 = 1, \qquad A = \frac{1 + f_e'}{1 + a_e}.$$

The quantity $A$ must be real in this case. Now if $A < 1$, then

$$f_e' < a_e,$$

and this condition is satisfied since $f_e' < 0$ and $a_e > 0$. On the other hand, if $A > -1$, then $f_e$ must satisfy the condition

$$1 + f_e' > -1 - a_e = -(1 + a_e),$$

which again is satisfied.

Hence, in summary, convergence to the root $X = X_e$ of the equation $f(X) = 0$, starting at a point in the phase plane close to the root, is guaranteed provided that

$$0 < a_e, \qquad -2 < f_e' < 0. \tag{11.22}$$

When these conditions are satisfied, in the linearisation about $X = X_e$, the equilibrium point is either a spiral point or a node in the phase plane spanned by axes $X$ and $Y \equiv \dot{X}$ if

$$\text{spiral point:} \quad (f_e' + 1)^2 + a_e^2 < 1,$$
$$\text{stable node:} \quad (f_e' + 1)^2 + a_e^2 \geq 1.$$

## 11.5 Exact correspondence between the differential and discretised equations

The definition of the function $f(X)$ in Equation (11.9) has been given as

$$f(X) \equiv \frac{1}{2}(\Delta t)^2 \gamma(X) F(X).$$

This function can be written in terms of a new function $\mu(X)$ where

$$f(X) = \mu(X) \frac{F(X)}{F'(X)}. \tag{11.23}$$

It follows that

$$f'(X) = \mu' \frac{F}{F'} + \mu \left( 1 - \frac{F}{F'^2} F'' \right)$$

which gives $f_e' = \mu_e \equiv \mu(X_e)$, since $F(X_e) = 0$. Hence

$$f(X) = \mu(X) \frac{F(X)}{F'(X)} = \mu_e \hat{\mu}(X) \frac{F(X)}{F'(X)} \tag{11.24}$$

where $\hat{\mu}(X)$ is a new function given by $\hat{\mu}(X) \equiv \mu(X)/\mu_e$, with $\hat{\mu}_e = 1$. Since the function $\gamma(X)$ is arbitrary subject to the conditions derived in Sect. 11.2, then the function $\hat{\mu}(X)$ is not strictly prescribed. In the simplest case, we may take $\hat{\mu}(X) \equiv 1$, giving

$$f(X) = \mu_e \frac{F(X)}{F'(X)} = f'_e \frac{F(X)}{F'(X)}. \tag{11.25}$$

Hence the iteration scheme given by Equation (11.11) becomes

$$X^{(l+1)} = \frac{2f'_e \frac{F(X^{(l)})}{F'(X^{(l)})} + 2X^{(l)} - (1 - a(X^{(l)}))X^{(l-1)}}{1 + a(X^{(l)})}. \tag{11.26}$$

Since $f'_e < 0$ and $(\Delta t)^2 > 0$, it follows that $(\gamma F)'_e = 2f'_e/(\Delta t)^2 < 0$. Hence the conditions in Equation (11.6) which are necessary for a stable solution are satisfied.

## 11.6  A one-variable example

A programme was written to implement the iterative solution based on Equation (11.26). For this particular example, the function $F(X)$ took the form

$$F(X) = \sin(X - 0.1),$$

so that one of the solutions of $F(X_e) = 0$ is $X_e = 0.1$. The values of $a$ and $f'_e$ were set (using $w \equiv -f'_e$). Since we are solving a second order difference equation, two starting values must be taken. These were set as $X^{(1)} = 2.0$ and $X^{(2)} = 1.1 \times X^{(1)}$. The quantity $F'(X^{(l)})$ was evaluated using the central difference. The program was scheduled to stop after 1000 iterations, but was stopped early when the difference between consecutive solutions was less than a specified quantity, or when a specified maximum number of iterations had been reached.

The output of the program contains the iterated values of $X$. The program was run twice, with different values of $a_e$. The output was:

$w = 1.0,\ a_e = 0.8$
Roots are complex
$1/10000, X = 0.11016$
$2/10000, X = 0.11$
$3/10000, X = 0.0988708$
$4/10000, X = 0.0988889$
$5/10000, X = 0.100125$
$6/10000, X = 0.100123$
$7/10000, X = 0.0999861$
$8/10000, X = 0.0999863$
$9/10000, X = 0.100002$
$10/10000, X = 0.100002$

11/10000, $X = 0.0999998$
12/10000, $X = 0.0999998$
13/10000, $X = 0.1$
14/10000, $X = 0.1$

$w = 1.0, a_e = 1.0$
Roots are identical
1/10000, $X = 0.100244$
2/10000, $X = 0.1$
3/10000, $X = 0.1$

It can be seen that the correct choice of the parameter $a_e$ is crucial: the second run produced convergence in only three iterations, while the first run needed fourteen iterations for the same level of convergence. However, there is no way of determining in advance how the optimal values of $f_e'$ and $a_e$ should be chosen.

## 11.7 The phaseplane equations for several variables

For the remainder of this Chapter, the notation $\partial_i \equiv \partial/\partial x_i$ will be used. The previous analysis can now be extended to the case of solving $M + 1$ simultaneous equations

$$F_i(\mathbf{X}_e) = 0, \quad (i = 0, \ldots, M) \tag{11.27}$$

where the functions $F_i$ are generally nonlinear functions of the $M + 1$ variables $\mathbf{X} = (X_0, \ldots, X_M)^T$. In order to proceed, the ordering on the suffices must be taken such that

$$\frac{\partial F_i}{\partial X_i} \neq 0, \quad (i = 0, \ldots, M). \tag{11.28}$$

We then form the set of $M + 1$ differential equations

$$\ddot{X}_i + k_i(\mathbf{X})\dot{X}_i = \gamma_i(\mathbf{X})F_i(\mathbf{X}), \quad (i = 0, \ldots, M) \tag{11.29}$$

where the functions $\gamma_i(\mathbf{X})$ are all non-zero. An analysis similar to that in Sect. 11.2 shows that the conditions for a stable solution at the equilibrium point $\mathbf{X} = \mathbf{X}_e$ are

$$k_{ie} > 0 \quad \text{and} \quad \partial_i(\gamma_i F_i)_e < 0. \tag{11.30}$$

Further, the required stability conditions are:

$$\text{complex roots:} \quad k_{ie} > 0 \text{ and } 4\partial_i(\gamma_i F_i)_e < -k_{ie}^2$$
$$\text{real roots:} \quad k_{ie} > 0 \text{ and } -k_{ie}^2 \leq 4\partial_i(\gamma_i F_i)_e < 0. \tag{11.31}$$

The discretisation of the differential equations follows as for the one-variable case. However, when producing equations equivalent to Equations (11.11) for each variable, we must decide whether or not to use variables replaced with their updated values in later equations of the sequence. In order to describe this replacement process, define the value $r$ such that

$$r = \begin{cases} 1 & \text{for replacement} \\ 0 & \text{for no replacement,} \end{cases} \qquad (11.32)$$

and then define the $(M + 1)$-tuple

$$\mathbf{W}_{i(>0)}^{(l)} \equiv \left( r X_0^{(l+1)} + (1 - r) X_0^{(l)}, r X_1^{(l+1)} + (1 - r) X_1^{(l)}, \right.$$
$$\left. \ldots, r X_{i-1}^{(l+1)} + (1 - r) X_{i-1}^{(l)}, X_i^{(l)}, X_{i+1}^{(l)}, \ldots, X_M^{(l)} \right),$$
$$\mathbf{W}_0^{(l)} \equiv \left( X_0^{(l)}, X_1^{(l)}, \ldots, X_M^{(l)} \right). \qquad (11.33)$$

Hence

$$\mathbf{W}_{i(>0)}^{(l)} = (X_0^{(l+1)}, X_1^{(l+1)}, \ldots, X_{i-1}^{(l+1)}, X_i^{(l)}, X_{i+1}^{(l)}, \ldots, X_M^{(l)}) \quad \text{for replacement}$$
$$= (X_0^{(l)}, X_1^{(l)}, \ldots, X_{i-1}^{(l)}, X_i^{(l)}, X_{i+1}^{(l)}, \ldots, X_M^{(l)}) \quad \text{for no replacement.}$$
$$(11.34)$$

This definition may be re-written in the following way. Define the $(M+1) \times (M+1)$ matrices $\mathbf{U}_{M,i}$ $(i = -1, 0, 1, \ldots, M)$ by

$$(\mathbf{U}_{M,i})_{pq} = \delta_{pq} \quad \text{for } q \leq i$$
$$= 0 \quad \text{otherwise,} \qquad (11.35)$$
$$\text{and} \quad \mathbf{U}_{M,-1} = \mathbf{0}_{(M+1) \times (M+1)}.$$

Hence $\mathbf{U}_{M,i}$ is the diagonal matrix with unit elements in the top $i$ rows and zeros elsewhere. Further, it follows that $\mathbf{U}_{M,i}^T = \mathbf{U}_{M,i}$. Then

$$\mathbf{W}_i^{(l)} = \mathbf{X}^{(l)} + r\mathbf{U}_{M,i-1}(\mathbf{X}^{(l+1)} - \mathbf{X}^{(l)}), \quad (i = 0, \ldots, M) \qquad (11.36)$$

for $r = 0$ or 1. Further, for any $(M + 1) \times (M + 1)$ matrix $\mathbf{A}$, define the new $(M + 1) \times (M + 1)$ matrix $\mathbf{A}_L$ such that

$$(\mathbf{A}_L)_{ik} \equiv \sum_{j=0}^{M} A_{ij}(\mathbf{U}_{M,i-1})_{jk} = \sum_{j=0}^{i-1} A_{ij}\delta_{jk} = \begin{cases} A_{ik} & \text{for } i > k \\ 0 & \text{for } i \leq k \end{cases}. \qquad (11.37)$$

Hence $\mathbf{A}_L$ is the lower left part of the matrix $\mathbf{A}$, below and not including the diagonal. Now, for $i = 0, \ldots, M$, define the quantities

$$a_i(\mathbf{W}_i^{(l)}) \equiv \frac{1}{2}(\Delta t) k_i(\mathbf{W}_i^{(l)}) \quad \text{and} \quad f_i(\mathbf{W}_i^{(l)}) \equiv \frac{1}{2}(\Delta t)^2 \gamma_i(\mathbf{W}_i^{(l)}) F_i(\mathbf{W}_i^{(l)}).$$

The multi-variable analogue of Equation (11.10) is then

$$(1 + a_i(\mathbf{W}_i^{(l)}))X_i^{(l+1)} - 2X_i^{(l)} - 2f_i(\mathbf{W}_i^{(l)}) + (1 - a_i(\mathbf{W}_i^{(l)}))X_i^{(l-1)} = 0. \quad (11.38)$$

The analysis follows that for the one-variable case, leading to the scheme

$$X_i^{(l+1)} = \frac{2(\partial_i f_i)_e \frac{F_i(\mathbf{W}_i^{(l)})}{\partial_i F_i(\mathbf{W}_i^{(l)})} + 2X_i^{(l)} - (1 - a_i(\mathbf{W}_i^{(l)}))X_i^{(l-1)}}{1 + a_i(\mathbf{W}_i^{(l)})}. \quad (11.39)$$

A programme was written to implement the iteration solution based on Equations (11.39), for the case of two variables and two functions $F_0(X, Y)$ and $F_1(X, Y)$. In this particular example, the functions took the form

$$F_0(X, Y) = 3X^3 - 2Y^5, \qquad F_1(X, Y) = X^3 - 2Y^5 + 4.$$

The solutions of the equations $F_0(X, Y) = 0$ and $F_1(X, Y) = 0$ are $X_e = 2^{\frac{1}{3}} = 1.25992$ and $Y_e = 3^{\frac{1}{5}} = 1.24573$. The values of $a_{0e}, a_{1e}, f'_{0e}$ and $f'_{1e}$ were set (using $w_{0e} \equiv -f'_{0e}$ and $w_{1e} \equiv -f'_{1e}$). Since we were solving a second order difference equation, two starting values must be taken for $X$ and for $Y$. These were set as $X^{(1)} = 2.0$, $X^{(2)} = 1.1 \times X^{(1)}$, and $Y^{(1)} = 2.0$, $Y^{(2)} = 1.1 \times Y^{(1)}$. The partial derivatives $\partial_i F_0(X, Y)$ and $\partial_i F_1(X, Y)$ were evaluated using central differences. Replacement was used—as soon as the variable $X$ had been iterated, the new value was used in the iteration of the next variable $Y$ in the list. The program stoped when the difference between consecutive solutions was less than a specified quantity, or when a specified maximum number of 1000 iterations had been reached.

The output of the program containing the iterated values of $X$ and $Y$ was as follows:

Roots for equation 0 are identical
Roots for equation 1 are identical
1/10000, $X = 3.83289$, $Y = 2.01745$
2/10000, $X = 3.06079$, $Y = 1.8112$
3/10000, $X = 2.50286$, $Y = 1.63183$
4/10000, $X = 2.07905$, $Y = 1.48861$
5/10000, $X = 1.76183$, $Y = 1.38372$
6/10000, $X = 1.53771$, $Y = 1.31527$
7/10000, $X = 1.39506$, $Y = 1.2766$
8/10000, $X = 1.31719$, $Y = 1.25793$
9/10000, $X = 1.28156$, $Y = 1.25015$
10/10000, $X = 1.26754$, $Y = 1.24726$
11/10000, $X = 1.26251$, $Y = 1.24625$
12/10000, $X = 1.26079$, $Y = 1.2459$
13/10000, $X = 1.26021$, $Y = 1.24579$
14/10000, $X = 1.26002$, $Y = 1.24575$
15/10000, $X = 1.25995$, $Y = 1.24574$
16/10000, $X = 1.25993$, $Y = 1.24573$

17/10000, $X = 1.25992$, $Y = 1.24573$
18/10000, $X = 1.25992$, $Y = 1.24573$

## 11.8 Connection with the Newton and SOR/SUR schemes

It can now be shown that the phaseplane scheme is really a generalisation of the Newton iterative scheme, and of the SOR/SUR iterative schemes.

*Connection with the Newton scheme*

To illustrate the connection with the Newton scheme, the analysis will be restricted to the case of one variable. Whereas $\xi^{(l)}$ was defined as the difference between $X^{(l)}$ and the equilibrium value $X_e$, now define the quantity $\delta^{(l)}$ as the difference between the $l$th and $(l + 1)$th iterates:

$$\delta^{(l)} \equiv X^{(l+1)} - X^{(l)}, \qquad \delta^{(l-1)} \equiv X^{(l)} - X^{(l-1)}.$$

In terms of these new quantities, Equation (11.10) becomes

$$(1 + a(X^{(l)}))(X^{(l)} + \delta^{(l)}) - 2X^{(l)} + (1 - a(X^{(l)}))(X^{(l)} - \delta^{(l-1)}) = 2f(X^{(l)}),$$

or

$$(1 + a(X^{(l)}))\delta^{(l)} - (1 - a(X^{(l)}))\delta^{(l-1)} = 2f(X^{(l)}). \qquad (11.40)$$

In particular, if the function $a$ is chosen such that $a(X^{(l)}) = 1$, then Equation (11.40) becomes

$$\delta^{(l)} = f(X^{(l)}) = f_e' \frac{F(X^{(l)})}{F'(X^{(l)})},$$

which gives

$$X^{(l+1)} = X^{(l)} + f_e' \frac{F(X^{(l)})}{F'(X^{(l)})}. \qquad (11.41)$$

Remembering that $-2 < f_e' < 0$, this is just the Newton algorithm given in Equation (10.6) with factor $(-f_e')$ to avoid overshoot.

*Connection with the SOR/SUR scheme*

Suppose we have a set of $M + 1$ *linear* equations

$$G_i(\mathbf{X}^0) \equiv b_{i0}X_0^{(l)} + b_{i1}X_1^{(l)} + \cdots + b_{ii}X_i^{(l)} + \cdots + b_{iM}X_M^{(l)} = 0,$$
$$(i = 0, \ldots, M).$$

where the $b_{ij}$ are real constants. The equations can be ordered such that $b_{ii} \neq 0$, $(i = 0, \ldots, M)$, and then it follows that

$$X_i^{(l)} = \frac{-G_i(\mathbf{X}^{(l)}) + b_{ii}X_i^{(l)}}{b_{ii}}.$$

Hence the SOR/SUR iterative scheme in this case will be

$$X_i^{(l+1)} = (1-\omega)X_i^{(l)} + \omega\left(\frac{-G_i(\mathbf{X}^{(l)}) + b_{ii}X_i^{(l)}}{b_{ii}}\right)$$

$$= X_i^{(l)} - \omega\frac{G_i(\mathbf{X}^{(l)})}{b_{ii}},$$

where $0 < \omega < 1$ for SUR and $1 < \omega < 2$ for SOR. However, $b_{ii} = \partial_i G_i(\mathbf{X}^{(l)})$, so that

$$X_i^{(l+1)} = X_i^{(l)} - \omega\frac{G_i(\mathbf{X}^{(l)})}{\partial_i G_i(\mathbf{X}^{(l)})}, \qquad (0 < \omega < 2). \qquad (11.42)$$

This result is the analogue of Equation (11.26) if the value $a = 1$ is taken in the equation, since $\omega = -f_e'$ with $-2 < f_e' < 0$.

## 11.9  Error analysis of the phase plane method

The error at the $l$th iteration step is $\xi^{(l)} = X^{(l)} - X_e$, and it should be expected that the magnitude of this error will decrease as the iteration number $l$ increases. It will now be shown how the errors in the solution develop with the iteration. This development can be given in terms of the exact derivatives evaluated at the equilibrium solution (which is not possible in practice since it is this solution which is being sought), or in terms of derivatives evaluated at the current timestep using finite differences.

### 11.9.1  One-variable case using exact derivatives

Using the substitution $X^{(l)} = X_e + \xi^{(l)}$, Equation (11.10) becomes

$$(1 + a(X^{(l)}))\xi^{(l+1)} = 2f(X^{(l)}) + 2\xi^{(l)} - (1 - a(X^{(l)}))\xi^{(l-1)},$$

which expands to

$$(1 + a_e + a_e'\xi^{(l)} + \cdots)\xi^{(l+1)} = 2f_e + 2f_e'\xi^{(l)} + f_e''(\xi^{(l)})^2 + \cdots + 2\xi^{(l)}$$
$$- (1 - a_e - a_e'\xi^{(l)} - \cdots)\xi^{(l-1)}$$
$$= 2(f_e' + 1)\xi^{(l)} - (1 - a_e)\xi^{(l-1)} + f_e''(\xi^{(l)})^2$$
$$+ a_e'\xi^{(l)}\xi^{(l-1)} + \cdots \qquad (11.43)$$

Hence convergence will be quadratic if the first two terms on the right hand side are zero:

$$(f'_e = -1 \text{ and } a_e = 1) \quad \Rightarrow \quad \text{quadratic convergence.} \tag{11.44}$$

### 11.9.2 One-variable case using central difference derivatives

The convergence can also be analysed when the derivative $F'(X^{(l)})$ is evaluated using a central difference formula. Since

$$F(X^{(l)} + \Delta) = F(X_e + \xi^{(l)} + \Delta) = F_e + (\xi^{(l)} + \Delta)F'_e + \frac{1}{2}(\xi^{(l)} + \Delta)^2 F''_e + \cdots$$

and

$$F(X^{(l)} - \Delta) = F(X_e + \xi^{(l)} - \Delta) = F_e + (\xi^{(l)} - \Delta)F'_e + \frac{1}{2}(\xi^{(l)} - \Delta)^2 F''_e + \cdots,$$

then the central difference formula using a positive increment $\Delta$ becomes

$$\frac{F(\xi^{(l)} + \Delta) - F(\xi^{(l)} - \Delta)}{2\Delta} = F'_e + F''_e \xi^{(l)}$$

—this is independent of the value of $\Delta$ to the first order in $\xi^{(l)}$. Hence the convergence property is the same as that given by Equation (11.44). It can be seen from the output listed in Sect. 11.6 that the solution has converged much more rapidly for the optimal case of $a_e = 1.0$ than for the case of $a_e = 0.8$.

### 11.9.3 Multi-variable case using exact derivatives

This analysis follows from Equation (11.38). Using Equation (11.36), we may define

$$\Xi_i^{(l)} \equiv W_i^{(l)} - X_e = X^{(l)} - X_e + rU_{N,i-1}(X^{(l+1)} - X_e + X_e - X^{(l)})$$
$$= \xi^{(l)} + rU_{N,i-1}(\xi^{(l+1)} - \xi^{(l)}). \tag{11.45}$$

On expanding to the second order in the $\xi$, it follows that

$$a_i(W_i^{(l)}) = a_i(X_e + \Xi_i^{(l)})$$
$$= a_{ie} + \sum_{m=0}^{M}(\partial_m a_i)_e(\Xi_i^{(l)})_m + \frac{1}{2}\sum_{m=0}^{M}\sum_{n=0}^{M}(\partial_m \partial_n a_i)_e(\Xi_i^{(l)})_m(\Xi_i^{(l)})_n,$$

and

$$f_i(\mathbf{W}_i^{(l)}) = \sum_{m=0}^{M} (\partial_m f_i)_e (\Xi_i^{(l)})_m + \frac{1}{2} \sum_{m=0}^{M} \sum_{n=0}^{M} (\partial_m \partial_n f_i)_e (\Xi_i^{(l)})_m (\Xi_i^{(l)})_n,$$

since $f_{ie} = 0$. Hence to the second order in $\boldsymbol{\xi}$, Equation (11.38) becomes

$$\left(1 + a_{ie} + \sum_{m=0}^{M} (\partial a_i)_e (\Xi_i^{(l)})_m\right) \xi_i^{(l+1)}$$

$$\approx 2\xi_i^{(l)} - \left(1 - a_{ie} - \sum_{m=0}^{M} (\partial a_i)_e (\Xi_i^{(l)})_m\right) \xi_i^{(l-1)}$$

$$+ 2 \sum_{m=0}^{M} (\partial_m f_i)_e (\Xi_i^{(l)})_m$$

$$+ \sum_{m=0}^{M} \sum_{n=0}^{M} (\Xi_i^{(l)})_m (\Xi_i^{(l)})_n. \tag{11.46}$$

To the first order in $\boldsymbol{\xi}$, this becomes

$$(1 + a_{ie})\xi_i^{(l+1)} \approx 2\xi_i^{(l)} - (1 - a_{ie})\xi_i^{(l-1)} + 2 \sum_{m=0}^{M} (\partial_m f_i)_e (\Xi_i^{(l)})_m$$

$$= 2\xi_i^{(l)} - (1 - a_{ie}) \xi_i^{(l-1)}$$

$$+ 2 \sum_{m=0}^{M} (\partial_m f_i)_e \left(\xi_m^{(l)} + r \sum_{k=0}^{M} (\mathbf{U}_{N,i-1})_{mk} (\xi_k^{(l+1)} - \xi_k^{(l)})\right). \tag{11.47}$$

To simplify the notation, define the $(M + 1) \times (M + 1)$ matrix $\phi$ by $\phi_{ij} \equiv (\partial_j f_i)_e$. Then equation (11.47) becomes

$$(1 + a_{ie})\xi_i^{(l+1)} = 2\xi_i^{(l)} - (1 - a_{ie})\xi_i^{(l-1)} + 2(\phi\xi^{(l)})_i$$
$$+ 2r(\phi_L\xi^{(l+1)})_i - 2r(\phi_L\xi^{(l)})_i, \tag{11.48}$$

where $\phi_L$ is the lower left triangular part of $\phi$ as defined in Equation (11.37). On defining the diagonal matrix $\mathbf{A}$ by

$$(\mathbf{A})_{ij} \equiv a_{ie}\delta_{ij},$$

Equation (11.48) can be written in the matrix form

$$(\mathbf{I} + \mathbf{A} - 2r\phi_L)\xi^{(l+1)} = 2(\mathbf{I} + \phi - r\phi_L)\xi^{(l)} - (\mathbf{I} - \mathbf{A})\xi^{(l-1)}$$
$$+ \text{higher order terms} \tag{11.49}$$

for $r = 0$ or 1.

Note that the $(M + 1)^2$ elements of $\phi$ are not all independent quantities, but are fixed by at most $M + 1$ independent parameters. This can be seen in the following way. On writing $F_i \equiv F_i(\mathbf{W}_i)$ and using the result

$$f_i = (\partial_i f_i)_e \frac{F_i}{\partial_i F_i},$$

it follows that

$$\partial_j f_i = (\partial_i f_i)_e \left( \frac{\partial_j F_i}{\partial_i F_i} - F_i \frac{\partial_j \partial_i F_i}{(\partial_i F_i)^2} \right).$$

Hence

$$\phi_{ij} \equiv (\partial_j f_i)_e = (\partial_i f_i)_e \frac{(\partial_j F_i)_e}{(\partial_i F_i)_e} \tag{11.50}$$

since $(F_i)_e = 0$. Then defining the two $(M + 1) \times (M + 1)$ matrices $\boldsymbol{\Phi}$ and $\mathbf{f}'$ by

$$\boldsymbol{\Phi}: \quad \Phi_{ij} \equiv \frac{(\partial_j F_i)_e}{(\partial_i F_i)_e} \tag{11.51}$$

$$\mathbf{f}': \quad f'_{ij} \equiv (\partial_i f_i)_e \delta_{ij} \tag{11.52}$$

it follows that

$$\phi = \mathbf{f}' \boldsymbol{\Phi}. \tag{11.53}$$

For a given problem, the elements of $\boldsymbol{\Phi}$ are fixed since the functions $F_i$ are specified. Hence the only values which must be chosen are the $M + 1$ elements of $\mathbf{f}'$. Since it also follows that $\phi_L = \mathbf{f}' \boldsymbol{\Phi}_L$, Equation (11.49) becomes

$$(\mathbf{I} + \mathbf{A} - 2r\mathbf{f}'\boldsymbol{\Phi}_L)\xi^{(l+1)} = 2(\mathbf{I} + \mathbf{f}'(\boldsymbol{\Phi} - r\boldsymbol{\Phi}_L))\xi^{(l)} - (\mathbf{I} - \mathbf{A})\xi^{(l-1)}$$
$$+ \text{ higher order terms} \tag{11.54}$$

for $r = 0$ or 1. We are then left to choose the $2(M + 1)$ independent parameters in the matrices $\mathbf{A}$ and $\mathbf{f}'$.

This result clearly simplifies for the case of a single variable. In this case, the value $r = 0$ must be taken for no replacement, and all matrices become scalar quantities: $\boldsymbol{\Phi}$ is replaced by $+1$, $\mathbf{f}'$ is replaced by $f'_e$, and $\mathbf{A}$ is replaced by $a_e$. Result (11.54) will then reduce to

$$(1 + a_e)\xi^{(l+1)} = 2(1 + f'_e)\xi^{(l)} - (1 - a_e)\xi^{(l-1)},$$

which is the linear part of Equation (11.43). In this single-variable case, it was possible to make the convergence quadratic by taking $f'_e = -1$ and $a_e = +1$. However, in the multi-variable case, we are not able to make the convergence linear by taking $\mathbf{A} = \mathbf{I}$ and $r = 0$ because the relation

$$\xi^{(l+1)} = (\mathbf{I} + \mathbf{f}'\boldsymbol{\Phi})\xi^{(l)}$$

still remains. Even taking $\mathbf{f}' = -\mathbf{I}$ still gives

$$\xi^{(l+1)} = (\mathbf{I} - \boldsymbol{\Phi})\xi^{(l)}$$

with $\boldsymbol{\Phi} \neq \mathbf{I}$.

## 11.9.4 Multi-variable case using central difference derivatives

Define the $(M+1)$-component vector $\mathbf{D}_i$ to have a positive value $\Delta_i$ in the $i$th place and zeros elsewhere:

$$(\mathbf{D}_i)_j \equiv \Delta_i \delta_{ij} \quad (i, j = 0, \dots, M). \tag{11.55}$$

Then the central difference approximation for the partial derivative of the function $F_i$ is

$$\partial_i F_i(\mathbf{W}_i^{(l)}) \approx \frac{F_i(\mathbf{W}_i^{(l)} + \mathbf{D}_i) - F_i(\mathbf{W}_i^{(l)} - \mathbf{D}_i)}{2\Delta_i}. \tag{11.56}$$

Now

$$
\begin{aligned}
F_i(\mathbf{W}_i^{(l)} \pm \mathbf{D}_i) &= F_i(\mathbf{X}_e + \boldsymbol{\Xi}_i^{(l)} \pm \mathbf{D}_i) \\
&= \sum_{m=0}^{M} (\partial_m F_i)_e (\boldsymbol{\Xi}_i^{(l)} \pm \mathbf{D}_i)_m \\
&\quad + \frac{1}{2} \sum_{m=0}^{M} \sum_{n=0}^{M} (\partial_m \partial_n F_i)_e (\boldsymbol{\Xi}_i^{(l)} \pm \mathbf{D}_i)_m (\boldsymbol{\Xi}_i^{(l)} \pm \mathbf{D}_i)_n \\
&= \sum_{m=0}^{M} (\partial_m F_i)(\boldsymbol{\Xi}_i^{(l)})_m \pm (\partial_i F_i)_e \Delta_i \pm \sum_{m=0}^{M} (\partial_i \partial_m F_i)_e (\boldsymbol{\Xi}_i^{(l)})_m \Delta_i \\
&\quad + \frac{1}{2} \sum_{m=0}^{M} \sum_{n=0}^{M} (\partial_m \partial_n F_i)_e (\boldsymbol{\Xi}_i^{(l)})_m (\boldsymbol{\Xi}_i^{(l)})_n + \frac{1}{2} (\partial_i \partial_i F_i)_e \Delta_i^2.
\end{aligned}
\tag{11.57}
$$

Hence it follows that

$$
\begin{aligned}
\partial_i F_i(\mathbf{W}_i^{(l)}) &\approx \frac{F_i(\mathbf{W}_i^{(l)} + \mathbf{D}_i) - F_i(\mathbf{W}_i^{(l)} - \mathbf{D}_i)}{2\Delta_i} \\
&\approx (\partial_i F_i)_e + \sum_{m=0}^{M} (\partial_i \partial_m F_i)_e (\boldsymbol{\Xi}_i^{(l)})_m
\end{aligned}
\tag{11.58}
$$

which is independent of $\Delta_i$ to the first order in $\mathbf{Z}$.

The expression for $F_i(\mathbf{W}_i^{(l)})$ is found from Equation (11.57) by formally putting $\Delta_i = 0$. Hence we have

$$\frac{F_i(\mathbf{W}_i^{(l)})}{\partial_i F_i(\mathbf{W}_i^{(l)})} = \frac{\sum_{m=0}^{M}(\partial_m F_i)_e(\Xi_i^{(l)})_m + \frac{1}{2}\sum_{m=0}^{M}\sum_{n=0}^{M}(\partial_m \partial_n F_i)_e(\Xi_i^{(l)})_m(\Xi_i^{(l)})_n}{(\partial_i F_i)_e + \sum_{m=0}^{M}(\partial_i \partial_m F_i)_e(\Xi_i^{(l)})_m}.$$

(11.59)

Remember that the elements $(i, \dots, M)$ of $\Xi_i^{(l)}$ are $\xi_j^{(l)}$ for $j = i, \dots, M$, while the first $i-1$ elements are either $\xi_j^{(l)}$ or $\xi_j^{(l+1)}$ depending on whether there is replacement ($r = 1$) or no replacement ($r = 0$).

Since the expression on the right hand side of Equation (11.59) is independent of the step $\Delta_i$ to the first order in $\xi$, it should be expected that substitution of this expression into Equation (11.39) leads to results which are identical to those of Equations (11.49)–(11.54). This may be seen as follows: to the first order in $\xi$, Equation (11.59) becomes

$$\frac{F_i(\mathbf{W}_i^{(l)})}{\partial_i F_i(\mathbf{W}_i^{(l)})} = \frac{\sum_{m=0}^{M}(\partial_m F_i)_e(\Xi_i^{(l)})_m}{(\partial_i F_i)_e} = \sum_{m=0}^{M}\left(\frac{\partial_m F_i}{\partial_i F_i}\right)_e(\Xi_i^{(l)})_m$$

$$= \sum_{m=0}^{M}\Phi_{im}(\Xi_i^{(l)})_m.$$

(11.60)

Equation (11.38) then becomes

$$(1 + a_{ie})\xi_i^{(l+1)'} \approx 2\xi_i^{(l)} - (1 - a_{ie})\xi_i^{(l-1)} + 2(\partial_i f_i)_e \sum_{m=1}^{N}\Phi_{im}(\Xi_i^{(l)})_m,$$

(11.61)

and Equations (11.49)–(11.54) then follow.

### 11.9.5 Multi-variable case using forward difference derivatives

Using Equation (11.57), the forward difference formula for the partial derivative is

$$\partial_i F_i(\mathbf{W}_i^{(l)}) \approx \frac{F_i(\mathbf{W}_i^{(l)} + \mathbf{D}_i) - F_i(\mathbf{W}_i^{(l)})}{\Delta_i}$$

$$\approx (\partial_i F_i)_e + \sum_{m=0}^{M}(\partial_i \partial_m F_i)_e(\Xi_i^{(l)})_m + \frac{1}{2}(\partial_i \partial_i F_i)_e \Delta_i,$$

(11.62)

and hence it follows that

$$\frac{F_i(\mathbf{W}_i^{(l)})}{\partial_i F_i(\mathbf{W}_i^{(l)})} = \frac{\sum_{m=0}^{M}(\partial_m F_i)_e(\mathbf{\Xi}_i^{(l)})_m + \frac{1}{2}\sum_{m=0}^{M}\sum_{n=0}^{M}(\partial_m \partial_n F_i)_e(\mathbf{\Xi}_i^{(l)})_m(\mathbf{\Xi}_i^{(l)})_n}{(\partial_i F_i)_e + \sum_{m=0}^{M}(\partial_i \partial_m F_i)_e(\mathbf{\Xi}_i^{(l)})_m + \frac{1}{2}(\partial_i \partial_i F_i)_e \Delta_i}. \tag{11.63}$$

To the first order, we again have

$$\frac{F_i(\mathbf{W}_i^{(l)})}{\partial_i F_i(\mathbf{W}_i^{(l)})} = \frac{\sum_{m=0}^{M}(\partial_m F_i)_e(\mathbf{\Xi}_i^{(l)})_m}{(\partial_i F_i)_e} = \sum_{m=0}^{M}\left(\frac{\partial_m F_i}{\partial_i F_i}\right)_e (\mathbf{\Xi}_i^{(l)})_m$$

$$= \sum_{m=0}^{M} \Phi_{im}(\mathbf{\Xi}_i^{(l)})_m. \tag{11.64}$$

Hence, to the first order, the corrections to the $\xi$ terms are given, as in the case of the central difference approximation, by Equation (11.61).

## 11.10 The phaseplane method applied to device modelling

The application of the phaseplane method will be illustrated by considering a HEMT with a two dimensional cross section. The modelling equations are

1. The Poisson equation

$$\nabla_{\mathbf{r}} \cdot (\varepsilon_0 \varepsilon_r \nabla \psi) = -q(N_D - n) \tag{11.65}$$

where $\psi$ is the electrostatic potential and $n$ is the electron density.
2. The continuity equation for electrons

$$\frac{\partial n}{\partial t} = \frac{1}{q}\nabla_{\mathbf{r}} \cdot \mathbf{J}, \tag{11.66}$$

where $\mathbf{J}$ is the current density. It is often more convenient to work with the *Fermi potential* $\phi$ defined by

$$\phi \equiv -\frac{1}{q}E_F, \tag{11.67}$$

rather than with the quasi Fermi level $E_F$. Then the argument $(E_F - E_c)/(k_B T)$ which appears in the associated Fermi integrals becomes $q(\psi - \phi)/(k_B T)$. The current density $\mathbf{J}$ can then be written

$$\mathbf{J} = -qn\mu\nabla_{\mathbf{r}}\phi + k_B \mu n \nabla_{\mathbf{r}} T. \tag{11.68}$$

3. The energy transport equation

$$\frac{\partial W}{\partial t} = \mathbf{J} \cdot \mathbf{E} - \frac{W - W_0}{\tau_e} - \nabla_{\mathbf{r}} \cdot \mathbf{s} \tag{11.69}$$

where $W$ is the total energy density, $\mathbf{E} \equiv -\nabla \psi$ is the electric field, and $\tau_e$ is the energy relaxation lifetime. The energy flux $\mathbf{s}$ will have the form

$$\mathbf{s} = -\mu W \nabla_\mathbf{r} \phi + 2 \frac{k_B}{q} \mu W \nabla_\mathbf{r} T. \tag{11.70}$$

The phaseplane solution method will be applied only to the Poisson equation, continuity equation, and the energy transport equation. It is not applied to the solution of the TISE, which is solved fully at each iteration step.

A non-uniform rectangular mesh is taken with the origin at one corner of the device, with the set of grid points being $((i, j) : i = 0, \ldots, N_x; j = 0, \ldots, N_y)$. The variable step lengths are $h_i \equiv x_{i+1} - x_i$ and $k_j \equiv y_{j+1} - y_j$.

1. The Poisson equation, Equation (11.65), becomes

$$
\begin{aligned}
0 = F_{(P)i,j} \\
\equiv \; & \frac{\varepsilon_{ri-\frac{1}{2},j} \frac{(\psi_{i,j}-\psi_{i-1,j})}{h_{i-1}} - \varepsilon_{ri+\frac{1}{2},j} \frac{(\psi_{i+1,j}-\psi_{i,j})}{h_i}}{\frac{1}{2}(h_{i-1}+h_i)} \\
& + \frac{\varepsilon_{ri,j-\frac{1}{2}} \frac{(\psi_{i,j}-\psi_{i,j-1})}{k_{j-1}} - \varepsilon_{ri,j+\frac{1}{2}} \frac{(\psi_{i,j+1}-\psi_{i,j})}{k_j}}{\frac{1}{2}(k_{j-1}+k_j)} \\
& + \frac{q}{\varepsilon_0}(N_{Di,j} - n_{i,j})
\end{aligned} \tag{11.71}
$$

where the relative permittivity $\varepsilon_r$ is evaluated at the half-points as $\varepsilon_{ri-\frac{1}{2},j} \equiv \frac{1}{2}(\varepsilon_{ri,j} + \varepsilon_{ri-1,j})$.

2. The continuity equation, Equation (11.66), and the expression for the current density in Equation (11.68), become

$$0 = F_{(C)i,j} \equiv \frac{(J_{xi+\frac{1}{2},j} - J_{xi-\frac{1}{2},j})}{\frac{1}{2}(h_{i-1}+h_i)} + \frac{(J_{yi,j+\frac{1}{2}} - J_{yi,j-\frac{1}{2}})}{\frac{1}{2}(k_{j-1}+k_j)} \tag{11.72}$$

and

$$J_{xi+\frac{1}{2},j} = \mu_{i+\frac{1}{2},j} n_{i+\frac{1}{2},j} \left( k_B \frac{(T_{i+1,j} - T_{i,j})}{h_i} - q \frac{(\phi_{i+1,j} - \phi_{i,j})}{h_i} \right), \tag{11.73}$$

with similar expressions for $J_{xi-\frac{1}{2},j}$, $J_{yi,j+\frac{1}{2}}$ and $J_{yi,j-\frac{1}{2}}$.

3. The energy transport equation, Equation (11.69), and the energy flux $\mathbf{s}$ become

$$
\begin{aligned}
0 = F_{(E)i,j} \equiv \; & (\mathbf{J} \cdot \mathbf{E})_{i,j} - \frac{(W_{i,j} - W_{0i,j})}{\tau_{ei,j}} \\
& - \frac{(s_{xi+\frac{1}{2},j} - s_{xi-\frac{1}{2},j})}{\frac{1}{2}(h_{i-1}+h_i)} - \frac{(s_{yi,j+\frac{1}{2}} - s_{yi,j-\frac{1}{2}})}{\frac{1}{2}(k_{j-1}+k_j)}
\end{aligned} \tag{11.74}
$$

with

$$(\mathbf{J} \cdot \mathbf{E})_{i,j} = \frac{1}{2}(E_{xi+\frac{1}{2},j}J_{xi+\frac{1}{2},j} + E_{xi-\frac{1}{2},j}J_{xi-\frac{1}{2},j})$$

$$+ \frac{1}{2}(E_{yi,j+\frac{1}{2}}J_{yi,j+\frac{1}{2}} + E_{yi,j-\frac{1}{2}}J_{yi,j-\frac{1}{2}}),$$

$$E_{xi+\frac{1}{2},j} = -\frac{(\psi_{i+1,j} - \psi_{i,j})}{h_i}, \tag{11.75}$$

$$S_{xi+\frac{1}{2},j} = -\mu_{i+\frac{1}{2},j}W_{i+\frac{1}{2},j}\frac{(\phi_{i+1,j} - \phi_{i,j})}{h_i}$$

$$+ 2\frac{k_B}{q}\mu_{i+\frac{1}{2},j}W_{i+\frac{1}{2},j}\frac{(T_{i+1,j} - T_{i,j})}{h_i}.$$

There will be similar expressions at the half points $(i - \frac{1}{2}, j)$, $(i, j + \frac{1}{2})$ and $(i, j - \frac{1}{2})$.

The unknown quantities to be solved are the $3(N_x - 1)(N_y - 1)$ internal values

$$\{\psi_{i,j}, \phi_{i,j}, T_{i,j} : i = 1, \dots, N_x - 1; j = 1, \dots, N_y - 1\}$$

taken in some order. The equations which are to be solved are

$$F_{(P)i,j} = 0, \quad F_{(C)i,j} = 0, \quad F_{(E)i,j} = 0,$$
$$(i = 1, \dots, N_x - 1; j = 1, \dots, N_y - 1), \tag{11.76}$$

taken in some order over the internal grid points of the device.

A damped equation similar to that in Equation (11.29) must be assigned to each unknown by choosing an appropriate function for each equation. Each of the Equations (11.76) at the grid point $(i, j)$ depends, either implicitly or explicitly, on all of the quantities $\psi_{i,j}$, $\phi_{i,j}$ and $T_{i,j}$. In principle, therefore, it does not matter which of these unknowns is assigned to which equation. In practice, however, there is a natural order in the assignment which can be made.

1. Since $\psi$ is the only variable of the set $(\psi, \phi, T)$ which appears explicitly in the equation $F_{(P)i,j} = 0$, a natural choice will be to assign this unknown to this equation, and to solve a damped equation

$$\ddot{\psi}_{i,j} + k_{(P),i,j}\dot{\psi}_{i,j} = \gamma_{(P),i,j}F_{(P)i,j} \tag{11.77}$$

at each internal grid point $(i, j)$. The values of $k_{(P),i,j}$ and $\gamma_{(P),i,j}$ have yet to be chosen. This assignment will then remain valid in one-dimensional cases for which there is no transport, and in which only the Poisson equation is solved with the TISE (Cole et al. 1997).

2. Both the unknowns $\phi_{i,j}$ and $T_{i,j}$ appear explicitly in both the equations $F_{(C)i,j} = 0$ and $F_{(E)i,j} = 0$. However, in order to consider models in which there is no energy transport (for which $T_{i,j}$ is independent of the grid point), then it is natural to assign the unknown $\phi_{i,j}$ with the equation $F_{(C)i,j} = 0$ and $T_{i,j}$ with the equation

$F_{(E)i,j} = 0$:

$$\ddot{\phi}_{i,j} + k_{(C),i,j}\dot{\phi}_{i,j} = \gamma_{(C),i,j}F_{(C)i,j} \tag{11.78}$$

$$\ddot{T}_{i,j} + k_{(E),i,j}\dot{T}_{i,j} = \gamma_{(E),i,j}F_{(E)i,j}. \tag{11.79}$$

The values of $\psi_{i,j}^{(l)}$, $\phi_{i,j}^{(l)}$ and $T_{i,j}^{(l)}$ will be updated at each grid point $(i, j)$ to the values $\psi_{i,j}^{(l+1)}$, $\phi_{i,j}^{(l+1)}$ and $T_{i,j}^{(l+1)}$ using Equation (11.39):

$$\psi_{i,j}^{(l+1)} = \frac{2(f'_{(P)i,j})e^{\frac{F_{(P)i,j}^{(l)}}{(\partial F_{(P)i,j}^{(l)})/(\partial \psi_{i,j}^{(l)})}} + 2\psi_{i,j}^{(l)} - (1 - a_{(P)i,j})\psi_{i,j}^{(l-1)}}{1 + a_{(P)i,j}}, \tag{11.80}$$

$$\phi_{i,j}^{(l+1)} = \frac{2(f'_{(C)i,j})e^{\frac{F_{(C)i,j}^{(l)}}{(\partial F_{(C)i,j}^{(l)})/(\partial \phi_{i,j}^{(l)})}} + 2\phi_{i,j}^{(l)} - (1 - a_{(C)i,j})\phi_{i,j}^{(l-1)}}{1 + a_{(C)i,j}}, \tag{11.81}$$

$$T_{i,j}^{(l+1)} = \frac{2(f'_{(E)i,j})e^{\frac{F_{(E)i,j}^{(l)}}{(\partial F_{(E)i,j}^{(l)})/(\partial T_{i,j}^{(l)})}} + 2T_{i,j}^{(l)} - (1 - a_{(E)i,j})T_{i,j}^{(l-1)}}{1 + a_{(E)i,j}}. \tag{11.82}$$

We must choose the values of the parameters

$$f'_{(P)i,j}, \qquad a_{(P)i,j}, \qquad f'_{(C)i,j}, \qquad a_{(C)i,j}, \qquad f'_{(E)i,j}, \qquad a_{(E)i,j}$$

which appear in the Equations (11.80)–(11.82) for each of the $3(N_x - 1)(N_y - 1)$ grid points $(i, j)$. It will be an impossible task to assign $6(N_x - 1)(N_y - 1)$ separate values, and it will be more convenient to assume that they are all independent of the grid point $(i, j)$; only the six values $a_{(P)}$, $(f'_{(P)})e$, $a_{(C)}$, $(f'_{(C)})e$, $a_{(E)}$, and $(f'_{(E)})e$ will then need to be assigned. That is, each of the equations with $\psi$ as the value to be solved will use the same parameters $a_{(P)}$ and $(f'_{(P)})e$, with similar assignments for the other two sets of equations. These assignments must be made within the restrictions imposed by Equation (11.22).

When this assignment has been made, the ordered steps in the iteration process are:

- The derivatives in the numerator are calculated using the central difference formula of Equation (11.56).
- The boundary values are then updated using appropriate boundary conditions.
- At each time step, Equation (11.80) is iterated once and then the updated value of $\psi_{i,j}$ is used to calculate the function $F_{(C)i,j}$.
- Equation (11.81) is iterated once and then the updated values of $\psi_{i,j}$ and $\phi_{i,j}$ are used to calculate the function $F_{(E)i,j}$.
- Equation (11.82) is iterated once, and then the whole process is moved to the next grid point.
- The boundary values are updated after all of the internal gridpoints have been iterated.

Hence the sequence of unknowns $(X_0, X_1, \ldots, X_M)$ in the general Equation (11.27) is taken in the order

$$(\psi_{1,1}, \phi_{1,1}, T_{1,1}, \psi_{1,2}, \phi_{1,2}, T_{1,2}, \ldots, \psi_{(N_x-1),(N_y-1)}, \phi_{(N_x-1),(N_y-1)},$$
$$T_{(N_x-1),(N_y-1)}).$$

## 11.11 Case study: a four-layer four-contact HEMT

The results of the preceding section were applied to the steady-state solution of the equations relating to a four-layer HEMT with a fieldplate. The device has four contacts—the source, gate, fieldplate, and drain, and two recesses. The overall width of the device is 3.8 μm. Starting from the edge containing the contacts, the device has layers (with thicknesses) of Si-GaAs (0.1 μm), Si-AlGaAs (0.03 μm), Si-GaAs (0.1 μm) and GaAs (0.5 μm).

Equations (11.71)–(11.75) were used for the solution on the internal points of the device. Neumann boundary conditions were applied on the free surfaces and Dirichlet conditions on the contacts (Cole et al. 1998; Hussain et al. 2003). In addition, subsidiary equations were used to specify the electron density $n$ and the energy density $W$. The TISE of Equation (6.64) was solved in one-dimensional sections along the $y$-direction perpendicular to the layer interfaces. The exchange correlation energy of Equation (6.63) was not included in this application. The TISE was solved numerically for the eigenvalues using a QL algorithm with implicit shifts (Press et al. 2002), and then a small number of corresponding eigenfunctions were found by back substitution. The energy eigenvalues $\lambda_\nu$ and eigenfunctions $\xi_\nu$ were used to calculate the electron density $n$ and the energy density $W$ whose forms are given in Equations (5.50) and (5.51).

A number of simulations were taken for this four-layer HEMT, at a number of different bias points. A non-uniform grid of $129 \times 146$ grid points was used in all cases. At each bias point, ranges of different values of the $a_0$ and $f_0'$ were searched, in order to find the particular values which corresponded to the fastest convergence, with stopping based on the magnitudes of the residuals. In each case, it was found that the values

$$a_{(P)} = a_{(C)} = a_{(E)} = 1.55, \qquad (f_{(P)}')_e = -0.6, \qquad (f_{(C)}')_e = -0.5,$$
$$(f_{(E)}')_e = -0.2$$

produced the fastest convergence. These same values were found to give the fastest convergence for a separate simulation of an eleven-layer HEMT with a fieldplate.

The device equations were also solved using the Newton method. There, it was found that a very small value for the overshoot factor had to be taken in the early stages of the iteration in order that the iterative process did not collapse. The starting value of this factor was taken as 0.001, rising to 0.4 after 500 iterations. It was found that a value for this overshoot factor larger than 0.4 caused the solution to become unstable. Using the phase plane method, however, stable solutions were obtained using a wide variation of the parameters $a_0$ and $f_0'$. Not all of the choices will give as rapid convergence as the Newton method, but stability is preserved. However, it was found that the values of $a_0$ and $f_0'$ shown above produced a speed-up of approximately 1.8 over the Newton method.

# Chapter 12
# Solution of equations: the multigrid method

The equations governing the modelling of many semiconductor devices, including the HEMT, are differential equations with associated boundary conditions. These equations are discretised on a grid which is fine enough to preserve all of the necessary physical detail. In an iterative solution of the equations on this fine grid, the high frequency components of the errors are rapidly eliminated, but the convergence then becomes slow. In the multigrid method, the equations and their partial solutions are moved up and down through a succession of grids, from the initial fine grid to a very coarse grid, and different error frequencies are eliminated on different grids. This method allows the equations to be solved on a fine grid with a significant decrease in computing time. Details of the multigrid method can be found in the literature (Bodine et al. 1993; Bramble 1993; Brandt 1977; Brandt et al. 1983; Briggs and McCormick 2000; Dick et al. 1999; Grinstein et al. 1983; Press et al. 2002; Zhu and Cangellaris 2006).

In the application of the multigrid method to the HEMT simulation described later in this Chapter, the following approach will be adopted: solve the Poisson, current continuity, and energy transport equations using the multigrid method, but solve the Schrödinger equation on each grid using a non-multigrid QL algorithm. An important consideration which must be taken into account in using this method in device modelling is that the coarsest grid must be coarse enough in order for an exact solution to be obtained rapidly on this grid, while at the same time preserving the essential physical detail of the device.

## 12.1 Description of the multigrid method

The multigrid method is most simply explained in terms of a problem involving only one dependent variable $X$. Consider the problem of solving the equation

$$F(X) = 0 \tag{12.1}$$

E.A.B. Cole, *Mathematical and Numerical Modelling of* 283
*Heterostructure Semiconductor Devices: From Theory to Programming*,
DOI 10.1007/978-1-84882-937-4_12, © Springer-Verlag London Limited 2009

on a two-dimensional region $A$, where $F$ is a differential operator (generally non-linear), and $X(x, y)$ is a single function whose solution is to be found when given boundary conditions are applied. On taking a discrete grid $(g)$, this equation will be discretised in the form

$$F_{(g)}(X_{(g)}) = 0. \tag{12.2}$$

For example, the Poisson equation must be solved in a HEMT simulation on the region $A$, in which Dirichlet conditions apply on the part of the boundary $\partial A_1$ and Neumann conditions apply on the remainder $\partial A_2$ of the boundary. The equation and boundary conditions are

$$\nabla_{\mathbf{r}} \cdot (\varepsilon(x, y)\nabla_{\mathbf{r}}\psi(x, y)) + \rho(x, y) = 0 \quad \text{internally}$$
$$\psi(x, y) - G(x, y) = 0 \quad \text{on } \partial A_1 \tag{12.3}$$
$$\mathbf{n} \cdot \nabla_{\mathbf{r}}\psi(x, y) = 0 \quad \text{on } \partial A_2$$

where $\psi$ is the electrostatic potential which takes the given functional values $G(x, y)$ on $\partial A_1$, and $\mathbf{n}$ is the normal vector to the boundary. Let $\psi_{(g),i,j}$ be the *exact* solution of the discretised equation on the discrete grid $(g)$. Then on a uniform grid with mesh spacings $h$ and $k$ in the $x$ and $y$ directions respectively, the internal discretised Poisson equation at the grid point $(i, j)$ using the five-point formula is

$$\frac{1}{h^2}\varepsilon_{i+\frac{1}{2},j}\psi_{(g),i+1,j} + \frac{1}{h^2}\varepsilon_{i-\frac{1}{2},j}\psi_{(g),i-1,j} + \frac{1}{k^2}\varepsilon_{i,j+\frac{1}{2}}\psi_{(g),i,j+1}$$
$$+ \frac{1}{k^2}\varepsilon_{i,j-\frac{1}{2}}\psi_{(g),i,j-1}$$
$$- \left( \frac{1}{h^2}(\varepsilon_{i+\frac{1}{2},j} + \varepsilon_{i-\frac{1}{2},j}) + \frac{1}{k^2}(\varepsilon_{i,j+\frac{1}{2}} + \varepsilon_{i,j-\frac{1}{2}}) \right) \psi_{(g),i,j} + \rho_{i,j} = 0 \tag{12.4}$$

with a similar discretisation on the boundaries. The exact solution $\psi(x, y)$ of Equation (12.3) and the exact solution of Equation (12.4) will not be the same at the gridpoints since the discretised equation is an approximation to the original differential equation. In the multigrid method, it is important to make clear distinctions between the exact solution of the differential equation, the exact solution of the discretised equation, and the approximate solution of the discretised equation.

In the multigrid method, the problem must be broadened into that of solving the associated problem

$$F(X) = f$$

where $f$ is some given right hand side. The discretised equation can be iterated using the Newton method which has been described previously in Chapter 10. Let $X^{(k)}$ be the solution after $k$ iterations. Then the correction $\delta X^{(k)}$ to be added to $X^{(k)}$ is

$$\delta X^{(k)} = J^{(k)-1} \left( f^{(k)} - F(X^{(k)}) \right),$$

where $J^{(k)} \equiv dF/dX^{(k)}$ is the Jacobian (which is a scalar quantity in the one variable problem). It follows that

$$X^{(k+1)} = X^{(k)} + a^{(k)}\delta X^{(k)}$$
$$= X^{(k)} + a^{(k)} J^{(k)-1} \left( f^{(k)} - F(X^{(k)}) \right) \qquad (12.5)$$

where $a^{(k)}$ is a factor with $(0 < a^{(k)} \leq 1)$ which is introduced to avoid overshoot. As described earlier in Chapter 10, this factor need not be constant, but can be made to vary as the iterations progress. Finally, a mixture of this Newton method and the successive relaxation method may be used by taking

$$X^{(k+1)} = (1 - \omega^{(k)})X^{(k)} + \omega^{(k)} \left( X^{(k)} + a^{(k)} J^{(k)-1} \left( f^{(k)} - F(X^{(k)}) \right) \right) \quad (12.6)$$

where $0 < \omega^{(k)} < 1$ for SUR and $1 \leq \omega^{(k)} < 2$ for SOR. Again, as this notation implies, the factor $\omega^{(k)}$ can be made to vary as the iterations progress.

Normally, the problem is to be solved on as fine a mesh as possible. However, the high frequency components of the correction terms are eliminated after relatively few iterations on this grid, and the relaxation process slows up after this point. If the iteration process can be transferred to a coarser grid, then other frequency components of the correction can be eliminated very rapidly. In the *Full Approximation Storage* method (FAS) of Brandt (1977), an exact solution of the discretised equation is determined on a very coarse grid with very little effort. This solution is moved up (and down) through a succession of finer grids, with a small number of iterations being performed on each grid, until the required solution on the finest grid is obtained.

In describing the multigrid method, the discussion will be restricted to uniform two-dimensional grids. Extension to one-dimensional and three-dimensional grids is straightforward. The notation will be such that we have a succession of $G$ uniform grids $\{(g) : g = 1, 2, \ldots, G\}$ with grid $(g) = (1)$ being the coarsest and grid $(g) = (G)$ the finest. Each grid, apart from the coarsest, is obtained from the previous coarse grid by halving the mesh length in each direction. Let grid $(g)$ have $c_{(g)x}$ internal nodes along the $x$-direction, and $c_{(g)y}$ nodes along the $y$-direction. Then

$$c_{(g+1)x} = 2c_{(g)x} + 1 \quad \text{and} \quad c_{(g+1)y} = 2c_{(g)y} + 1. \qquad (12.7)$$

Again, since there are $(c_{(g)x} + 1)$ and $(c_{(g)y} + 1)$ intervals on each grid $(g)$ along the $x$- and $y$-directions respectively, and this number doubles each time the grid is refined, it follows that

$$c_{(g)x} = (c_{(1)x} + 1)2^{g-1} - 1, \quad \text{and}$$
$$c_{(g)y} = (c_{(1)y} + 1)2^{g-1} - 1. \qquad (12.8)$$

## 12.2 Moving between grids: restriction and prolongation

Before describing the processes involved in moving between grids, we must first show how quantities defined on a grid $(g - 1)$ can be prolongated (or interpolated) onto the finer grid $(g)$, or quantities defined on a grid $(g)$ can be restricted onto the coarser grid $(g - 1)$. These two processes must complement each other in a way that preserves the fidelity of the solution. It is not only the solution $X$ of Equation (12.1) which must be treated in this way, but also all quantities which appear in the equations—for example, the charge density $\rho$ which appears in the Poisson equation. Generally, we will use $v(x, y)$, or $v_{i,j}$ in its discretised form, to denote any quantity which has to be treated in this way.

Let $v_{(g)}$ and $v_{(g-1)}$ represent the values of the quantity $v$ on grids $(g)$ and $(g - 1)$ respectively. The *prolongation operator* $P$ takes the values of $v_{(g-1)}$ into the values of $v_{(g)}$:

$$v_{(g)} = P v_{(g-1)}. \tag{12.9}$$

It is found that linear prolongation works well for many multigrid implementations (Briggs and McCormick 2000). On a two-dimensional region, this operation is given by

$$v_{(g)2i,2j} = v_{(g-1)i,j} \quad \text{internally and on boundaries,}$$

$$v_{(g)2i,2j+1} = \frac{1}{2} \left( v_{(g-1)i,j} + v_{(g-1)i,j+1} \right) \quad \text{internally and on } y \text{ boundaries,}$$

$$v_{(g)2i+1,2j} = \frac{1}{2} \left( v_{(g-1)i,j} + v_{(g-1)i+1,j} \right) \tag{12.10}$$

$$\text{internally and on } x \text{ boundaries,}$$

$$v_{(g)2i+1,2j+1} = \frac{1}{4} \left( v_{(g-1)i+1,j} + v_{(g-1)i,j} + v_{(g-1)i+1,j+1} + v_{(g-1)i,j+1} \right)$$

$$\text{internally.}$$

The *restriction operator* $R$ takes the values of $v_{(g)}$ into the values of $v_{(g-1)}$:

$$v_{(g-1)} = R v_{(g)}. \tag{12.11}$$

The simplest restriction method is that of straight injection, given by

$$v_{(g-1)i,j} = v_{(g)2i,2j}, \tag{12.12}$$

while the *full weighting* method on internal points is given by

$$v_{(g-1)i,j} = \frac{1}{4} v_{(g)2i,2j} + \frac{1}{8} \left( v_{(g)2i,2j+1} + v_{(g)2i,2j-1} + v_{(g)2i+1,2j} + v_{(g)2i-1,2j} \right)$$

$$+ \frac{1}{16} \left( v_{(g)2i+1,2j+1} + v_{(g)2i+1,2j-1} + v_{(g)2i-1,2j+1} + v_{(g)2i-1,2j+1} \right),$$

$$\tag{12.13}$$

and on the boundaries by

$$v_{(g-1)i,j} = v_{(g)2i,2j}. \tag{12.14}$$

The linear prolongation operator $P$ and the full weighting restriction operator $R$ are adjoint operators (Press et al. 2002). In the work of Briggs and McCormick (2000), it is shown how the full weighting restriction method is the most suitable inverse of the linear prolongation operator, and produces relatively smooth oscillatory modes in its operation. Another popular choice of the prolongation operator is the *half weighting* formula given internally by

$$v_{(g-1)i,j} = \frac{1}{2}v_{(g)2i,2j} + \frac{1}{8}\left(v_{(g)2i,2j+1} + v_{(g)2i,2j-1} + v_{(g)2i+1,2j} + v_{(g)2i-1,2j}\right) \tag{12.15}$$

and on the boundaries by

$$v_{(g-1)i,j} = v_{(g)2i,2j}. \tag{12.16}$$

Note that the three restriction operators $R$ defined above are linear, in the sense that

$$R\left(v_{1(g)} + v_{2(g)}\right) = Rv_{1(g)} + Rv_{2(g)}. \tag{12.17}$$

## 12.3 Implementation of the multigrid method

Let us restate the problem. We are seeking to find the exact continuous solution to the problem $F(X) = 0$. This generally being impossible, we then look to find an exact solution to the discretised problem $F_{(G)}(X_{(G)}) = 0$ on the finest grid $(G)$ at our disposal. In the multigrid method, we look for solutions of the expanded equation

$$F_{(g)}(X_{(g)}) = f_{(g)}, \quad (g = 1, \ldots, G) \tag{12.18}$$

where the $f_{(g)}$ are functions which are generated as follows. Let the quantity $X'_{(g)}$ represent only an *approximate* solution of the discretised Equation (12.18), which is obtained after only a small number of iterations using the Newton method. Then the functions $f_{(g)}$ are generated using the recurrence relation

$$f_{(g-1)} = F_{(g-1)}\left(RX'_{(g)}\right) + R\left(f_{(g)} - F_{(g)}(X'_{(g)})\right). \tag{12.19}$$

Therefore the function $f_{(g-1)}$ is defined using the restriction operator $R$ on the way down from the finest grid $(g) = (G)$ to the coarsest grid $(g) = (1)$. In order to show how this result is derived, let $\delta X_{(g)}$ be the correction to be added to the approximate solution $X'_{(g)}$ to give the exact solution $X_{(g)}$ on grid $(g)$. Then

$$X_{(g)} = X'_{(g)} + \delta X_{(g)},$$

and it follows that

$$F_{(g)}(X'_{(g)} + \delta X_{(g)}) = f_{(g)}.$$

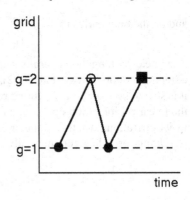

Then

$$F_{(g)}(X'_{(g)} + \delta X_{(g)}) - F_{(g)}(X'_{(g)}) = f_{(g)} - F_{(g)}(X'_{(g)}).$$

Applying the restriction operator $R$ to both sides, this becomes

$$R\left(F_{(g)}(X'_{(g)} + \delta X_{(g)}) - F_{(g)}(X'_{(g)})\right) = R\left(f_{(g)} - F_{(g)}(X'_{(g)})\right).$$

On applying the linearity property of the operator $R$, this result becomes

$$f_{(g-1)} - F_{(g-1)}(RX'_{(g)}) = F_{(g-1)}(X_{(g-1)}) - RF_{(g)}(X'_{(g)})$$
$$= R\left(f_{(g)} - F_{(g)}(X'_{(g)})\right),$$

and then Equation (12.19) follows.

We will first describe in detail how the multigrid method will apply in the simple case of two grids, that is, with $G = 2$. We therefore wish to find the exact solution of the discretised equation $F_{(2)}(X_{(2)}) = 0$. The steps taken in moving between grids are shown in Fig. 12.1, and are described as follows:

1. Set $f_{(g)} = 0$ on all grids $(g) = (1)$ and $(g) = (2)$.
2. Find the *exact* solution of Equation (12.18) on the coarsest grid $(g) = (1)$, remembering that $f_{(1)} = 0$.
3. Prolong this solution up to grid $(g) = (2)$ and take a small number of *pre-smoothing* iterations on grid $(g) = 2$; this produces the approximate solution $X'_{(2)}$.
4. Restrict the solution $X'_{(2)}$ back onto grid $(g) = (1)$, and find the new left hand side $f_{(1)}$ using the equation

$$f_{(1)} = F_{(1)}\left(RX'_{(2)}\right) + R\left(f_{(2)} - F_{(2)}(X'_{(2)})\right).$$

5. Solve Equation (12.18) *exactly* on the coarsest grid, using the newly generated function $f_{(1)}$.

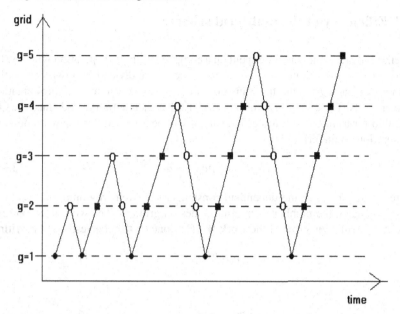

**Fig. 12.2** The steps involved in a five-grid FAS scheme. The transitions between grids are to be read from left to right. Exact solutions on the coarsest grid are represented by black circles, pre-smoothing stages are represented by white circles, and post-smoothing stages are represented by black squares.

6. Prolong this solution back up to grid $(g) = (2)$ and take a small number of *post-smoothing* iterations on grid $(g) = (2)$ to get the final solution.

There are three main comments to make regarding this scheme.

- The exact solution on the coarsest grid $(g) = (1)$ should be obtained without too much work; for example, in the one dimensional device simulation described later, the coarse grid has only $c_{(1)} = 24$ internal nodes.
- The small numbers of pre-smoothing and post-smoothing iterations are chosen in advance; in the device simulations described later, both these values are taken as ten, although the numbers do not have to be the same.
- The Newton scheme described in Equation (12.6) includes successive relaxation. Either SOR or SUR can be used to obtain the exact solutions on the coarsest grid, but SOR must not be used at the pre- and post-smoothing stages, since this will counteract the smoothing of the high frequency components.

The extension of this scheme to use more than two grids is straightforward, and can be described recursively (Briggs and McCormick 2000). Fig. 12.2 describes the steps in a five-grid process, and shows the stages involving the exact solution calculation on the coarsest grid, the pre-smoothing and the post-smoothing stages. However, extra cycles can be inserted at various stages. The device simulations which follow will use uniform grids, but the multigrid scheme can also be applied to non-uniform grids (Brandt 1977).

## 12.4 Efficiency of the multigrid scheme

A method of predicting the time spent on a complete multigrid operation will now be described. The operations of the multigrid process are divided into two parts—the iterations on each grid, and the transition processes of restriction and prolongation between grids. The quantum solution is involved at each iteration, and since the QL algorithm for tridiagonal matrices is $O(n)$, it will be assumed that the work done in one iteration on the grid $g$ is

$$W_{its}(g) = pc_{(g)} + qc_{(g)}^2 \qquad (12.20)$$

where $p$ and $q$ are machine-dependent constants for a given programming structure, and $c_{(g)}$ is again the number of internal nodes on grid $(g)$. The work done on grid $(g)$ can be written in terms of the work $W_{its}(1)$ done on the coarsest grid by writing

$$W_{its}(g) = \sigma(c_{(g)}, c_{(1)}) W_{its}(1), \quad (g = 1, \ldots, G) \qquad (12.21)$$

with

$$\sigma(c_{(g)}, c_{(1)}) \equiv \frac{pc_{(g)} + qc_{(g)}^2}{pc_{(1)} + qc_{(1)}^2}$$

$$= A_1 \left( \frac{c_{(g)}}{c_{(1)}} \right) + (1 - A_1) \left( \frac{c_{(g)}}{c_{(1)}} \right)^2, \qquad (12.22)$$

where the quantity $A_1$ is a constant which is given by

$$A_1 \equiv \frac{\frac{p}{c_{(1)}}}{(\frac{p}{c_{(1)}} + q)}. \qquad (12.23)$$

The value of $A_1$ will have the machine-dependent properties of the quantities $p$ and $q$. Most of the work performed in the remaining processes is in the descents in which the new right hand sides $f_{(g)}$ must be calculated, again using the quantum solution. It can be seen from Fig. 12.2 that ascents and descents are grouped in pairs, and the number of ascents from grid $(g)$ to grid $(g + 1)$ will equal the number of descents from grid $(g + 1)$ to grid $(g)$. It will therefore be assumed that the work $W_{trans}(g)$ expended in a single pairing of ascent and descent has the form

$$W_{trans}(g) = B_1 \left( \frac{c_{(g)}}{c_{(1)}} \right), \quad (g = 2, \ldots, G) \qquad (12.24)$$

where $B_1$ is also a constant which is machine-dependent.

It will be seen from the results in Sect. 12.5.1 that the assumptions (12.20) and (12.24) provide an excellent basis for predicting the total work done in a complete multigrid operation. Finding an expression for this total work amounts to counting the processes involved in the scheme shown in Fig. 12.2. When there are $G$ grids in total, there will be $G$ visits to the coarsest grid. Let $v_i$ $(i = 1, \ldots, G)$ be the number of iterations performed on the coarsest grid at the $i$th visit. Let the numbers of pre-smoothing and post-smoothing iterations on grid $(g)$ $(g = 2, \ldots, G)$ be denoted by $a_{(g)}$ and $b_{(g)}$ respectively. Then on each of the non-coarsest grids $(g)$, there will be $2(G - g + 1)$ visits involving $a_{(g)}$ pre-smoothing iterations, and $b_{(g)}$ post-smoothing iterations. In addition, there will be a total of $(G - g + 1)$ of ascents to, and descents from, grid $(g)$. Let $W_{its}$ and $W_{trans}$ be the *total* components of the work done in the complete multigrid operation in the iteration and transition stages respectively. Then the total work done in the multigrid solution will be

$$W = W_{its} + W_{trans}$$

$$= \left( \sum_{g=1}^{G} v_g + \sum_{g=2}^{G} (a_{(g)} + b_{(g)})(G - g + 1)R(c_{(g)}, c_{(1)}) \right) W_{its}(1)$$

$$+ B_1 \sum_{g=2}^{G} (G - g + 1)\left( \frac{c_{(g)}}{c_{(1)}} \right). \tag{12.25}$$

Now define the function $\gamma_G(x)$ by

$$\gamma_G(x) \equiv \sum_{g=2}^{G} (G - g + 1)x^{g-1}$$

$$= \begin{cases} \frac{x^2(x^{G-1}-1)}{(x-1)^2} - (G-1)\frac{x}{x-1} & \text{for } x \neq 1 \\ \frac{1}{2}G(G-1) & \text{for } x = 1. \end{cases} \tag{12.26}$$

Some relevant values of $\gamma_G(x)$ are given in Table 12.1. Using the results of Equations (12.8), (12.21) and (12.24), and assuming that the numbers of iterations $a_{(g)}$ and $b_{(g)}$ are independent of the grid number, so that $a_{(g)} \equiv a$ and $b_{(g)} \equiv b$, it can be shown that the total work done in the iteration stages and transition stages are

$$W_{its} = \left( \sum_{g=1}^{G} v_g + (a + b)\left[ (1 - A_1(1 + c_{(1)}^{-1})^2 \gamma_G(4) \right. \right.$$

$$+ (1 + c_{(1)}^{-1})(A_1 - 2(1 - A_1)c_{(1)}^{-1})\gamma_G(2)$$

$$\left. \left. + c_{(1)}^{-1}(c_{(1)}^{-1}(1 - A_1) - A_1)\gamma_G(1) \right] \right) W_{its}(1), \tag{12.27}$$

and

$$W_{trans} = B_1 \left( (1 + c_{(1)}^{-1})\gamma_G(2) - c_{(1)}^{-1}\gamma_G(1) \right). \tag{12.28}$$

The numbers of pre and post-smoothing numbers $a$ and $b$ must be specified in the multigrid solution. The values $v_1$ of the numbers of iterations on the coarsest grid, which are necessary to get an exact solution on that grid, can be specified in

**Table 12.1** Values of the functions $\gamma_G(1)$, $\gamma_G(2)$ and $\gamma_G(3)$ for grids $(G)$ in the range $(1)$–$(6)$.

| $G$ | $\gamma_G(1)$ | $\gamma_G(2)$ | $\gamma_G(4)$ |
|---|---|---|---|
| 1 | 0 | 0 | 0 |
| 2 | 1 | 2 | 4 |
| 3 | 3 | 8 | 24 |
| 4 | 6 | 22 | 108 |
| 5 | 10 | 52 | 448 |
| 6 | 15 | 114 | 1812 |

advance, but they are more likely to change during the multigrid process if convergence criteria are used to stop the iterations. In this more likely case, the total $\sum_{g=1}^{G} \nu_g$ can be used to provide an upper limit on the total work done on the lowest grid. When calculating the values of $W_{its}$ and $W_{trans}$ through Equation (12.27) and Equation (12.28), it is necessary to determine the values of $A_1$, $B_1$ and $W_{its}(1)$; it will be shown in Sect. 12.5.1 how this can be achieved by performing a relatively small number of runs.

The values of the constants $A_1$ and $B_1$ will depend on the size of the coarse grid $(g) = (1)$, but what if a different coarse grid $(g) = (1')$ is to be used? Simple algebra performed on Equations (12.23) and (12.24) will give the results

$$A_{1'} = \frac{A_1}{A_1 + \frac{c_{(1')}}{c_{(1)}}(1 - A_1)} \quad \text{and} \tag{12.29}$$

$$B_{1'} = \frac{c_{(1')}}{c_{(1)}} B_1. \tag{12.30}$$

Hence if $A_1$ and $B_1$ have been determined on a coarse grid $(1)$, the corresponding values $A_{1'}$ and $B_{1'}$ can be found for any other coarse grid $(1')$.

## 12.5 Multigrids applied to device modelling

The multigrid technique will now be described in its application to device modelling. Two case studies will be presented. The first is a study of a one dimensional device, in which only the coupled Poisson and Schrödinger equations are solved; there is no current or energy transport. The second case is a study of a two dimensional HEMT (Cole et al. 1997, 1998).

The choices of the coarsest and finest grids which are imposed on the device are crucial. Any grid that is used must not be so coarse as to lose the important physical detail. For example, it would not be satisfactory to have a grid which included a thin material layer contained entirely between two adjacent grid lines. However, we need only be careful in this way when constructing the coarsest grid, since this will not be coarsened further, and refinements will only add extra grid lines without moving the original lines.

**Fig. 12.3** A device with four layers and no recess structure. The only contact is the gate at the end $y = 0$ and a free surface at the end $y = Y$. There is no current and energy transport, and a one dimensional simulation is suitable for this device.

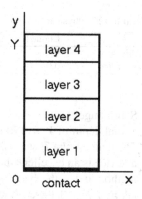

**Table 12.2** Composition of the layer structure, with the layers numbered from the end at the gate.

| Layer | Thickness (nm) | Al fraction | $N_D$ (m$^{-3}$) |
|---|---|---|---|
| 1 | 30 | 0.3 | $1.5 \times 10^{24}$ |
| 2 | 20 | 0.0 | $1.0 \times 10^{20}$ |
| 3 | 20 | 0.3 | $1.5 \times 10^{24}$ |
| 4 | 100 | 0.0 | $1.0 \times 10^{20}$ |

Another point to consider is the placement of the grid lines, especially in relation the physical layer structure in the $y$-direction. We wish to put the grid lines at the layer interfaces, but if we were to use a uniform mesh then a great burden could be imposed on the grid structure if we tried to put a grid line at each interface; in effect, depending on the physical make-up of the layer structure, we could end up with a coarse grid that was too fine. This would be unsatisfactory since, in the multigrid method, we must find *exact* solutions on the coarsest grid with a relatively small amount of computational effort. Similar considerations apply in the $x$-direction when considering the positions of the recess boundaries.

If a uniform mesh is used, then a simple approach would be to adjust the physical boundaries to the grid lines. If the grid is too coarse, then there will be a danger that the device configuration which is solved is different from the one we set out to solve. Hence the choice of the coarsest grid must be a compromise between this consideration and the requirement that it be as coarse as we can make it. One solution would be to impose strict minimum and maximum limits on the number $c_{(1)}$ of internal points of the coarsest grid, and then to choose the value of $c_{(1)}$ within these limits in order to minimise the average displacement of the boundaries.

### 12.5.1 Case study 1: application to a one-dimensional device

Consider the device with four layers which is shown in Fig. 12.3. It has a gate at the end $y = 0$, and a free surface at the end $y = Y$. It has no recess structure, no current continuity, and no energy transport; we need only consider the coupled Poisson–

**Table 12.3** Physical and adjusted interface distances, measured in nm, from the gate end.

|          | Contact | Interface 1 | Interface 2 | Interface 3 | $Y$   |
|----------|---------|-------------|-------------|-------------|-------|
| Physical | 0.0     | 30.0        | 50.0        | 70.0        | 170.0 |
| Adjusted | 0.0     | 27.2        | 47.6        | 68.0        | 170.0 |

Schrödinger problem in one dimension. In this particular simulation, the device has a total depth of $Y = 170$ nm, and is modelled using the data shown in Table 12.2. The temperature $T_e$ was taken as the constant value 300 K. The consideration of such a device is an excellent starting point for becoming familiar with both the multigrid method and with obtaining a numerical solution of the Schrödinger equation. On taking $E_F = 0$, the equations to be solved are a subset of those detailed in Sect. 6.3:

$$\rho = q(N_D - n),$$

$$\frac{d}{dy}\left(\varepsilon_0\varepsilon_r\frac{d\psi}{dy}\right) + \rho = 0 \quad \text{internally},\tag{12.31}$$

$$\psi(0) - E_h + V_g - \phi_b = 0 \quad \text{on the Schottky gate},\tag{12.32}$$

$$\frac{d\psi}{dy} = 0 \quad \text{on } y = Y, \quad \text{and}\tag{12.33}$$

$$n = n_1 + n_2,\tag{12.34}$$

where the expressions for $n_2$ and $n_3$ inside and outside the quantum wells are given in Sect. 5.4.

### Device results

A programme, which was written specifically for this one dimensional simulation, was run for the case of a coarse grid with $c_{(1)} = 24$ and a fine grid with $c_{(4)} = 199$ running through 4 grids. The physical interfaces were first adjusted to fit to the nearest coarse grid lines. Table 12.3 shows the outcome of this adjustment. A minimum number of 11 iterations was imposed on the coarsest grid in order to allow the Newton update factor $\alpha^{(k)}$ to reach a reasonably large value. Automatic stopping was then used on this grid. It was found that the number of iterations on the coarse grid at successive visits were 48, 11, 11 and 11, Ten pre-smoothing and ten post-smoothing iterations were used on the higher grids. A constant value of $w^{(k)} = 1.8$ was taken for the SOR iterations in order to calculate the exact solution of the Poisson equation on the coarsest grid. Only 4 eigensolutions were used to calculate the contribution $n_2$ to the total electron concentration. The results file produced the following output:

```
--------------------------------------------------------------------
DEVICE PARAMETERS
=================
Number of layers=4
Layer 1: Ymin=0.000000e+000
         Al_1=0.300000    In_1=0.000000    Nd_1=1.500000e+024
Layer 2: interface12=3.000000e-008
         Al_2=0.000000    In_2=0.000000    Nd_2=1.000000e+020
Layer 3: interface23=5.000000e-008
```

```
         Al_3=0.300000     In_3=0.000000     Nd_3=1.500000e+024
Layer 4: interface34=7.000000e-008
         Al_4=0.000000     In_4=0.000000     Nd_4=1.000000e+020
Ymax=1.700000e-007
```

Electron temperature T_e = 300K = constant.
The thermal energy is (k_B)(T_e)=2.585108e-002eV

```
Physical interfaces at 3.000000e-008  5.000000e-008  7.000000e-008
Adjusted interfaces at 2.720000e-008  4.760000e-008  6.800000e-008
```

Solving initially on coarsest grid 24 internal points....dun.

Vg=0.0

COARSE GRID RESULTS:
====================
Results on grid 24 internal grid points:-

The first 10 eigenvalues are:-
```
-1.945636e-002
-2.457071e-003   (difference=1.699929e-002)
 3.121697e-003   (difference=5.578768e-003)
 1.497405e-002   (difference=1.185236e-002)
 1.943534e-002   (difference=4.461291e-003)
 3.227811e-002   (difference=1.284276e-002)
 4.301998e-002   (difference=1.074188e-002)
 4.674606e-002   (difference=3.726082e-003)
 4.966077e-002   (difference=2.914704e-003)
 6.451633e-002   (difference=1.485556e-002)
```

Number of eigenstates summed=4

```
Processing time for coarse solution   =     0.05seconds
Processing time for multigrid section =     0.00seconds
Total processing time                 =     0.05seconds
```

MULTIGRID CYCLES
================
Running between 4 grids....coarse grid 24, fine grid 199 internal points.
NPRE=10,  NPOST=10,  V_cycles=1

```
 1
   2  pre
 1
   2  post
     3  pre
   2  pre
 1
   2  post
     3  post
       4  pre
     3  pre
   2  pre
 1
   2  post
     3  post
       4  post
```

FINE GRID RESULTS:
==================
Results on grid 1x201

The thermal energy is (k_B)(T_e)=2.585108e-002eV

The first 10 eigenvalues are:-

```
-3.021521e-002
-1.831933e-002   (difference=1.189587e-002)
-7.920708e-003   (difference=1.039863e-002)
 1.962266e-002   (difference=2.754337e-002)
 3.038128e-002   (difference=1.075862e-002)
 3.426651e-002   (difference=3.885234e-003)
 5.496256e-002   (difference=2.069604e-002)
 6.282752e-002   (difference=7.864964e-003)
 7.493377e-002   (difference=1.210625e-002)
 8.576004e-002   (difference=1.082627e-002)

Processing time for coarse solution    =    0.05seconds
Processing time for multigrid section =    0.99seconds
Total processing time                  =    1.04seconds
------------------------------------------------------------------
```

**Table 12.4** The first ten eigenvalues on a fine grid of 199 internal points. Four grids were used, with the coarsest grid having 24 internal points. The second column shows the difference between consecutive eigenvalues, and can be compared with the thermal energy $k_B T_e = 2.585 \times 10^{-2}$ eV.

| Eigenvalue (eV) | Difference (eV) |
|---|---|
| $-3.021521 \times 10^{-2}$ | |
| $-1.831933 \times 10^{-2}$ | $1.189587 \times 10^{-2}$ |
| $-7.920708 \times 10^{-3}$ | $1.039863 \times 10^{-2}$ |
| $1.962266 \times 10^{-2}$ | $2.754337 \times 10^{-2}$ |
| $3.038128 \times 10^{-2}$ | $1.075862 \times 10^{-2}$ |
| $3.426651 \times 10^{-2}$ | $3.885234 \times 10^{-3}$ |
| $5.496256 \times 10^{-2}$ | $2.069604 \times 10^{-2}$ |
| $6.282752 \times 10^{-2}$ | $7.864964 \times 10^{-3}$ |
| $7.493377 \times 10^{-2}$ | $1.210625 \times 10^{-2}$ |
| $8.576004 \times 10^{-2}$ | $1.082627 \times 10^{-2}$ |

Table 12.4 shows the values of the ten lowest eigenvalues found in the final solution on the finest grid, together with differences between consecutive eigenvalues. This Table shows that some of the differences between consecutive eigenvalues fall below the value $k_B T_e = 2.585 \times 10^{-2}$ eV. This is due to the existence of the double well, and indicates that we should not rely on the method of stopping to include eigensolutions in the sum for $n_2$ as soon as the difference falls below $k_B T_e$. Results for the conduction band profile $E_c$, together with the first three eigenvalues, are shown in Figs. 12.4 and 12.5 for the coarsest and finest grids respectively. The quantum wells can be clearly seen in the Figures.

*Multigrid performance*

Eleven separate runs were made on different combinations of coarse and fine grids in order to test the performance of the multigrid method. For purposes of comparison all the results were obtained for a common gate voltage of 0.0 V. Various combinations of coarse and fine grid sizes were investigated. Timings for the separate runs are given in Table 12.5, and are described below. Automatic stopping was applied in run 1 on the solution on the coarsest grid, while a fixed number of 250 iterations on

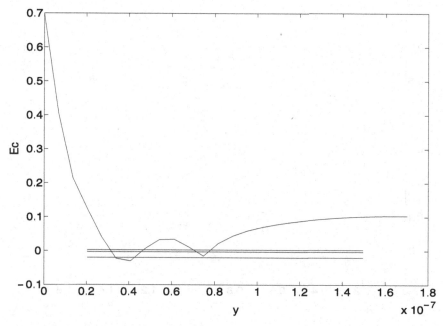

**Fig. 12.4** The conduction band profile $E_c$, showing the quantum wells and the first three energy levels on the coarse grid of 24 internal points.

the coarsest grid was used in runs 2–11. In all cases, 10 pre-smoothing and 10 post-smoothing iterations were used on the higher grids. The values in columns A and B are the numbers of internal points on each grid. The values in columns D and E are the times for the initial coarse solution and the total multigrid times respectively. Values in brackets are values predicted using Equations (12.25)–(12.28).

- *Run 1* This is the run which produced the results shown in Figs. 12.4 and 12.5. Four grids were used, with the coarsest grid having 24 internal points and the finest grid having 199 internal points. In order to compare timings on the separate runs, the basic time unit was defined to be the time taken to produce the results on the coarse grid on this run, and all other runtimes are measured in terms of this unit. The total time taken to obtain the final result in this run was 21.32 units.
- *Runs 2–5* These runs show the times achieved for a fixed fine grid but starting at successively coarser coarse grids. Run 2 is the non-multigrid version in which the coarse and fine grids are identical. This run used 250 iterations, in a time of 103.69 units to achieve the same level of convergence as in run 1. Thus the multigrid process produced a speedup of 4.9 over the non-multigrid method, while producing identical results. The total time taken in the multigrid process increased as the size of the coarse grid increased; this increase was caused by the increase in work done in obtaining the exact solution on the coarse grid. The times taken to move through the grids, found by subtracting column D from column E, increased much more slowly.

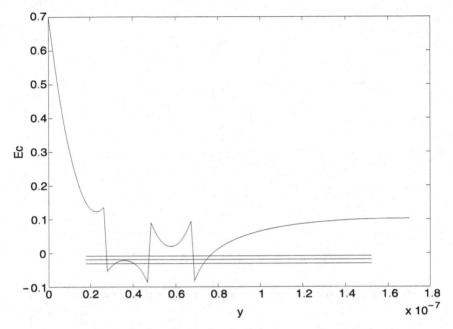

**Fig. 12.5** The conduction band profile $E_c$, showing the quantum wells and the first three energy levels on the finest grid of 199 internal points.

**Fig. 12.6** The cross section of the simple non-recessed two-layer HEMT, considered in Case study 2, showing the source, gate, drain and $n^+$ regions around the source and drain. Results shown in later Figures are viewed looking at the corner A.

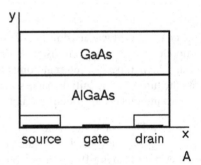

- *Runs 6–10* These runs all start on a coarse grid of 24 internal points, but each run has a different finest grid.
- *Run 11* This run is a 2-grid process running from 199 to 399 internal points. This run is used to provide data so that times may be predicted using Equations (12.25)–(12.28).

In making predictions about the work done in one of the multigrid processes, only the data from runs 5 and 11 are needed. The time taken will be used as a measure of the work done in a process. The results of run 5 show that $W_{its}(1) = 5.32/250 = 0.02128$ units, since 250 iterations were performed on this coarsest grid. Taking $G = 4$, $c_{(1)} = 24$, $a = b = 10$, and $v_g = 250$ for run 5, and calculating the

**Table 12.5** Results of various runs with different coarse and fine grids. The last two columns show the times in basic time units for the solution on the coarse grid and the total time taken in this unit for the complete multigrid solution. The values in brackets are predicted values. In run 1, automatic stopping was used on the coarse grid, In runs 2–11, there were 250 iterations imposed on the coarsest grid.

| Run | Fine (A) | Coarse (B) | No. of grids (C) | Coarse time (D) | Total time (E) |
|-----|------|--------|-------------|-------------|------------|
| 1 | 199 | 24 | 4 | 1.00 (1.01) | 21.32 (21.71) |
| 2 | 199 | 199 | 1 | 103.69 | 103.69 |
| 3 | 199 | 99 | 2 | 35.16 (34.66) | 79.95 (79.84) |
| 4 | 199 | 49 | 3 | 12.84 (12.96) | 56.24 (56.25) |
| 5 | 199 | 24 | 4 | 5.32 | 40.90(41.28) |
| 6 | 49 | 24 | 2 | 5.32 | 11.84 (11.96) |
| 7 | 99 | 24 | 3 | 5.32 | 21.61 (21.92) |
| 8 | 199 | 24 | 4 | 5.32 | 40.90 (41.28) |
| 9 | 399 | 24 | 5 | 5.32 | 90.18 (90.43) |
| 10 | 799 | 24 | 6 | 5.32 | 242.96(242.90) |
| 11 | 399 | 199 | 2 | 103.69 | 237.15 |

**Table 12.6** Composition of the layers and depletion regions of the simple two dimensional structure considered in Case study 2. The layers are numbered from the contact end.

| Layer | Thickness (nm) | Al fraction | $N_D$ (m$^{-3}$) |
|-------|----------------|-------------|------------------|
| 1 | 100 | 0.3 | $5.0 \times 10^{23}$ |
| 2 | 300 | 0.0 | $1.0 \times 10^{19}$ |

value $W_{its} = 103.69/250 = 0.4148$ units from run 11, the value of $A_1$ can be found by solving Equation (12.27). The value of $B_1$ is found as follows: consider run 11 with coarsest grid (1′) consisting of 199 internal points. Equation (12.29) provides the value of $A_{1'}$. The value of $B_{1'}$ can be calculated using the appropriate values in Equations (12.25)–(12.28) with $W = 237.15/250$. The value of $B_1$ can then be calculated using Equation (12.30). Hence knowing the values of $A_1$ and $B_1$ for one coarse grid, they may be found for any other coarse grid. Using the values found above, Equations (12.25)–(12.28) can be used to predict the work done in any multigrid process for a particular implementation. The numbers in brackets in Table 12.5 were predicted in this way by using the values found above. Some entries are not followed by bracketed values, since they were used in the prediction process. It can be seen that there is very close agreement between the bracketed values and the observed values.

## 12.5.2  Case study 2: application to a two-dimensional device

The simple two layer device shown in Fig. 12.6 was simulated. This device has no recess structure, and no fieldplate. $n^+$ regions surround the source and drain.

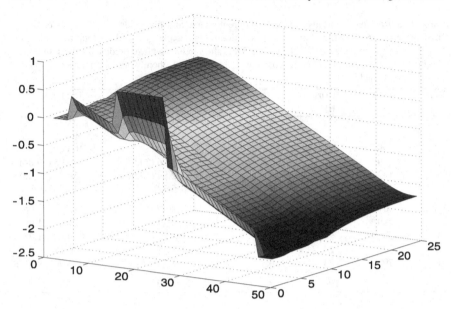

**Fig. 12.7** The conduction band $E_c$, for the two dimensional structure, on the coarsest grid of $48 \times 24$ internal points. The plot is viewed from the corner $A$ in Fig. 12.6.

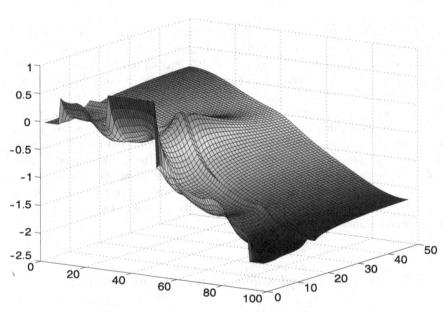

**Fig. 12.8** The conduction band $E_c$, for the two dimensional structure, on the finest grid of $97 \times 49$ internal points. The plot is viewed from the corner $A$ in Fig. 12.6.

The electron temperature was held constant at $T_e = 300$ K. Hence the simulation was run without energy transport. Three eigensolutions were used in the calculation of $n_2$. The multigrid routine used 20 pre-smoothing iterations and 6 post-smoothing iterations. A fixed number of 200 iterations were used to calculate the exact solution on the coarsest grid. A multigrid simulation involving two grids was used, with the coarse grid having $48 \times 24$ internal points and the fine grid having $97 \times 49$ internal points. The layer structure is shown in Table 12.6. SOR and SUR were not used. Fig. 12.7 shows the conduction band profile for the coarse grid, viewed from the corner $A$ in Fig. 12.6, and Fig. 12.8 gives the corresponding plot on the fine grid. Voltages of 0.0 V, $-0.1$ V and 2.0 V were taken on the source, gate and drain respectively.

# Chapter 13
# Approximate and numerical solutions of the Schrödinger equation

In practically all realistic physical situations, the hamiltonian operator is such that it is not possible to specify the potential as a single function. In this case, it is not possible to calculate the exact eigensolutions of the Schrödinger equation. Even when the potential is specified as a functional form, it is possible to solve the Schrödinger equation only for a very limited number of artificial potentials. Many analytical and numerical approximation methods are available for the solution of the equation, and in this chapter some of the main techniques will be described which will be of relevance to device modelling. These will include the WKB approximation, both time independent and time dependent perturbation theory, the variational method, and numerical methods.

## 13.1 The WKB approximation

The classical approximation to quantum mechanics arises in the formal limit $\hbar \to 0$. In the *Wentzel-Kramers-Brillouin* (WKB) method, the wave function is expanded in powers of $\hbar$, and this method provides a useful approximation in cases in which the potential function varies slowly with position. The description of this method will be limited to the one dimensional situation, in which the potential changes along the $x$-axis. Unless otherwise stated, the particle mass $m$ will be treated as a constant in this section. A prime on any function denotes the derivative of that function with respect to its argument.

### 13.1.1 The basis of the WKB method

The solutions of the TISE

$$u'' + \frac{2m}{\hbar^2}(E - V(x))u = 0 \tag{13.1}$$

E.A.B. Cole, *Mathematical and Numerical Modelling of Heterostructure Semiconductor Devices: From Theory to Programming*, DOI 10.1007/978-1-84882-937-4_13, © Springer-Verlag London Limited 2009

will be expanded in powers of $\hbar$. Writing the solution $u(x)$ in terms of a new function $w(x)$ given by

$$u \equiv e^{\frac{i}{\hbar}w(x)}, \tag{13.2}$$

then Equation (13.1) can be written

$$i\hbar w'' = w'^2 - 2m(E - V(x)). \tag{13.3}$$

On making the expansion

$$w = w_0 + \hbar w_1 + \cdots, \tag{13.4}$$

then Equation (13.3) becomes

$$i\hbar(w_0'' + \hbar w_1'' + \cdots) = (w_0' + \hbar w_1' + \cdots)^2 - 2m(E - V),$$

and equating like powers of $\hbar$ leads to the set of equations

$$w_0'^2 = 2m(E - V(x)), \tag{13.5}$$
$$i w_0'' = 2w_0' w_1', \tag{13.6}$$
$$\cdots \quad \cdots$$

The WKB approximation stops after Equation (13.6).

The regions of $x$ for which $E > V(x)$ and $E < V(x)$ must be considered separately:

- For $E > V(x)$, define the function

$$k_1(x) \equiv \sqrt{\frac{2m(E - V(x))}{\hbar^2}}. \tag{13.7}$$

Then Equations (13.5) and (13.6) give

$$w_0'^2 = \hbar^2 k_1^2, \qquad w_0' = \pm\hbar k_1(x), \qquad w_0 = \pm\hbar \int^x k_1(\xi)d\xi,$$

$$w_1' = \frac{i w_0''}{2 w_0'} = \frac{i}{2}\frac{k_1'}{k_1}, \quad \text{and} \quad w_1(x) = \frac{1}{2}i \ln k_1(x).$$

Hence the function $u(x)$ is given, from Equation (13.2) by

$$u(x) = \exp\frac{i}{\hbar}\left(\pm\hbar \int^x k_1(\xi)d\xi + i\frac{\hbar}{2}\ln k_1(x)\right) = \frac{1}{\sqrt{k_1(x)}}e^{\pm i\int^x k_1(\xi)d\xi}, \tag{13.8}$$

which leads to the general solution

$$u(x) = \frac{1}{\sqrt{k_1(x)}}\left(P_1 e^{i\int^x k_1(\xi)d\xi} + Q_1 e^{-i\int^x k_1(\xi)d\xi}\right), \tag{13.9}$$

where $P_1$ and $Q_1$ are arbitrary constants.

- For $E < V(x)$, define the function

$$k_2(x) \equiv \sqrt{\frac{2m(V(x) - E)}{\hbar^2}}.$$
(13.10)

Then in a similar way, the expression for $u(x)$ becomes

$$u(x) = \frac{1}{\sqrt{k_2(x)}} e^{\pm \int^x k_2(\xi)d\xi},$$
(13.11)

which leads to the general solution

$$u(x) = \frac{1}{\sqrt{k_2(x)}} \left( P_2 e^{\int^x k_2(\xi)d\xi} + Q_2 e^{-\int^x k_2(\xi)d\xi} \right)$$
(13.12)

where $P_2$ and $Q_2$ are arbitrary constants.


### 13.1.2 The limit of the approximation

For $E > V(x)$, then the TISE Equation (13.1) is

$$u'' + k_1(x)^2 u = 0.$$
(13.13)

On writing

$$u_\pm(x) = \frac{1}{\sqrt{k_1(x)}} e^{\pm i \int^x k_1(\xi)d\xi},$$
(13.14)

it is found that $u_\pm$ satisfies the equation

$$u''_\pm + \left[ k_1(x)^2 - \frac{3}{4} \left( \frac{k'_1}{k_1} \right)^2 + \frac{k''_1}{2k_1} \right] u_\pm = 0.$$

This is a good approximation to Equation (13.13) if

$$|k_1(x)|^2 \gg \left| \frac{3}{4} \left( \frac{k'_1}{k_1} \right)^2 - \frac{k''_1}{2k_1} \right|,$$

and this condition is satisfied when

$$\left| \frac{k'_1}{k_1^2} \right| \ll 1 \quad \text{and} \quad \left| \frac{k''_1}{k_1^3} \right| \ll 1.$$
(13.15)

In other words, $k_1$ must not be too small for the approximation to apply.

The physical interpretation of this result can be seen as follows. The de Broglie wavelength is

**Fig. 13.1** The accessible and inaccessible regions are denoted as Ac and In respectively. The point $x = a_1$ is a turning point with the inaccessible region on the left and the accessible side on the right.

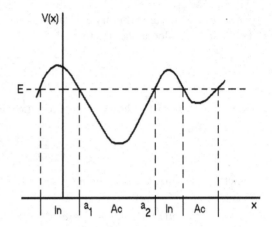

$$\lambda = \frac{\hbar}{p} = \frac{2\pi\hbar}{\sqrt{2m(E - V)}} = \frac{2\pi}{k_1(x)}. \qquad (13.16)$$

Hence the first condition in Equation (13.15) gives

$$\left|\frac{d}{dx}\left(\frac{1}{k_1}\right)\right| \ll 1, \quad \text{or} \quad \left|\frac{d\lambda}{dx}\right| \ll 1.$$

That is, the de Broglie wavelength is slowly varying. Further, the spatial derivative of $k_1$ is

$$k_1' = \frac{d}{dx}\sqrt{\frac{2m(E - V)}{\hbar^2}} = -\frac{m}{\hbar\sqrt{2m(E - V)}}V' = \frac{mF}{\hbar p}$$

where $F = -dV/dx$ is the classical force. Hence

$$\left|\frac{k_1'}{k_1^2}\right| = \left|\frac{m\hbar F}{p^3}\right| \ll 1,$$

from which it follows that the momentum must not be too small. Note that this physical interpretation cannot be applied in the case $E < V(x)$, since this case will not apply classically.

### 13.1.3 The connection formulae

In general, the region along the $x$-axis will be divided into two types: the classically accessible regions for which $E > V(x)$, and the classically inaccessible regions for which and $E < V(x)$. The points at which $V(x) = E$ are called the *turning points*. This situation is illustrated in Fig. 13.1. Near a turning point, the value of $k_1(x)$ which was defined in Equation (13.7) is small, and so the approximation breaks down near these points. However, the solution of the TISE should be well behaved at these turning points. We now consider the problem of finding the *connection formulae* for the WKB approximation: can we find the WKB approximation on one side of a turning point at $x = a$ if we are given the approximation on the other side of the point? If it is assumed that $V'(a) \neq 0$, then the result is given by the following theorem:

**Theorem 13.1.** *Connection formulae. The arrows indicate the directions to follow from the given expression to the derived expression:*

*inaccessible side*          *accessible side*

$$\frac{1}{2\sqrt{k_2(x)}}e^{-|\int_a^x k_2(\xi)d\xi|} \longrightarrow \frac{1}{\sqrt{k_1(x)}}\cos\left(\left|\int_a^x k_1(\xi)d\xi\right| - \frac{\pi}{4}\right) \qquad (13.17)$$

$$\frac{\sin\eta}{\sqrt{k_2(x)}}e^{|\int_a^x k_2(\xi)d\xi|} \longleftarrow \frac{1}{\sqrt{k_1(x)}}\cos\left(\left|\int_a^x k_1(\xi)d\xi\right| - \frac{\pi}{4} + \eta\right) \qquad (13.18)$$

*where the quantity $\eta/\pi$ must not be close to an integer.*

*Proof.* The following outline proof of these connection formulae follows that of Landau and Lifshitz (1981).

*Proof of the first connection formula*

Let $x = a_1$ be a turning point with the inaccessible region on the left and the accessible side on the right. Then it can be seen from Fig. 13.1 that $V'(a_1) < 0$. For values of $x$ near to this point, then

$$-k_2(x)^2 = k_1(x)^2 = \frac{2m}{\hbar^2}(E - V(x)) \approx C(x - a_1) \qquad (13.19)$$

where $C \equiv -2mV'(a_1)/\hbar^2 > 0$ is a constant. It follows that, near $x = a_1$, the equation $u'' + k_1(x)^2 u = 0$ becomes

$$u'' + C(x - a_1)u = 0, \quad \text{or} \quad \frac{d^2u}{dw^2} - wu = 0$$

where $w \equiv -C^{\frac{1}{3}}(x - a_1)$. The solution of this equation which is finite for all $w$ is the *Airy function*

$$\Phi(w) \equiv \frac{1}{\sqrt{\pi}}\int_0^\infty \cos\left(\frac{1}{3}t^3 + tw\right)dt,$$

which has the property that

$$\Phi(w) \to \frac{1}{2}w^{-\frac{1}{4}}e^{-\frac{2}{3}w^{\frac{3}{2}}} \quad \text{as } w \to \infty$$

$$\to |w|^{-\frac{1}{4}}\cos\left(\frac{2}{3}|w|^{\frac{3}{2}} - \frac{1}{4}\pi\right) \quad \text{as } w \to -\infty.$$

Hence near $x = a_1$, it follows that

$$u(x) = u_1(x) \equiv A\Phi\left(-C^{\frac{1}{3}}(x - a_1)\right)$$

where $A$ is a constant. Now suppose that this linear potential can be used at some point $x < a_1$, which is sufficiently far enough from $a_1$ so that these asymptotic limits apply. Then at this value of $x$, we have

$$u_1(x) \approx A \cdot \frac{1}{2}\left(C^{\frac{1}{3}}(a_1 - x)\right)^{-\frac{1}{4}} e^{-\frac{2}{3}\left(C^{\frac{1}{3}}(a_1 - x)\right)^{\frac{3}{2}}}$$

$$= AC^{\frac{1}{6}}\frac{1}{2\sqrt{k_2(x)}}e^{-\left|\int_{a_1}^{x} k_2(\xi)d\xi\right|},$$

where Equation (13.19) has been used. Again, suppose that this linear potential can be used at some point $x > a_1$, which is sufficiently far enough from $a_1$ so that these asymptotic limits apply. Then at this value of $x$, we have

$$u_1(x) \approx A \cdot \left|-C^{\frac{1}{3}}(x - a_1)\right|^{-\frac{1}{4}} \cos\left(\frac{2}{3}\left| - C^{\frac{1}{3}}(x - a_1)\right|^{\frac{3}{2}} - \frac{\pi}{4}\right)$$

$$= AC^{\frac{1}{6}}\frac{1}{\sqrt{k_1(x)}} \cos\left(\int_{a_1}^{x} k_1(\xi)d\xi - \frac{\pi}{4}\right).$$

The first connection formula follows on taking $A = C^{-\frac{1}{6}}$. A similar analysis holds at a turning point such as $x = a_2$, at which the inaccessible region lies to its right: this time, $C < 0$.

*Proof of the second connection formula*

Let

$$u_2(x) \approx \frac{1}{\sqrt{k_1(x)}} \cos\left(\int_{a_1}^{x} k_1(\xi)d\xi - \frac{\pi}{4} + \eta\right)$$

be another exact solution on the accessible side $x > a_1$. The solution for $x < a_1$ will be

$$u_2(x) \approx \frac{B}{\sqrt{k_2(x)}}e^{\int_x^{a_1} k_2(\xi)d\xi} + \text{negligible potential}.$$

On evaluating the Wronskian $W[u_1, u_2]$ on both sides of $x = a_1$, it is found that $W[u_1, u_2] = -B = -\sin\eta$. Hence the second connection formula is proved.

These connection formulae may also be proved using Bessel's equation of order $1/3$, since the Airy function can be expressed in terms of these Bessel functions (Schiff 1955). ☐

When using these formulae, care must be taken not to use them in the reverse order. In the first formula, any slight error in the phase of the cosine term would multiply the left hand side by an exponential which would swamp the original exponential. If the second formula were to be run backwards, then even a very small addition to

**Fig. 13.2** A potential barrier, with the turning points at $x = a_1$ and $x = a_2$. A particle approaches it from the left. Regions I and III are accessible, while region II is inaccessible.

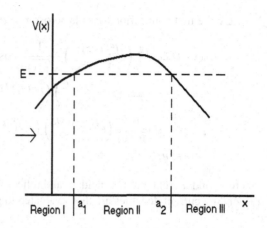

the exponent on the left hand side would result in a large change in the argument of the cosine term.

### 13.1.4 Examples

*1. Penetration through a barrier*

In the classical limit, there is no penetration, and then the transmission coefficient $T = 0$. In the quasi-classical approximation, $T$ is expected to be very small, and so the barrier must be wide.

Consider the situation shown in Fig. 13.2. The turning points are at $x = a_1$ and $x = a_2$, with the inaccessible region II in the interval $a_1 < x < a_2$. A particle is approaching from the left. In the accessible region III we have

$$u_{III}(x) \approx \frac{1}{\sqrt{k_1(x)}} e^{i \int_{a_2}^{x} k_1(\xi)d\xi}$$

$$= \frac{1}{\sqrt{k_1(x)}} \cos\left( \int_{a_2}^{x} k_1(\xi)d\xi \right) + \frac{i}{\sqrt{k_1(x)}} \cos\left( \int_{a_2}^{x} k_1(\xi)d\xi - \frac{1}{2}\pi \right).$$

$$(13.20)$$

The WKB approximation in region II is found by applying the second connection formula to both terms, taking $\eta = \frac{1}{4}\pi$ and $\eta = -\frac{1}{4}\pi$ respectively:

$$u_{II}(x) \approx \frac{1}{\sqrt{2}} \frac{1}{\sqrt{k_2(x)}} \left( e^{\int_{x}^{a_2} k_2(\xi)d\xi} \right) - \frac{1}{\sqrt{2}} \frac{i}{\sqrt{k_2(x)}} \left( e^{\int_{x}^{a_2} k_2(\xi)d\xi} \right)$$

$$= \frac{2e^{-i\frac{\pi}{4}}}{2\sqrt{k_2(x)}} \left( e^{\int_{a_1}^{a_2} k_2(\xi)d\xi} \right) \left( e^{-\int_{a_1}^{x} k_2(\xi)d\xi} \right).$$

$$(13.21)$$

Now use the first connection formula across $x = a_1$:

$$u_I(x) \approx 2e^{-i\frac{\pi}{4}} \left(e^{\int_{a_1}^{a_2} k_2(\xi)d\xi}\right) \frac{1}{\sqrt{k_1(x)}} \cos\left(\int_x^{a_1} k_1(\xi)d\xi - \frac{1}{4}\pi\right)$$

$$= \frac{1}{\sqrt{k_1(x)}} \left(e^{\int_{a_1}^{a_2} k_2(\xi)d\xi}\right) \left(e^{-i\int_x^{a_1} k_1(\xi)d\xi}\right)$$

$$- \frac{i}{\sqrt{k_1(x)}} \left(e^{\int_{a_1}^{a_2} k_2(\xi)d\xi}\right) \left(e^{i\int_x^{a_1} k_1(\xi)d\xi}\right)$$

$$\equiv u_i + u_r, \tag{13.22}$$

where $u_i$ and $u_r$ relate to the incident and reflected solutions respectively. The corresponding current densities are then evaluated to give

$$s_i = \frac{\hbar}{m}|u_i|^2 \frac{d}{dx}(arg(u_i)) = \frac{\hbar}{m}e^{2\int_{a_1}^{a_2} k_2(\xi)d\xi}$$

$$s_t = \frac{\hbar}{m} \quad \text{and}$$

$$s_r = -\frac{\hbar}{m}e^{2\int_{a_1}^{a_2} k_2(\xi)d\xi}.$$

Hence the transmission and reflection coefficients are

$$T = \left|\frac{s_t}{s_i}\right| = e^{-2\int_{a_1}^{a_2} k_2(\xi)d\xi} \tag{13.23}$$

$$R = 1 - T = 1 - e^{-2\int_{a_1}^{a_2} k_2(\xi)d\xi}. \tag{13.24}$$

It is easily verified that a direct calculation of $R$ produces the result $R = 1$: this is why it is evaluated as $R = 1 - T$. Strictly speaking, this calculation cannot be applied to finding $R$ and $T$ for a square well potential, since the quantity $E - V(x)$ is not linear close to the edges. However, we do get the correct answer if we try it this way!—prove this.

2. *Discontinuous barrier at $x = a_1$*

Fig. 13.3 shows a discontinuous barrier at $x = a_1$ between regions I and II. In regions I, II, and III, we have respectively

$$u_{III} \approx \frac{1}{\sqrt{k_1(x)}} e^{i\int_{a_2}^x k_1(\xi)d\xi},$$

$$u_{II} \approx \frac{e^{-i\frac{\pi}{4}}}{\sqrt{k_2(x)}} \left(e^{\int_{a_1}^{a_2} k_2(\xi)d\xi}\right) \left(e^{-\int_{a_1}^x k_2(\xi)d\xi}\right), \quad \text{and}$$

$$u_I \approx \frac{1}{\sqrt{k_1(x)}} \left(Ae^{-i\int_x^{a_1} k_1(\xi)d\xi} + Be^{i\int_x^{a_1} k_1(\xi)d\xi}\right).$$

Continuity of the pairs $u_I$ and $u_{II}$, and $u_I'$ and $u_{II}'$, at $x = a_1$ gives the two relations

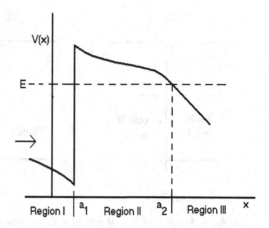

**Fig. 13.3** A discontinuous barrier at $x = a_1$. Regions I and III are accessible, while region II is inaccessible.

$$\frac{1}{\sqrt{k_1(a_1)}}(A + B) = \frac{e^{-i\frac{\pi}{4}}}{\sqrt{k_2(a_1)}}\left(e^{\int_{a_1}^{a_2} k_2(\xi)d\xi}\right)$$

$$i\sqrt{k_1(a_1)}(A - B) = -e^{-i\frac{\pi}{4}}\sqrt{k_2(a_1)}\left(e^{\int_{a_1}^{a_2} k_2(\xi)d\xi}\right),$$

from which it follows that

$$A = e^{-i\frac{\pi}{4}}\frac{(k_1(a_1) + ik_2(a_1))}{2\sqrt{k_1(a_1)k_2(a_1)}}\left(e^{\int_{a_1}^{a_2} k_2(\xi)d\xi}\right),$$

$$s_i = \frac{\hbar}{m}|A|^2,$$

$$s_t = \frac{\hbar}{m}, \quad \text{and} \tag{13.25}$$

$$T = \left|\frac{s_t}{s_i}\right| = \frac{1}{|A|^2} = \frac{4k_1(a_1)k_2(a_1)}{k_1(a_1)^2 + k_2(a_1)^2}\left(e^{-2\int_{a_1}^{a_2} k_2(\xi)d\xi}\right).$$

### 3. Cold emission of electrons from a metal

A metal has a surface in the plane perpendicular to the $x$-axis. The electrons inside the metal have an energy $-W$, and move in a potential $-V_0$. An electric field $F > 0$ is applied in the positive $x$-direction, perpendicular to the metal surface. The energy $W$ that must be supplied by the field in order that electrons are ejected from the metal is the work function. Fig. 13.4 show the situations in which $F = 0$ and $F \neq 0$. The potential is given by

$$V(x) = \begin{cases} -V_0 & \text{for } x < 0 \\ -qFx & \text{for } x > 0. \end{cases}$$

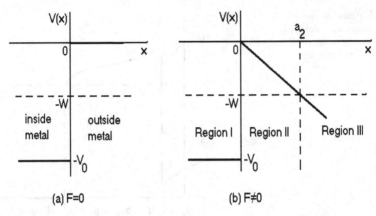

**Fig. 13.4** An electric field $F > 0$ applied in the positive $x$-direction, perpendicular to the metal surface which lies in the plane $x = 0$. In (a), the field is zero. In (b), the field is not zero, and enables the value of $a_2$ to be calculated.

As before, we have regions I, II, and III, with $a_1 = 0$. The value of $a_2$ is given by $-W = -qFa_2$, or $a_2 = W/(qF)$. Hence

$$k_1(x) = \sqrt{\frac{2m(-W - (-V_0))}{\hbar^2}} = \sqrt{\frac{2m(V_0 - W)}{\hbar^2}} = \text{const.} \quad \text{for } x \leq 0$$

$$k_2(x) = \sqrt{\frac{2m(-qFx + W)}{\hbar^2}} = \sqrt{\frac{2m(W - qFx)}{\hbar^2}} \quad \text{for } 0 \leq x \leq a_2,$$

which gives

$$k_1(0) = \sqrt{\frac{2m(V_0 - W)}{\hbar^2}}, \qquad k_2(0) = \sqrt{\frac{2mW}{\hbar^2}},$$

and

$$\int_{a_1}^{a_2} k_2(\xi)d\xi = \int_0^{\frac{W}{qF}} \sqrt{\frac{2m(W - qF\xi)}{\hbar^2}} d\xi$$

$$= \frac{2}{3} \frac{\sqrt{2m}}{\hbar} \frac{W^{\frac{3}{2}}}{qF}.$$

It follows that the transmission coefficient is given from Equation (13.25) by

$$T = \frac{4\sqrt{\frac{2m(V_0-W)}{\hbar^2}}\sqrt{\frac{2mW}{\hbar^2}}}{\frac{2m(V_0-W)}{\hbar^2} + \frac{2mW}{\hbar^2}} e^{-\frac{4}{3}\frac{\sqrt{2m}}{\hbar}\frac{W^{\frac{3}{2}}}{qF}}$$

$$= \frac{4\sqrt{W(V_0 - W)}}{V_0} e^{-\frac{4}{3}\frac{\sqrt{2m}}{\hbar}\frac{W^{\frac{3}{2}}}{qF}}. \tag{13.26}$$

## 13.2 Time independent perturbation theory

Suppose that an unperturbed hamiltonian function $H^{(0)}$ possesses a known complete set of orthonormal eigenfunctions $u_n^{(0)}$ and energy levels $E_n^{(0)}$, so that

$$H^{(0)} u_n^{(0)} = E_n^{(0)} u_n^{(0)}. \tag{13.27}$$

The actual hamiltonian of the physical system has the form

$$H = H^{(0)} + W$$

where $W \equiv W(\mathbf{r})$ is treated as a perturbation to the hamiltonian, which is independent of time $t$:

$$\frac{\partial}{\partial t} W(\mathbf{r}) = 0.$$

It is required to find the eigenfunctions $u_n$ and energy levels $E_n$ of the full hamiltonian using the equation

$$H u_n = E_n u_n. \tag{13.28}$$

For example, the hamiltonian of the helium atom is

$$H = -\frac{\hbar^2}{2m_e} \nabla_1{}^2 - \frac{\hbar^2}{2m_e} \nabla_2{}^2 - 2q^2 (1/r_1 + 1/r_2),$$

where $\nabla_1$ and $\nabla_2$ are the operators $\nabla_\mathbf{r}$ with respect to the coordinates $\mathbf{r}_1$ and $\mathbf{r}_2$ of the two electrons. In this case, the description

$$H^{(0)} = -\frac{\hbar^2}{2m_e} \nabla_1{}^2 - \frac{\hbar^2}{2m_e} \nabla_2{}^2$$
$$W = -2q^2 (1/r_1 + 1/r_2)$$

can be used. In order to proceed with the solution of Equation (13.28), a real parameter $\varepsilon$ with $0 \leq \varepsilon \leq 1$ is introduced by writing the full hamiltonian as

$$H = H^{(0)} + \varepsilon W. \tag{13.29}$$

The eigensolutions of this augmented hamiltonian are then expanded in powers of $\varepsilon$ in the form

$$u_n = u_n^{(0)} + \varepsilon u_n^{(1)} + \varepsilon^2 u_n^{(2)} + \cdots \tag{13.30}$$
$$E_n = E_n^{(0)} + \varepsilon E_n^{(1)} + \varepsilon^2 E_n^{(2)} + \cdots \tag{13.31}$$

where the $u_n^{(i)}$ and $E_n^{(i)}$ are to be determined for $i \geq 1$, but which are independent of $\varepsilon$. On substituting these into Equation (13.28), it follows that

$$(H^{(0)} + \varepsilon W)(u_n^{(0)} + \varepsilon u_n^{(1)} + \varepsilon^2 u_n^{(2)} + \cdots)$$

$$= (E_n^{(0)} + \varepsilon E_n^{(1)} + \varepsilon^2 E_n^{(2)} + \cdots)(u_n^{(0)} + \varepsilon u_n^{(1)} + \varepsilon^2 u_n^{(2)} + \cdots).$$

Collecting together the various powers of $\varepsilon$, this equation becomes

$$
\begin{aligned}
0 = {}& + \varepsilon^0 (H^{(0)} u_n^{(0)} - E_n^{(0)} u_n^{(0)}) \\
= {}& + \varepsilon (H^{(0)} u_n^{(1)} + W u_n^{(0)} - E_n^{(0)} u_n^{(1)} - E_n^{(1)} u_n^{(0)}) \\
& + \varepsilon^2 (H^{(0)} u_n^{(2)} + W u_n^{(1)} - E_n^{(0)} u_n^{(2)} - E_n^{(1)} u_n^{(1)} - E_n^{(2)} u_n^{(0)}) \\
& + \varepsilon^3 (\cdots) \\
& + \cdots .
\end{aligned}
\tag{13.32}
$$

Finally, on equating the coefficients of like powers of $\varepsilon$, we get

$$0 = H^{(0)} u_n^{(0)} - E_n^{(0)} u_n^{(0)} \tag{13.33}$$

$$0 = H^{(0)} u_n^{(1)} + W u_n^{(0)} - E_n^{(0)} u_n^{(1)} - E_n^{(1)} u_n^{(0)} \tag{13.34}$$

$$0 = H^{(0)} u_n^{(2)} + W u_n^{(1)} - E_n^{(0)} u_n^{(2)} - E_n^{(1)} u_n^{(1)} - E_n^{(2)} u_n^{(0)} \tag{13.35}$$

$$\cdots \quad \cdots$$

Note that Equation (13.27) is recovered in Equation (13.33).

### 13.2.1 The first order non-degenerate case

The unknown function $u_n^{(1)}$ will be expanded in terms of the complete orthonormal set $u_n^{(0)}$ as

$$u_n^{(1)} = \sum_r a_r^n u_r^{(0)}, \tag{13.36}$$

where the coefficients $a_r^n$ will need to be determined. On substituting this expression into Equation (13.34) and using Equation (13.33), it follows that

$$\sum_r a_r^n E_r^{(0)} u_r^{(0)} + W u_n^{(0)} - E_n^{(0)} \sum_r a_r^n u_r^{(0)} - E_n^{(1)} u_n^{(0)} = 0.$$

Now take the inner product with $u_s^{(0)}$ for any index $s$:

$$a_s^n E_s^{(0)} + (u_s^{(0)}, W u_n^{(0)}) - E_n^{(0)} a_s^n - E_n^{(1)} \delta_{sn} = 0. \tag{13.37}$$

In particular, on taking $s = n$, this result gives

$$E_n^{(1)} = (u_n^{(0)}, W u_n^{(0)}).$$

Now define the inner product $W_{sn}$ as

$$W_{sn} \equiv (u_s^{(0)}, W u_n^{(0)}). \tag{13.38}$$

Then to the first order in Equation (13.31), and putting $\varepsilon = 1$, it follows that

$$E_n = E_n^{(0)} + E_n^{(1)}$$
$$= E_n^{(0)} + W_{nn}. \tag{13.39}$$

Since the functions $u_n^{(0)}$ are known, this enables the first order correction to the energy levels to be calculated.

The corrections to the eigenfunctions to the first order may now be calculated by taking $s \neq n$ in Equation (13.37):

$$a_s^n E_s^{(0)} + W_{sn} - E_n^{(0)} a_s^n = 0,$$

which gives

$$a_s^n = \frac{W_{sn}}{E_n^{(0)} - E_s^{(0)}} \quad (s \neq n). \tag{13.40}$$

Note that the denominator is not zero since, for the moment, we are considering only the non-degenerate case. It is easily shown that $a_n^n = i\gamma_n$, where $\gamma_n$ is an arbitrary real quantity. Hence this quantity may be taken as zero, and the corrections to the eigenfunctions to the first order are given, from Equation (13.30) with $\varepsilon = 1$ and from Equation (13.36), by

$$u_n = u_n^{(0)} + u_n^{(1)}$$
$$= u_n^{(0)} + \sum_{r \neq n} \frac{W_{rn}}{E_n^{(0)} - E_r^{(0)}} u_r^{(0)}. \tag{13.41}$$

### 13.2.2  The second order non-degenerate case

The second order correction to the energy levels can now be calculated. On taking the expansion

$$u_n^{(2)} = \sum_r b_r^n u_r^{(0)}, \tag{13.42}$$

where the coefficients $b_r^n$ are as yet unknown, its substitution into Equation (13.35) gives

$$\sum_r b_r^n E_r^{(0)} u_r^{(0)} + W \sum_r a_r^n u_r^{(0)} - E_n^{(0)} \sum_r b_r^n u_n^{(0)} - E_n^{(1)} \sum_r a_r^n u_n^{(0)} - E_n^{(2)} u_n^{(0)} = 0.$$

Now take the inner product with $u_s^{(0)}$ for any index $s$:

$$b_s^n E_s^{(0)} + \sum_r a_r^n W_{sr} - E_n^{(0)} b_s^n - E_n^{(1)} a_s^n - E_n^{(2)} \delta_{ns} = 0. \tag{13.43}$$

In particular, on taking $s = n$, this result reduces to

$$E_n^{(2)} = \sum_r a_r^n W_{nr} = \sum_{r \neq n} \frac{W_{nr} W_{rn}}{E_n^{(0)} - E_r^{(0)}}. \tag{13.44}$$

Hence to the second order, and putting $\varepsilon = 1$, Equation (13.30) becomes

$$\begin{aligned} E_n &= E_n^{(0)} + E_n^{(1)} + E_n^{(2)} \\ &= E_n^{(0)} + W_{nn} + \sum_{r \neq n} \frac{W_{nr} W_{rn}}{E_n^{(0)} - E_r^{(0)}}. \end{aligned} \tag{13.45}$$

### 13.2.3 The degenerate case

The procedure involved in this case will be outlined only, without giving the full details. Suppose that the energy eigenvalue $E_n^{(0)}$ is $m$-fold degenerate, so that

$$E_n^{(0)} = E_{n+1}^{(0)} = E_{n+2}^{(0)} = \cdots = E_{n+m-1}^{(0)}.$$

Then the calculations which led to the result in Equation (13.45) will break down since the denominator of the quantity

$$a_i^n = \frac{W_{in}}{E_n^{(0)} - E_i^{(0)}}, \quad (i = n+1, \ldots, n+m-1)$$

is zero. Further, the functions $u_n^{(0)}, \ldots, u_{n+m-1}^{(0)}$ are not necessarily orthogonal. The difficulty is overcome by using the Schmidt Orthogonalisation Procedure to form a set of orthogonal functions $u_n^{(0)\prime}, \ldots, u_{n+m-1}^{(0)\prime}$ from the set of degenerate eigenfunctions such that

$$W_{kr}' \equiv (u_k^{(0)\prime}, W u_r^{(0)\prime}) = 0 \quad \text{whenever } E_k^{(0)} = E_r^{(0)}, \; k \neq r. \tag{13.46}$$

Hence $W_{in}' = 0$ whenever $E_n^{(0)} = E_i^{(0)}$, and the offending terms are removed from the summations.

### 13.2.4 Example: linear perturbation to the harmonic oscillator

A linear harmonic oscillator whose classical equation of motion is $m\ddot{x} = -\mu x$ is perturbed by a small extra potential $W = \gamma \mu x$ where $\gamma$ is a suitably "small" real constant. It is required to calculate the changes to the energy levels to both the first and second orders. In fact, it is easy to calculate the new energy levels exactly by making a translation on the $x$-axis, but the problem will be tackled using the perturbation method described above.

For this problem, the unperturbed solution is

$$E_n = \left(n + \frac{1}{2}\right)\hbar\omega_c$$

$$u_n^{(0)} = \left(\frac{\alpha}{2^n \sqrt{\pi n!}}\right)^{\frac{1}{2}} H_n(\alpha x) e^{-\frac{1}{2}\alpha^2 x^2}$$

where

$$\omega_c^2 \equiv \frac{\mu}{m}, \qquad \alpha^4 \equiv \frac{\mu m}{\hbar^2},$$

and the Hermite polynomials $H_n(\xi)$ satisfy the relations

$$H_n' = 2n H_{n-1} \quad \text{and} \quad \int_{-\infty}^{\infty} H_n(\xi) H_m(\xi) e^{-\xi^2} d\xi = \sqrt{\pi} 2^n n! \delta_{nm}.$$

On defining the set of constants

$$C_{nm} \equiv \frac{\gamma \mu \alpha}{\sqrt{\pi}} \left(\frac{1}{2^{n+m} n! m!}\right)^{\frac{1}{2}},$$

and using integration by parts, it is easily shown that the matrix elements of the perturbation $W$ are

$$\begin{aligned}
W_{nm} &= (u_n^{(0)}, \gamma \mu x u_m^{(0)}) \\
&= C_{nm} \int_{-\infty}^{\infty} e^{-\alpha^2 x^2} x H_n(\alpha x) H_m(\alpha x) dx \\
&= \frac{1}{\alpha^2} C_{nm} \int_{-\infty}^{\infty} e^{-\xi^2} \xi H_n(\xi) H_m(\xi) d\xi \\
&= \frac{1}{2\alpha^2} C_{nm} \int_{-\infty}^{\infty} e^{-\xi^2} (2n H_{n-1} H_m + 2m H_n H_{m-1}) d\xi \\
&= \frac{\gamma \mu}{\alpha} \left(\frac{n}{2}\right)^{\frac{1}{2}} \quad \text{for } m = n - 1 \\
&= \frac{\gamma \mu}{\alpha} \left(\frac{n+1}{2}\right)^{\frac{1}{2}} \quad \text{for } m = n + 1 \\
&= 0 \quad \text{otherwise.}
\end{aligned}$$

In particular, this result shows that $W_{nn} = 0$, and so there is no correction to the first order. To the second order, Equation (13.45) gives

$$\begin{aligned}
E_n &= \left(n + \frac{1}{2}\right)\hbar\omega_c + 0 + \frac{W_{n(n-1)} W_{(n-1)n}}{E_n^{(0)} - E_{n-1}^{(0)}} + \frac{W_{n(n+1)} W_{(n+1)n}}{E_n^{(0)} - E_{n+1}^{(0)}} \\
&= \left(n + \frac{1}{2}\right)\hbar\omega_c + \frac{\gamma^2 \mu^2 n}{2\hbar\omega_c \alpha^2} - \frac{\gamma^2 \mu^2 (n+1)}{2\hbar\omega_c \alpha^2}
\end{aligned}$$

$$= \left(n + \frac{1}{2}\right)\hbar\omega_c - \frac{1}{2}\gamma^2\mu. \tag{13.47}$$

Hence the change to the energy levels to the second order is a constant value $\frac{1}{2}\gamma^2\mu$. In fact, it can be shown that this is the exact change, there being zero contribution from the remaining infinite number of terms in Equation (13.31).

## 13.3 Time dependent perturbation theory

Suppose that the hamiltonian of a system has the form $H = H^{(0)} + W$ where $W \equiv W(\mathbf{r}, t)$ is a perturbation to the hamiltonian which is now a function of time $t$:

$$\frac{\partial}{\partial t}W(\mathbf{r}, t) \neq 0.$$

Suppose as before that the complete orthonormal set of eigenfunctions of the hamiltonian $H^{(0)}$ are known, and can be calculated exactly through the equation

$$H^{(0)}u_n^{(0)} = E_n^{(0)}u_n^{(0)}. \tag{13.48}$$

It is required to calculate the time dependence of the wave function $\Psi \equiv \Psi(\mathbf{r}, t)$ through the Schrödinger equation

$$i\hbar\frac{\partial}{\partial t}\Psi(\mathbf{r}, t) = (H^{(0)} + W(\mathbf{r}, t))\Psi(\mathbf{r}, t). \tag{13.49}$$

In the following discussion, we will write $W(t) \equiv W(\mathbf{r}, t)$ in order to ease the notation, but it must be remembered that $W(t)$ is also a function of the space coordinate $\mathbf{r}$.

Now expand the unknown solution in terms of the complete orthonormal set:

$$\Psi(\mathbf{r}, t) = \sum_k c_k(t)e^{-\frac{i}{\hbar}E_k^{(0)}t}u_k^{(0)} = \sum_k c_k(t)e^{-i\omega_k t}u_k^{(0)} \tag{13.50}$$

where

$$\omega_k \equiv \frac{E_k^{(0)}}{\hbar}. \tag{13.51}$$

Then Equation (13.49) becomes

$$i\hbar\sum_k \frac{d}{dt}\left(c_k(t)e^{-i\omega_k t}\right)u_k^{(0)} = \sum_k c_k(t)e^{-i\omega_k t}(H^{(0)} + W(t))u_k^{(0)},$$

which reduces to

$$i\hbar \sum_k \frac{dc_k}{dt} e^{-i\omega_k t} u_k^{(0)} = \sum_k c_k(t) e^{-i\omega_k t} W(t) u_k^{(0)}. \tag{13.52}$$

On taking the inner product with $u_j^{(0)}$ for any index $j$, and remembering that $(u_k^{(0)}, u_j^{(0)}) = \delta_{kj}$, then Equation (13.52) becomes

$$i\hbar \frac{dc_j}{dt} = \sum_k e^{i(\omega_j - \omega_k)t} (u_j^{(0)}, W(t) u_k^{(0)}) c_k, \quad (j = 0, 1, \ldots). \tag{13.53}$$

In particular, if the system is known to be in the state $u_L^{(0)}$ at time $t = 0$ for some index $L$, then $\Psi(\mathbf{r}, 0) = u_L^{(0)}$. Hence it can be seen from Equation (13.50) that the initial conditions on the coefficients $c_k$ are

$$c_k(0) = \delta_{kL}. \tag{13.54}$$

Equations (13.53) are a set of simultaneous first order linear differential equations whose coefficients are functions of $t$. They must be solved using the initial conditions in Equation (13.54).

### Approximation

There is no general solution of the set of Equations (13.53). However, an approximate solution can be obtained for "small" times during which the coefficients $c_i$ remain close to their initial values. In this case, Equations (13.53) reduce to

$$i\hbar \frac{dc_j}{dt} = (u_j^{(0)}, W(t) u_L^{(0)}) e^{i(\omega_j - \omega_L)t}, \quad (j = 0, 1, \ldots),$$

which have solutions

$$c_j(t) = c_j(0) - \frac{i}{\hbar} \int_0^t (u_j^{(0)}, W(t') u_L^{(0)}) e^{i(\omega_j - \omega_L)t'} dt'$$

$$= c_j(0) - \frac{i}{\hbar} \int_0^t W_{jL}(t') e^{i(\omega_j - \omega_L)t'} dt', \tag{13.55}$$

where

$$W_{ij}(t) \equiv (u_i^{(0)}, W(t) u_j^{(0)}). \tag{13.56}$$

For example, consider an electron in a uniform magnetic field $\mathbf{B}_0$ which is perturbed by an oscillating plane polarised magnetic field $\mathbf{B} \sin(\omega t)$ perpendicular to $\mathbf{B}_0$, where $\omega > 0$. Then $W(t) = -\mathbf{m}_\mu \cdot \mathbf{B} \sin(\omega t) \equiv A \sin(\omega t)$ where $\mathbf{m}_\mu$ is the magnetic moment vector, and $A \equiv \mathbf{B} \cdot \mathbf{m}_\mu$. Equation (13.55) then becomes

$$c_j(t) = c_j(0) - \frac{i}{\hbar} A_{jL} \int_0^t e^{i(\omega_j - \omega_L)t'} \sin(\omega t') dt'$$

$$= c_j(0) + \frac{i}{2\hbar} A_{jL} \left( \frac{e^{i(\omega_j - \omega_L + \omega)t} - 1}{\omega_j - \omega_L + \omega} - \frac{e^{i(\omega_j - \omega_L - \omega)t} - 1}{\omega_j - \omega_L - \omega} \right) \quad (13.57)$$

for $|\omega_j - \omega_L| \neq |\omega|$. Hence by POSTULATE 5 of quantum mechanics which was introduced in Sect. 2.8, the probability of finding the system in state $j \neq L$ after a short time $t$ has elapsed is

$$\left| c_k(t) e^{-\frac{i}{\hbar} E_j^{(0)} t} \right|^2 = \frac{|A_{jL}|^2}{4\hbar^2} \left| \frac{e^{i(\omega_j - \omega_L + \omega)t} - 1}{\omega_j - \omega_L + \omega} - \frac{e^{i(\omega_j - \omega_L - \omega)t} - 1}{\omega_j - \omega_L - \omega} \right|^2 . \quad (13.58)$$

This quantity is large in comparison with all the other quantities $|c_k(t)|$ if either $\omega_j \approx \omega_L - \omega$ or $\omega_j \approx \omega_L + \omega$. If $\omega_j \approx \omega_L + \omega$, then $\omega_j > \omega_L$, or $E_j^{(0)} > E_L^{(0)}$, and the system gains energy by absorbing a photon from the magnetic field. On the other hand, if $\omega_j \approx \omega_L - \omega$, then $\omega_j < \omega_L$, or $E_j^{(0)} < E_L^{(0)}$, and the system loses energy to the magnetic field. A photon is emitted in this case, and the situation is that of *stimulated emission*.

## 13.4 The Variational Principle

The Variational Principle allows us to make a fairly accurate estimate of the ground-state energy by making an informed guess for the groundstate eigenfunction. This guess is called the *trial function*. Of course, the guess will not be completely accurate, but the method is fairly lenient in that it allows us to make a rough guess while producing a good estimate of the groundstate energy.

Let the set $\{u_i : i = 0, 1, 2, \ldots\}$ be a complete set of orthonormal eigenfunctions of the operator $H$ corresponding to the exact (and unknown) energy eigenvalues $E_0 \leq E_1 \leq E_2 \leq \cdots$. These will satisfy the equation

$$H u_i = E_i u_i, \quad i = 0, 1, 2, \ldots. \quad (13.59)$$

Let $\tau$ be any chosen trial function which is not necessarily normalised, and which can be expanded in terms of the set of functions $u_i$:

$$\tau = \sum_i c_i u_i. \quad (13.60)$$

Then

$$(\tau, H\tau) = \left( \sum_i c_i u_i, H \sum_j c_j u_j \right)$$

$$= \sum_i \sum_j \overline{c_i} c_j (u_i, H u_j)$$

$$= \sum_i \sum_j \overline{c}_i c_j E_j(u_i, u_j)$$

$$= \sum_i |c_i|^2 E_i$$

$$\geq \sum_i |c_i|^2 E_0 = E_0 \sum_i |c_i|^2. \tag{13.61}$$

Further, it is easily verified that $(\tau, \tau) = \sum_i |c_i|^2$. Hence the groundstate energy $E_0$ satisfies the condition

$$E_0 \leq \frac{(\tau, H\tau)}{(\tau, \tau)} \tag{13.62}$$

for all trial functions $\tau$. This result is used by choosing a "reasonable" trial function $\tau$ which is a function of a number $P$ of parameters $p_r$ $(r = 1, \ldots, P)$, and then minimising the function $(\tau, H\tau)/(\tau, \tau)$ by taking

$$\frac{\partial}{\partial p_r}\left[\frac{(\tau, H\tau)}{(\tau, \tau)}\right] = 0, \quad (r = 1, \ldots, P). \tag{13.63}$$

This will give an upper limit on the groundstate energy $E_0$ through the result of Equation (13.62). In practice, it turns out that a reasonable guess will return an upper limit which is very close, if not equal to, the groundstate energy.

As an example, consider the problem of finding the groundstate energy for the linear harmonic oscillator whose classical equation of motion is $m\ddot{x} = -\mu x$. The hamiltonian operator for this problem is

$$H = -\frac{\hbar^2}{2m}\frac{d^2}{dx^2} + \frac{1}{2}\mu x^2.$$

Now choose a one-parameter trial function

$$\tau = e^{-px^2} \tag{13.64}$$

where the single parameter $p$ is real and positive. Note that this choice of function is based on the fact that it is real, symmetric, and decreases to zero for large $|x|$. Then

$$(\tau, \tau) = \int_{-\infty}^{\infty} e^{-2px^2} dx, \quad \text{and}$$

$$(\tau, H\tau) = \int_{-\infty}^{\infty} e^{-px^2}\left(-\frac{\hbar^2}{2m}\frac{d^2}{dx^2} + \frac{1}{2}\mu x^2\right)e^{-px^2}$$

$$= -\frac{\hbar^2}{2m}\int_{-\infty}^{\infty}(-2p + 4p^2 x^2)e^{-2px^2} dx + \frac{1}{2}\mu\int_{-\infty}^{\infty} x^2 e^{-2px^2} dx$$

$$= \frac{p\hbar^2}{m}\int_{-\infty}^{\infty} e^{-2px^2} dx + \left(\frac{1}{2}\mu - \frac{2p^2\hbar^2}{m}\right)\int_{-\infty}^{\infty} x^2 e^{-2px^2} dx.$$

On using the relations

$$\int_{-\infty}^{\infty} e^{-2px^2} dx = \sqrt{\frac{\pi}{2p}} \quad \text{and} \quad \int_{-\infty}^{\infty} x^2 e^{-2px^2} dx = \frac{1}{4p}\sqrt{\frac{\pi}{2p}},$$

it follows that

$$E^{(p)} \equiv \frac{(\tau, H\tau)}{(\tau, \tau)} = \frac{p\hbar^2}{m} + \frac{1}{4p}\left(\frac{1}{2}\mu - \frac{2p^2\hbar^2}{m}\right). \tag{13.65}$$

The stationary value of this quantity is given by

$$0 = \frac{dE^{(p)}}{dp} = \frac{\hbar^2}{m} - \frac{\mu}{8p^2} - \frac{\hbar^2}{2m},$$

or

$$p = \frac{1}{2\hbar}\sqrt{m\mu}.$$

It is easily verified that this value of $p$ corresponds to a minimum of the quantity $E^{(p)}$. Inserting this optimal value of $p$ into the expression for $E^{(p)}$ gives

$$E^{(p)}{}_{min} = \frac{1}{2}\hbar\sqrt{\frac{\mu}{m}} = \frac{1}{2}\hbar\omega_c,$$

where $\omega_c^2 \equiv \mu/m$. Hence

$$E_0 \le \frac{1}{2}\hbar\omega_c. \tag{13.66}$$

In fact, this value is the exact value of $E_0$.

## 13.5 Discretisation of the Schrödinger equation

The complexity of any HEMT device simulation means that the TISE

$$-\frac{\hbar^2}{2}\nabla_{\mathbf{r}}\cdot\left(\frac{1}{m_e}\nabla_{\mathbf{r}}u_n\right) + V(\mathbf{r})u_n = E_n u_n \quad (n = 0, 1, 2, \ldots). \tag{13.67}$$

can only be solved numerically. The grid chosen for this solution must be of the most suitable type to fit with the geometry of the particular device. In the case of the HEMT with a rectangular profile in the $x$-$y$ plane, the most suitable grid is a rectangular grid with axes parallel to the sides of the device. Further, most HEMT devices have the boundaries of the various material layers arranged parallel to one edge, which will be taken as the $x$ direction. Consequently, changes in material type will take place most rapidly along the $y$ direction, and it is then common practice to solve the Schrödinger equation in one dimensional slices along the $y$ direction.

The discretisation of the Schrödinger equation will be described for both the two and one dimensional cases, using finite differences on non-uniform grids.

## 13.5.1 Discretisation in two dimensions

On a two dimensional rectangular grid, Equation (13.67) can be written

$$-\frac{\partial}{\partial x}\left(\frac{1}{m_e}\frac{\partial u_n}{\partial x}\right) - \frac{\partial}{\partial y}\left(\frac{1}{m_e}\frac{\partial u_n}{\partial y}\right) + \frac{2}{\hbar^2}V(x, y)u_n = \frac{2}{\hbar^2}E_n u_n, \quad (n = 0, 1, 2, \ldots).$$

(13.68)

This equation will be discretised on a nonuniform two dimensional rectangular grid which is composed of the set of points

$$\{(x_i, y_j) : i = 0, 1, \ldots, N_x; j = 0, 1, \ldots, N_y\},$$

with the mesh spacings being

$$h_i \equiv x_{i+1} - x_i, \qquad k_j \equiv y_{j+1} - y_j.$$

The kinetic energy term in Equation (13.67) has the form $\nabla_{\mathbf{r}} \cdot (a\nabla_{\mathbf{r}}\phi)$, where $a \equiv -\hbar^2/(2m_e)$ and $\phi \equiv u_n$. An expression has already been found for the discretisation of the quantity $\nabla_{\mathbf{r}} \cdot (a\nabla_{\mathbf{r}}\phi)$ on a two dimensional rectangular mesh in Equation (7.18). To ease the notation, suppress the index $n$ on the eigensolutions, and write $\mu \equiv 1/m_e$, $\mu_{i,j} \equiv \mu(x_i, y_j)$, $V_{i,j} \equiv V(x_i, y_j)$, and $u_{i,j} \equiv u(x_i, y_j)$. Using the result of Equation (7.18), the discretisation of the Equation (13.68) becomes

$$\frac{2}{\hbar^2}E u_{i,j} = -\frac{(\mu_{i-1,j} + \mu_{i,j})}{h_{i-1}(h_{i-1} + h_i)}u_{i-1,j} - \frac{(\mu_{i,j} + \mu_{i+1,j})}{h_i(h_{i-1} + h_i)}u_{i+1,j}$$
$$+ \frac{1}{h_{i-1} + h_i}\left(\frac{\mu_{i-1,j} + \mu_{i,j}}{h_{i-1}} + \frac{\mu_{i,j} + \mu_{i+1,j}}{h_i}\right)u_{i,j}$$
$$- \frac{(\mu_{i,j-1} + \mu_{i,j})}{k_{j-1}(k_{i,j-1} + k_j)}u_{i,j-1} - \frac{(\mu_{i,j} + \mu_{i,j+1})}{k_j(k_{j-1} + k_j)}u_{i,j+1}$$
$$+ \frac{1}{k_{j-1} + k_j}\left(\frac{\mu_{i,j-1} + \mu_{i,j}}{k_{j-1}} + \frac{\mu_{i,j} + \mu_{i,j+1}}{k_j}\right)u_{i,j}$$
$$+ \frac{2}{\hbar^2}V_{i,j}u_{i,j}.$$

(13.69)

In the case of a uniform mesh for which $h_i \equiv h$ and $k_j \equiv k$, this equation simplifies to

$$\frac{2}{\hbar^2}E u_{i,j} = -\frac{(\mu_{i-1,j} + \mu_{i,j})}{2h^2}u_{i-1,j} - \frac{(\mu_{i,j} + \mu_{i+1,j})}{2h^2}u_{i+1,j}$$
$$- \frac{(\mu_{i,j-1} + \mu_{i,j})}{2k^2}u_{i,j-1} - \frac{(\mu_{i,j} + \mu_{i,j+1})}{2k^2}u_{i,j+1}$$

$$+ \left( \frac{1}{2h^2} (\mu_{i-1,j} + 2\mu_{i,j} + \mu_{i+1,j}) \right.$$

$$+ \frac{1}{2k^2} (\mu_{i,j-1} + 2\mu_{i,j} + \mu_{i,j+1}) \left. \right) u_{i,j}$$

$$+ \frac{2}{\hbar^2} V_{i,j} u_{i,j}. \tag{13.70}$$

### 13.5.2 Discretisation in one dimension

Along the $y$ axis only, Equation (13.67) has the form the form

$$-\frac{d}{dy} \left( \frac{1}{m_e} \frac{du_n}{dy} \right) + \frac{2}{\hbar^2} V(x, y) u_n = \frac{2}{\hbar^2} E_n u_n, \quad (n = 0, 1, 2, \ldots). \tag{13.71}$$

The one dimensional grid is composed of the set of points

$$\{y_j : j = 0, 1, \ldots, N_y\}$$

and the mesh spacings will be

$$k_j = y_{j+1} - y_j.$$

In keeping with the layer structure, the effective mass $m_e$ will depend on $y$ only: $m_e \equiv m_e(y)$. Again, we write $\mu(y) \equiv 1/m_e(y)$, with $\mu_j \equiv \mu(y_j)$. Notice that the $x$ dependence has been retained in the potential $V$ in Equation (13.71). This is because the potential will depend on the electrostatic potential $\psi(x, y)$ which varies appreciably in both the $x$- and $y$-directions. Consequently, we will write $V_j \equiv V(x, y_j)$, while remembering that the numerical solution of the one dimensional equation refers to a one dimensional slice for a particular value of $x$. Discretisation of Equation (13.71) will then take the form

$$\frac{2}{\hbar^2} E u_j = -\frac{(\mu_{j-1} + \mu_j)}{k_{j-1}(k_{j-1} + k_j)} u_{j-1} - \frac{(\mu_j + \mu_{j+1})}{k_j(k_{j-1} + k_j)} u_{j+1}$$

$$+ \frac{1}{k_{j-1} + k_j} \left( \frac{\mu_{j-1} + \mu_j}{k_{j-1}} + \frac{\mu_j + \mu_{j+1}}{k_j} \right) u_j$$

$$+ \frac{2}{\hbar^2} V_j u_j. \tag{13.72}$$

Again in the case of a uniform mesh for which $k_j \equiv k$, this equation simplifies to

$$\frac{2}{\hbar^2} E u_j = -\frac{(\mu_{j-1} + \mu_j)}{2k^2} u_{j-1} - \frac{(\mu_j + \mu_{j+1})}{2k^2} u_{j+1}$$

$$+ \frac{1}{2k^2} (\mu_{j-1} + 2\mu_j + \mu_{j+1}) u_j + \frac{2}{\hbar^2} V_j u_j. \tag{13.73}$$

## 13.6 Numerical solution: the iteration method

The iteration method solves for the eigensolutions, in the order of increasing eigenvalue, starting from the groundstate. Each subsequent solution is found in terms of those which have been previously found, and are obtained by performing a relaxation iteration on the Schrödinger equation. The eigenvalue is re-calculated at each iteration using an equation similar to Equation (13.62), although trial functions are not involved in this process.

### 13.6.1 The basis of the iteration method

The TISE is written in the form

$$Hu_n = E_n u_n, \quad (n = 0, 1, 2, \ldots) \tag{13.74}$$

where the operator $H$ is given by

$$H \equiv -\frac{\hbar^2}{2} \nabla_{\mathbf{r}} \cdot \frac{1}{m_e} \nabla_{\mathbf{r}} + V(\mathbf{r}). \tag{13.75}$$

In solving for the $n$th level, the eigenfunction is calculated to be orthogonal to all of the eigenfunctions found on the previous levels $i = 0, 1, 2, \ldots, n - 1$. That is, whenever an iteration is made to find $u_n$, the replacement

$$u_n \rightarrow u_n - \sum_{i=0}^{n-1} (u_n, u_i) u_i \tag{13.76}$$

is made. This ensures that all eigenfunctions found previously, and the most recently iterated eigenfunction $u_n$, are orthogonal. The stages of the process are described for the first two eigensolutions, and then for the general eigensolution:

*Solution for $n = 0$.*

1. Make initial guesses for $E_0$ and $u_0$.
2. Iterate Equation (13.74) for $u_0$.
3. Update $E_0 = (u_0, Hu_0)/(u_0, u_0)$.
4. Repeat steps 2 and 3 until convergence.
5. Normalise $u_0$.

*Solution for $n = 1$.*

1. Make initial guesses for $E_1$ and $u_1$.
2. Iterate Equation (13.74) for $u_1$.

3. Replace $u_1 \to u_1 - (u_1, u_0)u_0$.
4. Update $E_1 = (u_1, Hu_1)/(u_1, u_1)$.
5. Repeat steps 2,3 and 4 until convergence.
6. Normalise $u_1$.

*Solution for a general value of n.*

1. Make initial guesses for $E_n$ and $u_n$.
2. Iterate Equation (13.74) for $u_n$.
3. Replace $u_n \to u_n - \sum_{i=0}^{n-1}(u_n, u_i)u_i$.
4. Update $E_n = (u_n, Hu_n)/(u_n, u_n)$.
5. Repeat steps 2,3 and 4 until convergence.
6. Normalise $u_n$.

### 13.6.2 Numerical implementation of the iteration method

Although a full programme for the implementation will not be presented, a description will be given of the main important functions which will be involved in such a programme. These functions are written for the special case in which the grid is uniform, and for the case in which the effective mass is a function of the $y$-coordinate only—this situation will apply in the case of a HEMT in which the $y$-direction is taken perpendicular to the layer structure. Initial declarations must be made of the type

```
int Nx,Ny; // numbers of grid spacings
double xmin,xmax,ymin,ymax; // limits of x and y
double h=(xmax-xmin)/Nx, k=(ymax-ymin)/Ny; // grid spacings (uniform)
int N_eig; // number of eigensolutions
double Eig_f[N_eig][Nx+1][Ny+1]; // eigenfunctions
double Eig_v[N_eig]; // eigenvalues
double V_pot[Nx+1][Ny+1];
int maxcount_S_int=1; // maximum internal iteration counter
int maxcount_S_ext=25; // maximum external iteration counter
double error_Sch=1.0e-6;
double w_Sch=1.4; // SOR factor for internal iteration
```

The coefficients which are found in the discretised Schrödinger equation (13.69) must be declared in the function set_fixedmatelems:

```
void set_fixedmatelems(double **s_0b, double **s_0f,
                double **s_b0, double **s_f0, double **s_00)
// setting fixed matrix elements for the Schrodinger equation.
// described for a uniform grid only.
// me_eff a function of y only.
{
  int i,j;
  double mb,m0,mf,mhf,mhb;
  double h2=h*h, k2=k*k;
  double S_cons=0.5*hPbar*hPbar;
  for(i=1;i<Nx;i++) {
    for (j=1;j<Ny;j++) {
      mb=me_eff(j-1);
      m0=me_eff(j);
      mf=me_eff(j+1);
```

```
        mhf=0.5*(m0+mf);   mhb=0.5*(mb+m0);

        s_0b[i][j]=S_cons/(k2*mhb);
        s_0f[i][j]=S_cons/(k2*mhf);
        s_b0[i][j]=S_cons/(h2*m0);
        s_f0[i][j]=S_cons/(h2*m0);
        s_00[i][j]=S_cons*((1.0/mhf+1.0/mhb)/k2+2.0/(h2*m0));
      }
    }
}
```

A total of N_eig eigensolutions are to be found, and these are initialised using the function init_eig():

```
void init_eig(void)
{
  int n_eig,i,j;
  // fixed boundary values:
  for(n_eig=0;n_eig<=N_eig-1;n_eig++) {
    for(i=0;i<=Nx;i++) Eig_f[n_eig][i][0]=Eig_f[n_eig][i][Ny]=0.0;
    for(j=0;j<=Ny;j++) Eig_f[n_eig][0][j]=Eig_f[n_eig][Nx][j]=0.0;
  }
  // initial guess:
  for(n_eig=0;n_eig<=N_eig-1;n_eig++) {
    Eig_v[n_eig]=0.010;   //0.01
    for(i=1;i<Nx;i++) for(j=1;j<Ny;j++) Eig_f[n_eig][i][j]=0.01;
    normalise(Eig_f[n_eig]);
  }
}
```

The eigensolution at the *n*th level is then found using the function calc_eig() which iterates all of the solutions from the groundstate solution:

```
void calc_eig(void)
{
  int i,j,n_eig,ne,counter_S_ext,counter_S_int;
  double eig_v_prev;
  double F[Nx+1][Ny+1];
  double ip;
  for(i=0;i<=Nx;i++) F[i][0]=F[i][Ny]=0.0;
  for(j=0;j<=Ny;j++) F[0][j]=F[Nx][j]=0.0;
  for(n_eig=0;n_eig<=N_eig-1;n_eig++) { //start of n_eig counter
    counter_S_ext=1;
    do {
      eig_v_prev=Eig_v[n_eig];
      // now iterate eigenfunction:
      counter_S_int=1;
      do {
        for(i=1;i<Nx;i++) {
          for(j=1;j<Ny;j++) {
            Eig_f[n_eig][i][j]=(1.0-w_Sch)*Eig_f[n_eig][i][j]
                       +w_Sch*(S_f0[i][j]*Eig_f[n_eig][i+1][j]
                         +S_b0[i][j]*Eig_f[n_eig][i-1][j]
                         +S_0f[i][j]*Eig_f[n_eig][i][j+1]
                         +S_0b[i][j]*Eig_f[n_eig][i][j-1])
                         /(S_00[i][j]+V_pot[i][j]-Eig_v[n_eig]);
          }
        }
  for(i=Nx-1;i>0;i--) {
    for(j=1;j<Ny;j++) {
      Eig_f[n_eig][i][j]=(1.0-w_Sch)*Eig_f[n_eig][i][j]
      +w_Sch*(S_f0[i][j]*Eig_f[n_eig][i+1][j]
      +S_b0[i][j]*Eig_f[n_eig][i-1][j]
      +S_0f[i][j]*Eig_f[n_eig][i][j+1]
      +S_0b[i][j]*Eig_f[n_eig][i][j-1])
      /(S_00[i][j]+V_pot[i][j]-Eig_v[n_eig]);
```

```
    }
  }
counter_S_int++;
  }while(counter_S_int<=maxcoun_S_int);
  normalise(Eig_f[n_eig]);  // just in case the thing blows up.
  if(n_eig!=0) {
for(ne=0;ne<n_eig;ne++) {
  ip=inner_prod(Eig_f[ne],Eig_f[n_eig]);
  for(i=1;i<Nx;i++) for(j=1;j<Ny;j++)
    Eig_f[n_eig][i][j]=Eig_f[n_eig][i][j]-ip*Eig_f[ne][i][j];
}
    }
    // now update eigenvalue:
    for(i=1;i<Nx;i++)
for(j=1;j<Ny;j++)
  F[i][j]=(S_00[i][j]+V_pot[i][j])*Eig_f[n_eig][i][j]
  -S_f0[i][j]*Eig_f[n_eig][i+1][j]
  -S_b0[i][j]*Eig_f[n_eig][i-1][j]
  -S_0f[i][j]*Eig_f[n_eig][i][j+1]
  -S_0b[i][j]*Eig_f[n_eig][i][j-1];
  Eig_v[n_eig]=inner_prod(Eig_f[n_eig],F)
  /inner_prod(Eig_f[n_eig],Eig_f[n_eig]);
    counter_S_ext++;
    }while((fabs((Eig_v[n_eig]-eig_v_prev)/Eig_v[n_eig]))>error_Sch)
            && (counter_S_ext<=maxcoun_S_ext));
  printf("   %d",counter_S_ext-1);
  // now finally normalise:
  normalise(Eig_f[n_eig]);
  }//end of n_eig counter
}
```

The function inner_prod is used to calculate the inner products:

```
double inner_prod(double **v1, double **v2)
{
  int i,j;
  double sum=0.0;
  for(i=0;i<=Nx;i++) for(j=0;j<=Ny;j++) sum+=v1[i][j]*v2[i][j];
  sum*=(h*k);
  return sum;
}
```

Finally, each eigenfunction is normalised using the function normalise:

```
void normalise(double **v)
{
  int i,j;
  double mag;
  mag=sqrt(inner_prod(v,v));
  for(i=0;i<=Nx;i++) for(j=0;j<=Ny;j++) v[i][j]/=mag;
}
```

## 13.7 Numerical solution: the trial function method

This is an extension of the variational method introduced in Sect. 13.4 for the groundstate energy. The required eigenfunctions $u_n$ which satisfy the equation

$$Hu_n = E_n u_n, \quad (n = 0, 1, 2, \ldots) \tag{13.77}$$

will be expanded in terms of a set of known trial functions. The coefficients in this expansion will depend on a set of parameters $p_n$. The method will be illustrated here in terms of a one dimensional spatial problem, although the extension to the case of more than one spatial dimension is straightforward.

### 13.7.1 The basis of the trial function method

The $n$th calculated (and therefore approximate) eigenfunction will be denoted by $u_n^{(p_n)}(x)$, where $p_n$ is a parameter, or a set of parameters, appropriate to the $n$th eigenfunction. The function $u_n^{(p_n)}(x)$ will be written as a linear combination of trial functions $v_{ni}^{(p_n)}(x)$:

$$u_n^{(p_n)}(x) = \sum_{i=0}^{n} A_{ni}^{(p_n)} v_{ni}^{(p_n)}(x) \tag{13.78}$$

where the functions $v_{ni}^{(p_n)}(x)$, $(i = 0, \ldots, n)$ are a chosen set of linearly independent functions which depend on a set of parameters $p_n$. The reasoning behind this particular choice will be discussed in Sect. 13.7.2. The coefficients $A_{ni}^{(p_n)}$ will generally be complex numbers.

The coefficients $A_{ni}$ have to be determined, but the trial functions can be ordered so that $A_{nn} \neq 0$ for each index $n$. Further, the $u_n^{(p_n)}(x)$ do not need to be normalised at this stage, and so without loss of generality we may impose the condition

$$A_{nn} = 1, \quad n = 0, 1, 2, \ldots. \tag{13.79}$$

However, the $u_n^{(p_n)}(x)$ will be orthogonal, with

$$\left( u_n^{(p_n)}, u_m^{(p_m)} \right) = 0 \quad (m \neq n),$$

or

$$\sum_{i=0}^{n} \sum_{j=0}^{m} \overline{A_{mi}^{(p_m)}} A_{ni}^{(p_n)} \left( v_{mj}^{(p_m)}, v_{ni}^{(p_n)} \right) = 0, \quad (m \neq n). \tag{13.80}$$

Since the functions $v_{ni}^{(p_n)}(x)$ are a chosen set, then the inner products in Equation (13.80) can be calculated in advance. Further, the $A_{ik}$ can be restricted to be real quantities. In practice, Equation (13.80) is used as follows: calculate the $A_{ik}$ for all of the eigensolutions up to the $(n-1)$th solution, and then use this equation to solve a set of linear equations for the corresponding values $A_{ni}$, $(i = 0, 1, \ldots, n-1)$. The following steps show how this iterative process applies to the solution of the first three eigensolutions.

*Solution for n = 0.*

For $n = 0$, Equation (13.78) becomes

$$u_0^{(p_0)}(x) = A_{00}^{(p_0)} v_{00}^{(p_0)}(x) = v_{00}^{(p_0)}(x)$$

since $A_{00}^{(p_0)} = 1$. Now write the approximate solution $u_0^{(p_0)}(x)$ as a linear combination of the complete set of the exact (and unknown) eigenfunctions $u_i(x)$:

$$u_0^{(p_0)}(x) = \sum_i c_i u_i(x)$$

where the coefficients $c_i$ will themselves be functions of the parameter $p_0$. Then

$$\left( u_0^{(p_0)}, H u_0^{(p_0)} \right) = \sum_{i,j} \overline{c_i} c_j E_j (u_i, u_j)$$

$$= \sum_i |c_i|^2 E_i$$

$$\geq E_0 \left( u_0^{(p_0)}, u_0^{(p_0)} \right).$$

Hence

$$E_0 \leq \frac{(u_0^{(p_0)}, H u_0^{(p_0)})}{(u_0^{(p_0)}, u_0^{(p_0)})}. \tag{13.81}$$

Having chosen a suitable trial function $v_{00}^{(p_0)}(x)$ in terms of a parameter $p_0$, the right hand side of Equation (13.81) can be calculated. The optimal value $\pi_0$ of $p_0$ can then be found to make this quantity a minimum. This will give an upper bound on the value of $E_0$ which, in the absence of further information, will be taken as the calculated value of $E_0$.

*Solution for n = 1.*

For $n = 1$, Equation (13.78) becomes

$$u_1^{(p_1)}(x) = A_{10}^{(p_1)} v_{10}^{(p_1)}(x) + A_{11}^{(p_1)} v_{11}^{(p_1)}(x). \tag{13.82}$$

Since $u_0^{(\pi_0)}(x)$ and $u_1^{(p_1)}(x)$ are orthogonal, the condition $(u_0^{(\pi_0)}, u_1^{(p_1)}) = 0$ gives

$$A_{10}^{(p_1)} (v_{00}^{(\pi_0)}, v_{10}^{(p_1)}) = -(v_{00}^{(\pi_0)}, v_{11}^{(p_1)})$$

since $A_{11} = 1$, and therefore it follows that

$$A_{10}^{(p_1)} = -\frac{\left( v_{00}^{(\pi_0)}, v_{11}^{(p_1)} \right)}{\left( v_{00}^{(\pi_0)}, v_{10}^{(p_1)} \right)}. \tag{13.83}$$

Again write the approximate solution $u_1^{(p_1)}(x)$ as a linear combination of the exact complete set of eigenfunctions:

$$u_1^{(p_1)}(x) = \sum_i d_i u_i(x) = d_0 u_0 + \sum_{i \neq 0} d_i u_i(x)$$

where the coefficients $d_i$ will themselves be functions of the parameter $p_1$. Now $d_0$ must be zero because $u_0$ and $u_1^{(p_1)}(x)$ are orthogonal. (That is not strictly true, because $u_1^{(p_1)}(x)$ is only an approximation to $u_1$.) Hence

$$\left(u_1^{(p_1)}, H u_1^{(p_1)}\right) = \sum_{i,j \neq 0} E_j \bar{d}_i d_j (u_i, u_j)$$

$$\geq E_1 \left(u_1^{(p_1)}, u_1^{(p_1)}\right),$$

or

$$E_1 \leq \frac{(u_1^{(p_1)}, H u_1^{(p_1)})}{(u_1^{(p_1)}, u_1^{(p_1)})}. \tag{13.84}$$

Hence for $n = 1$, the steps in the process are

1. Choose the trial functions $v_{10}^{(p_1)}$ and $v_{11}^{(p_1)}$ in terms of a parameter $p_1$.
2. Calculate $A_{10}^{(p_1)}$ using Equation (13.83), remembering that $A_{11} = 1$.
3. Evaluate the right hand side of Equation (13.84) using Equation (13.82).
4. Find the optimal value $\pi_1$ of $p_1$ which minimises the right hand side of Equation (13.84).
5. Use the minimum value of the right hand side of Equation (13.84) as an upper bound, and therefore as an estimate, of $E_1$.

*Solution for $n = 2$.*

The process for $n = 2$ will be described briefly in order to establish the pattern of the solution method. Equation (13.78) becomes

$$u_2^{(p_2)}(x) = A_{20}^{(p_2)} v_{20}^{(p_2)}(x) + A_{21}^{(p_2)} v_{21}^{(p_2)}(x) + A_{22}^{(p_2)} v_{22}^{(p_2)}(x).$$

Since this function and the already determined functions $u_0^{(\pi_0)}$ and $u_1^{(\pi_1)}$ are orthogonal, and remembering that we take $A_{22} = 1$, this gives the two equations

$$A_{20}^{(p_2)}(v_{00}^{(\pi_0)}, v_{20}^{(p_2)}) + A_{21}^{(p_2)}(v_{00}^{(\pi_0)}, v_{21}^{(p_2)}) = -(v_{00}^{(\pi_0)}, v_{22}^{(p_2)}),$$

$$A_{20}^{(p_2)} \left[ A_{10}^{(\pi_1)}(v_{10}^{(\pi_1)}, v_{20}^{(p_2)}) + (v_{11}^{(\pi_1)}, v_{20}^{(p_2)}) \right]$$

$$+ A_{21}^{(p_2)} \left[ A_{10}^{(\pi_1)}(v_{10}^{(\pi_1)}, v_{21}^{(p_2)}) + (v_{11}^{(\pi_1)}, v_{21}^{(p_2)}) \right]$$

$$= -A_{10}^{(\pi_1)}(v_{10}^{(\pi_1)}, v_{22}^{(p_2)}) + (v_{11}^{(\pi_1)}, v_{22}^{(p_2)}).$$

These are two simultaneous equations for the quantities $A_{20}^{(p_2)}$ and $A_{21}^{(p_2)}$, which can be solved to give $u_2^{(p_2)}(x)$ as a function of the parameter $p_2$. The optimal value $\pi_2$ of this parameter is then found which makes the quantity

$$\frac{(u_2{}^{(p_2)}, H u_2{}^{(p_2)})}{(u_2{}^{(p_2)}, u_2{}^{(p_2)})}$$

a minimum. This minimum value is then taken as an estimate for $E_2$.

### Solution for a general value of n.

The above solutions for the cases of $i = 0$, 1, and 2 show that each solution is built on the preceding ones. In order to get the solution for a general value of $i = n$, it is necessary to obtain the solutions for the earlier cases $i = 0, 1, 2, \ldots, n - 1$. Remembering that the $A_{ik}$ are taken to be real quantities, and that we are at liberty to take $A_{nn} = 1$, Equation (13.80) can be written

$$\sum_{i=0}^{n-1} \left[ \sum_{j=0}^{m} A_{mj}{}^{(\pi_m)} \left( v_{mj}{}^{(\pi_m)}, v_{ni}{}^{(p_n)} \right) \right] A_{ni}{}^{(p_n)}$$

$$= -\sum_{j=0}^{m} A_{mj}{}^{(\pi_m)} \left( v_{mj}{}^{(\pi_m)}, v_{ni}{}^{(p_n)} \right),$$

$$(m = 0, 1, 2, \ldots, n - 1).$$

This is a set of $n$ linear equations, from which the $n$ quantities $A_{n0}, A_{n1}, \ldots, A_{n(n-1)}$ can be solved in terms of the parameter $p_n$ and the optimal parameters $\pi_0, \ldots, \pi_{n-1}$ found previously at the lower levels. These equations have the structure

$$\sum_{i=0}^{n-1} B_{mi} A_{ni}{}^{(p_n)} = C_m, \quad (m = 0, 1, \ldots, n - 1)$$

where the matrix $\mathbf{B}$ and column vector $\mathbf{C}$ have elements

$$B_{mi} \equiv \sum_{j=0}^{m} A_{mj}{}^{(\pi_m)} \left( v_{mj}{}^{(\pi_m)}, v_{ni}{}^{(p_n)} \right), \tag{13.85}$$

$$C_m \equiv -\sum_{j=0}^{m} A_{mj}{}^{(\pi_m)} \left( v_{mj}{}^{(\pi_m)}, v_{nn}{}^{(p_n)} \right). \tag{13.86}$$

These equations are solved for the quantities $A_{n0}, A_{n1}, \ldots, A_{n(n-1)}$ which are then substituted into the function in Equation (13.78). Using this function, the optimal value $\pi_n$ of $p_n$ is found which minimises the quantity

$$\frac{(u_n{}^{(p_n)}, H u_n{}^{(p_n)})}{(u_n{}^{(p_n)}, u_n{}^{(p_n)})}. \tag{13.87}$$

This quantity is then taken as an estimate for $E_n$.

### 13.7.2  Choice of trial functions

One of the most commonly used trial function is the Fang-Howard variational function (Das Sarma et al. 1979; Ng and Khoie 1991; Norris et al. 1985; Stern and Das Sarma 1984). However, there are circumstances in which these forms are not accurate enough (Lehmann and Jasiukiewicz 2002; Valadares and Sheard 1993). These functions take the form

$$v_{ni}{}^{(p_n)}(x) \equiv x^{i+1}e^{-p_n x}, \quad (i = 0, 1, 2, \ldots) \tag{13.88}$$

and they have the properties

$$v_{ni}{}^{(p_n)}(0) = 0, \qquad \lim_{x \to \infty} v_{ni}{}^{(p_n)}(x) = 0. \tag{13.89}$$

The first derivative of the function $v_{ni}{}^{(p_n)}(x)$ is

$$\frac{d}{dx}v_{ni}{}^{(p_n)}(x) = x^i(i + 1 - p_n x)e^{-p_n x}. \tag{13.90}$$

It is possible to take a slightly more general form

$$v_{ni}{}^{(p_n)}(x) = x^{i+\alpha}e^{-p_n x}, \quad (i = 0, 1, 2, \ldots) \tag{13.91}$$

where $\alpha > 0$. Different values of $\alpha$ may be taken for different physical situations, but there is no method of determining in advance what the optimal value of $\alpha$ is for any particular application.

### 13.7.3  Numerical implementation of the trial function method

The processes of obtaining the solutions for $n = 0$, 1 and 2 which were described in Sect. 13.7.1 may appear cumbersome to implement. However, these processes can be done iteratively, and only one function is required in the numerical implementation. This function is listed as `calc_A` below. A programme was written for a one dimensional simulation, using the variational method, for the solution of the first few eigenvalues and eigenfunctions for a simple potential well. The eigenfunctions are considered to vanish at the values $x = x_{min}$ and $x = x_{max}$. The main declarations at the beginning of the programme are listed as follows:

```
// PARAMETERS:
int noofeigenvalues; // the specified number of eigensolutions
double xmin, xmax: //  values of x at which the eigenfunctions vanish
int gridx; // the number of grid points
// PROBLEM SPECIFICATION:
double V0; // a characteristic potential value
double wellmax, wellmin; // min and max x-values at the ends of the well
```

The shape of the potential well is specified in the function `double potential(double x)`. In this particular listing, a very simple square well potential is specified as follows:

```
double potential(double x)
{
    double temp;
    if ((x>=wellmin)&&(x<wellmax)) { temp=0.0; }
    else { temp=V0; }
    return temp;
}
```

The trial functions $v_j^{(p)}(x) = x^{j+1}e^{-px}$ are specified in the function

```
double v(int j, double x, double p)
{
    double temp;
    temp=pow(x,j+1)*exp(-p*x);
    return temp;
}
```

and their derivatives are specified in the function

```
double dv(int j, double x, double p)
{
    double temp,eps=1.0e-20;
    if(j==0) { temp=(1.0-p*x)*exp(-p*x); }
    else { temp=pow(x,j)*(j+1-p*x)*exp(-p*x); }
    return temp;
}
```

The functions $u_n^{(p_n)}(x)$ given by Equation (13.78) are specified in the function

```
double u(int n, double x, double p)
{
    int i;
    double temp=0.0;
    for(i=0;i<=n;i++) { temp+=A[n][i]*v(i,x,p); }
    return temp;
}
```

The set of linear equations for the quantities $A$ are solved in the function

```
void calc_A(int lev, double p)
{
    int mm,ii,jj;
    double d;
    double B[noofequations][noofequations];
    double C[noofequations];
    for(mm=0;mm<=lev-1;mm++) {
        C[mm]=0.0;
        for(jj=0;jj<=lev-1;jj++) {
            C[mm]-=A[lev][lev]*A[mm][jj]
                            *innerprod(jj,p_opt[mm],v,identity,lev,p,v);
        }
        for(ii=0;ii<=lev-1;ii++) {
            B[mm][ii]=0.0;
            for(jj=0;jj<=lev-1;jj++) {
                B[mm][ii]+=A[mm][jj]*innerprod(jj,p_opt[mm],v,identity,ii,p,v);
            }
        }
    }
    solve_leqts_LU(B,C,lev);
    for(ii=0;ii<=lev-1;ii++) A[lev][ii]=C[ii];
}
```

where the function `solve_leqts_LU` solves the linear equations using LU decomposition. The expressions in Equation (13.87) are calculated using the function

```
double etrial(int lev, double p)
{
    double hm= (0.5*hPbar_prob*hPbar_prob/me_prob);
    double temp;
    if(lev!=0) calc_A(lev,p);
    temp=(hm*innerprod(lev,p,du,identity,lev,p,du)
          +innerprod(lev,p,u,potential,lev,p,u))
         /innerprod(lev,p,u,identity,lev,p,u);
    return temp;
}
```

Both this function, and the calculation of the matrix elements of Equations (13.85) and (13.86), are calculated using the simple trapezium integration routine in the function

```
double innerprod(int n1, double p1,double (*fn1)(int,double,double),
double (*fn)(double),
int n2, double p2,double (*fn2)(int,double,double))
{
    double x,sum=0.0;
    int i;
    for(i=1;i<gridx;i++) {
      x=i*dx;
      sum+=(*fn1)(n1,x,p1)*(*fn)(x)*(*fn2)(n2,x,p2);
    }
    sum+=0.5*((*fn1)(n1,xmax,p1)*(*fn)(xmax)*(*fn2)(n2,xmax,p2)
          +(*fn1)(n1,xmin,p1)*(*fn)(xmin)*(*fn2)(n2,xmin,p2));
    sum*=dx;
    return sum;
}
```

However, this routine should be replaced with one based on Simpson's rule. This integration can be made more specific if the particular trial functions are hard-wired into the programme. For example, using the functions defined in Equation (13.88), it is easily shown that

$$\left(v_{ni}^{(p_n)}, v_{mj}^{(p_m)}\right) = \frac{(i+j+2)!}{(p_n + p_m)^{i+j+3}}, \tag{13.92}$$

with similar expressions found for the inner products of the derivatives of these functions.

The programme was run for the square well potential which was specified in the function `potential()`, using the values

```
noofeigenvalues= 5;
xmin=   0.0;
xmax=  10.0;
gridx= 500;
V0= 10.0;
wellmin= 0.50;
wellmax= 5.50;
me_prob= 1.0;
hPbar_prob= 1.0;
```

The output of the programme was as follows:

Calculating 5 eigenvalues ..
Energy level E[0]=0.739854 (optimal pi_0=0.66934)
Energy level E[1]=3.045310 (optimal pi_1=0.8092)
Energy level E[2]=6.258380 (optimal pi_2=0.9091)
Energy level E[3]=8.367810 (optimal pi_3=0.84916)
Energy level E[4]=8.669500 (optimal pi_4=0.8092)

## 13.8 Numerical solution: the matrix method

The discretised one dimensional Schrödinger equation can be written in the form of
a matrix eigenvalue problem. Equation (13.72), written as

$$
E u_j = -\frac{\hbar^2}{2} \frac{(\mu_{j-1} + \mu_j)}{k_{j-1}(k_{j-1} + k_j)} u_{j-1}
$$

$$
+ \left[ \frac{\hbar^2}{2} \frac{1}{k_{j-1} + k_j} \left( \frac{\mu_{j-1} + \mu_j}{k_{j-1}} + \frac{\mu_j + \mu_{j+1}}{k_j} \right) + V_j \right] u_j
$$

$$
- \frac{\hbar^2}{2} \frac{(\mu_j + \mu_{j+1})}{k_j(k_{j-1} + k_j)} u_{j+1}, \tag{13.93}
$$

has the structure

$$
a_j u_{j-1} + b_j u_j + c_j u_{j+1} = E u_j \quad (j = 0, 1, 2, \ldots, N_y) \tag{13.94}
$$

where, for the internal points $j = 1, 2, \ldots, N_y - 1$, the coefficients are given by

$$
a_j \equiv -\frac{\hbar^2}{2} \frac{(\mu_{j-1} + \mu_j)}{k_{j-1}(k_{j-1} + k_j)} \tag{13.95}
$$

$$
b_j \equiv \frac{\hbar^2}{2} \frac{1}{k_{j-1} + k_j} \left( \frac{\mu_{j-1} + \mu_j}{k_{j-1}} + \frac{\mu_j + \mu_{j+1}}{k_j} \right) + V_j \tag{13.96}
$$

$$
c_j \equiv -\frac{\hbar^2}{2} \frac{(\mu_j + \mu_{j+1})}{k_j(k_{j-1} + k_j)}. \tag{13.97}
$$

This set of equations can be written in matrix form

$$
\mathbf{A} \mathbf{u} = E \mathbf{u} \tag{13.98}
$$

where $\mathbf{u} \equiv (u_1, u_2, \ldots, u_{N_y-1})^T$ is a column vector, and $\mathbf{A}$ is the $(N_y - 1) \times (N_y - 1)$
matrix

$$A = \begin{pmatrix} b_1 & c_1 & 0 & 0 & 0 & \cdots & 0 \\ a_2 & b_2 & c_2 & 0 & 0 & \cdots & 0 \\ 0 & a_3 & b_3 & c_3 & 0 & \cdots & 0 \\ 0 & 0 & a_4 & \ddots & c_4 & \cdots & 0 \\ \vdots & \vdots & \vdots & \cdots & \vdots & \vdots & \vdots \\ 0 & \cdots & \cdots & \cdots & \cdots & \cdots & \cdots \end{pmatrix}. \tag{13.99}$$

This matrix is tridiagonal, but is only symmetric if $a_{j+1} = c_j$ ($j = 1, 2, \ldots, N_y - 2$), or

$$k_j + k_{j+1} = k_{j-1} + k_j.$$

This condition is clearly satisfied when the mesh is uniform, in which case the problem reduces to that of finding the eigenvalues of a real symmetric tridiagonal matrix.

Even when this matrix is tridiagonal but not symmetric, it is still possible to use the same technique as for a symmetric matrix. This is possible because it can be seen from Equations (13.95) and (13.97) that the condition $c_{j-1}a_j > 0$ holds. That is, the matrix $A$ is *quasi-symmetric*. It is therefore possible to transform $A$ into a real symmetric matrix using a similarity transform (Wilkinson 1965). In order to do this, define a new diagonal matrix $D$ by

$$d_{11} \equiv 1, \qquad d_{ii} \equiv \left( \frac{a_2 a_3 \ldots a_i}{c_1 c_2 \ldots c_{i-1}} \right)^{\frac{1}{2}}.$$

Then the new matrix $T \equiv D^{-1}AD$ has the properties

$$T_{ii} = b_i, \qquad T_{i,i+1} = T_{i+1,i} = (c_i a_{i+1})^{\frac{1}{2}}.$$

It then follows that

$$T(D^{-1}u) = D^{-1}ADD^{-1}u = D^{-1}(Eu) = E(D^{-1}u).$$

Hence the required eigenvalues $E$ are also the eigenvalues of the symmetric matrix $T$. Once the eigenfunctions $w \equiv D^{-1}u$ of the matrix $T$ have been found, the eigenfunctions $u$ of the physical problem are found from the inversion $u = Dw$.

The eigenvalues can be found using a QL algorithm with implicit shifts. If these eigenvalues are then put into ascending order, we only need to solve for the first $L$ of the corresponding eigenfunctions for use in evaluating the electron density $n$ and the energy density $W$ (Cole et al. 1997). This can be done by solving the equations using the Newton method with an initial guess

$$(u_n)_j = \sin \left( \frac{n \pi j}{N_y} \right), \qquad (n = 0, \ldots, L - 1; \, j = 0, \ldots, N_y).$$

# Chapter 14
# Genetic algorithms and simulated annealing

Genetic algorithms (GA) allow a function, or set of functions, to be optimised using random processes to "breed" generations of solutions which should converge to the optimal solution (Man et al. 1999; Michalewicz 1996; Mitchell 1998). Many applications require either the maximisation or minimisation of a function. For example, in many fields of theoretical physics, a sum of least squares must be minimised, or the energy of a system must be minimised.

Many of the standard numerical techniques which exist for the optimisation of functions apply to functions which can be specified in a relatively straightforward manner. However, many of the functions which arise in the modelling of semiconductor devices cannot be specified in this manner, and some of these functions can only be evaluated using many subsidiary functions and integrals. The example which will be given in Sect. 14.7 regarding the approximation of the associated Fermi integrals is a case in point, for which the approach of the genetic algorithm method is the most appropriate one (Cole 2007).

It must be said that there is a widespread distaste for the use of the GA method. This is mainly due to the fact that the method relies on random throws of dice in order to produce good results—or to produce bad results on bad days. I regard the method as a poor substitute in situations in which other standard tried and tested methods are available. It certainly cannot be used to produce complete device simulations. However, there are times when tricky subsidiary optimisation problems cannot be dealt with using standard methods, and then the GA method comes into its own.

Very often, it is required to maximise a set of functions, or to find the values of the arguments which make the functions zero. Such a problem can be transformed into one of a minimisation process. For example, the function $f(x)$ can be maximised by minimising the function $-f(x)$, and the zero of the function $g(x)$ can be found by minimising the function $|g(x)|$, or by minimising the function $g(x)^2$. Further, when the topic of simulated annealing (SA) is discussed in Sect. 14.6, it will be required that a function falls into the lowest possible "groundstate". Whatever the problem, the resulting function to be minimised will be called the *fitness function*.

E.A.B. Cole, *Mathematical and Numerical Modelling of
Heterostructure Semiconductor Devices: From Theory to Programming*,
DOI 10.1007/978-1-84882-937-4_14, © Springer-Verlag London Limited 2009

Hence without loss of generality, only the minimisation process will be described in this Chapter.

Since the computer implementation of both GA and SA requires the generation of random numbers, it is unwise to rely on just one run of the programme. Several runs should be made, and the optimal fitness found from those of each run.

## 14.1 How genetic algorithms work

Suppose that it is required to find the optimal value of a function $f(x_1, \ldots, x_{num\_vars})$ of a set of variables $x_1, \ldots, x_{num\_vars}$. Each variable will represent a *gene* and, as described below, a particular value of that variable will be represented by a string of the bits 0 and 1. The genes are arranged in a particular order to form a *chromosome* and, in the context of our optimisation problem, a chromosome will represent a concatenated string of 0s and 1s formed from the individual genes. Hence a chromosome must contain the information regarding the ordering of the individual genes. In the case in which $f$ is a function of one variable $x_1$ only, then the chromosome is the gene itself.

A brief outline of the genetic approach will be illustrated with a simple example. A detailed explanation and justification will be left to later sections. Suppose that it is required find the value $x_{opt}$ in the range $\pi/2 < x_{opt} < \pi$ which minimises the function

$$f(x) = -x \sin x. \tag{14.1}$$

It is easily verified that the solution is $x_{opt} = 2.028757$, giving a minimum value of $f(x_{opt}) = -1.819706$.

The steps in the solution using a GA are as follows:

1. Decide on a precision, that is, the number of decimal places after the decimal point, for the solution. In the particular example discussed here, a precision of six decimal places will be required.
2. Decide on the search range. This is any range which is known to contain the solution, but the search for a solution will be more effective if the range is as small as possible. It will be shown in a later section how the range may be refined automatically. In the particular example discussed here, the search range $\pi/2 < x < \pi$ will be used.
3. Construct a binary representation based on the search range and the required precision. This will consist of a string of 0s and 1s, such that any combination will correspond to a value of $x$ in the search range for a given precision. For example, it will be shown how the chromosome

$$110111010011111110010$$

will correspond to the value $x = 2.249573$ with a precision of six decimal places in the search range $\pi/2 < x < \pi$, giving $f(2.249573) = -1.750936$. The construction will ensure that any random string of 0s and 1s of this length will correspond to a unique value of $x$ in the given search range. Note that the chromosome contains twenty one bits—this is its *length*. The length of a chromosome will depend on the required precision and on the span of the search range.

4. Construct an initial population of such chromosomes at random. This will be our initial *generation*. In our example, a low population of twenty chromosomes will be used, and this will remain the population size as subsequent generations are bred. Each chromosome will represent a different value of $x$ to six decimal places in the given range. In practice, a population of fifty—one hundred will always be more suitable.

5. This generation is now allowed to *breed* a new generation of twenty chromosomes. This is done using three genetic operators: *roulette wheel selection, crossover*, and *mutation*. First, in the roulette wheel selection process, a new generation of twenty chromosomes is selected at random from the original population, with the probability of selection of an individual chromosome being higher if the function value (fitness) of this chromosome is low. Hence, at the end of this stage, a new population has been selected which will probably contain copies of some individuals, and not contain others that were in the original population. Second, chromosomes are selected with a given fixed probability (the *crossover probability*). These selected chromosomes are paired randomly (having ensured that an even number are chosen), and each pair is allowed to produce a new pair by randomly choosing a common breakpoint and allowing them to exchange sub-chromosomes after this point. For example, a crossover after the sixth bit of the pair

$$001000010100101100001, 101011101101011011001$$

will produce a new pair

$$001000101101011011001, 101011010100101100001$$

Third, each bit of each chromosome is allowed to mutate, with a given fixed probability, (the *mutation probability*) into its inverse, so that a 1 can mutate into a 0, and a 0 into a 1. For example, mutations in the fourth and ninth bits of the chromosome

$$001101011000000110110$$

will produce the chromosome

$$001001010000000110110$$

These three processes go to make up the complete breeding step from one generation to the next. Some of the original chromosomes will have come through unchanged, some will have undergone crossover, and some will have had individual bits mutated. Each chromosome in the population will correspond uniquely to a value of $x$, and hence will correspond to a function value $f(x)$, or fitness. Successive generations are bred in this way, until some stopping criterion is met. For example, a maximum number of generations may be imposed, or the breeding process may be allowed to stop when the average fitness of a generation is

**Table 14.1** An initial population of twenty randomly generated chromosomes. The chromosomes are listed in column 2, the corresponding value of $x$ is shown in column 3, and the corresponding function value is shown in column 4.

| Number | Chromosome | $x$ | $f(x)$ |
|---|---|---|---|
| 1 | 00100100110101011011 | 1.796815 | −1.751116 |
| 2 | 01101110100111110010 | 2.249573 | −1.750936 |
| 3 | 00000001010001101100 | 1.574712 | −1.574700 |
| 4 | 00001001011000111100 | 1.628414 | −1.625711 |
| 5 | 01010110001111000101 1 | 2.099932 | −1.812753 |
| 6 | 10100010001111111111 | 2.566350 | −1.396194 |
| 7 | 01000001001010101011 1 | 1.970655 | −1.815202 |
| 8 | 00100001010010110000 1 | 1.775080 | −1.738170 |
| 9 | 10101110110101101 1001 | 2.643595 | −1.262758 |
| 10 | 00011011111101000000 0 | 1.742315 | −1.716749 |
| 11 | 11011000001010110010 0 | 2.897190 | −0.701053 |
| 12 | 01000011100010011110 0 | 1.985208 | −1.817167 |
| 13 | 11000111011110101001 1 | 2.794784 | −0.949942 |
| 14 | 00101101001101101001 1 | 1.848222 | −1.777553 |
| 15 | 11011110111001001001 | 2.938779 | −0.591947 |
| 16 | 01011111000110100010 0 | 2.154336 | −1.797831 |
| 17 | 01110100101100011010 0 | 2.286821 | −1.725226 |
| 18 | 00110101100000011011 0 | 1.899109 | −1.797673 |
| 19 | 11010010011011110101 1 | 2.862010 | −0.789786 |
| 20 | 10111100000010100011 1 | 2.724595 | −1.103507 |

close to the lowest fitness taken over preceding generations. In any event, it is always wise to impose a longstop maximum number of iterations.

A programme was run using the initial population of twenty randomly generated chromosomes shown in Table 14.1. In this table, the chromosomes are listed in column 2, the corresponding value of $x$ is shown in column 3, and the corresponding function value is shown in column 4. The first number column plays no part in the genetic process: it has been included in order to make reference to specific chromosomes. The result of breeding for five hundred generations is shown in Table 14.2. Note how all the function values have decreased so that they are much closer to the minimum value. Note that chromosome 7 has produced the smallest function value $f(2.027634) = -1.819704$ of this final generation. However, there is no guarantee that this is the smallest value that has been encountered through the generations. In fact, the even lower function value $f(2.028757) = -1.819706$ was encountered in generation 168: this is why the minimum value should be stored as the generations are bred.

The average function value has decreased from $-1.474799$ in the initial generation to $-1.731864$ in the final generation. In this sense, the "fittest" chromosomes have been bred using the genetic operators. In later sections, it will be seen that a breeding programme based on the average fitness of each generation will guarantee that an optimum population will be bred.

At first sight, it would seem that the processes of crossover and mutation would spoil the progress to an optimal solution. After all, what is the point of selecting the

**Table 14.2** The resulting population of twenty chromosomes after 500 generations.

| Number | Chromosome | $x$ | $f(x)$ |
|--------|------------|-----|--------|
| 1 | 01011000001100111111 | 2.112003 | −1.810172 |
| 2 | 00011000001011110111 | 1.719202 | −1.700305 |
| 3 | 01101010001011010111 | 2.222324 | −1.767099 |
| 4 | 01001010001001101101 | 2.025799 | −1.819694 |
| 5 | 00111000001011010111 | 1.915528 | −1.802830 |
| 6 | 00101001001111111101 | 1.823901 | −1.765791 |
| 7 | 01001010011100111111 | 2.027634 | −1.819704 |
| 8 | 00101000001101111111 | 1.817575 | −1.762510 |
| 9 | 01011010001011111101 | 2.124178 | −1.807149 |
| 10 | 01011000001001101111 | 2.111686 | −1.810245 |
| 11 | 01001000001100111111 | 2.013829 | −1.819405 |
| 12 | 10011101001001001110 | 2.535058 | −1.445043 |
| 13 | 01011010001001111101 | 2.123986 | −1.807200 |
| 14 | 01011001001011010111 | 2.118014 | −1.808732 |
| 15 | 01001000011011111111 | 2.015459 | −1.819467 |
| 16 | 00001000001011111111 | 1.621225 | −1.619164 |
| 17 | 00001010001001111101 | 1.633112 | −1.629942 |
| 18 | 00001000001100111101 | 1.621128 | −1.619075 |
| 19 | 00001010001011010111 | 1.633275 | −1.630089 |
| 20 | 00000000111011111111 | 1.573672 | −1.573665 |

fittest members at the selection process, only to introduce unfit ones? The problem is that an uninterrupted march to an optimal solution may lead us to a *local* optimum value, and not to the *global* one; the processes of crossover and mutation will lead us out of that danger.

There is no guarantee that a complete run will give anything like the optimum solution, since the whole process is governed by chance. A second run may produce a better optimum result. It very often happens that the optimum solution of a run is achieved after a relatively few generations. Code should be written so that a relatively small number of generations is taken in each run, and more than one run should be used. The overall optimum solution should then be taken as the optimum of the optimums of all of the runs.

The numerical processing of the algorithms involve applying the genetic operators to many chromosomes over many generations. This process is most efficiently done using parallel computing. However, here we demonstrate the coding of the algorithms using non-parallel methods.

## 14.2 Chromosome representation

A chromosome consists of a fixed-length string consisting of the bits 0 and 1. The string will uniquely represent a value of $x$ lying in the given search range for the variable. The length of the string is determined by the precision required of the variable, and the length of the search range. In the example of Sect. 14.1, the precision was specified as six decimal places and the search range ($\pi/2 < x < \pi$) had length $\pi - \pi/2 = \pi/2$. The length of the chromosome was calculated as *chromosome_length* = 21.

The chromosome structure is constructed so that *any* combination of 0s and 1s will give a value of $x$ to the given precision in the given spanning range. This construction is given as follows. In general, suppose that the search range of $x$ is specified as $x_{min} < x < x_{max}$, and that the required precision is $P$. Then the range must be split into, at least, a number $N$ intervals, where

$$N \geq N' \equiv (x_{max} - x_{min}) \times 10^P.$$

Therefore the integer length $L$ of the chromosome must be found such that

$$2^{L-1} < N' \leq 2^L.$$

In our example in which $x_{min} = \pi/2$, $x_{max} = \pi$ and $P = 6$, then $N' = 1570796.4$, and

$$2^{20} = 1048576 < 1570796.4 \leq 2097152 = 2^{21}.$$

Hence the chromosome length is $L = 21$. Any binary string of length $L$ will then correspond to a value of $x$ in the specified range to the specified precision.

All genetic operations are performed on the binary strings themselves, and not on the corresponding real values of $x$. Hence a function value must be found for any binary string. This is done in the following way. Let $C_{(2)}$ represent a binary string of length $L$. First, convert the binary string from base 2 into an integer $C_{(10)}$ in base ten. Then calculate the corresponding value of $x$ as

$$x = x_{min} + \frac{C_{(10)}}{2^L - 1}(x_{max} - x_{min}). \tag{14.2}$$

The function value will be calculated with this value of $x$. In our example, the conversion would be

$$x = \frac{\pi}{2} + \frac{C_{(10)}}{2097151}\frac{\pi}{2}.$$

Hence, for example, for the chromosome

$$C_{(2)} \equiv 010010101010001011010$$

would correspond to the value

$$C_{(10)} = 611418$$

giving a value of $x = 2.028757$ to 6 decimal places. The corresponding function value is then $f(2.028757) = -1.819706$.

The following function codes the process of finding the chromosome length $L$ when $x_{min}$, $x_{max}$ and *precision* are specified:

```
int chromosome_length;
unsigned two_mult;
```

```
void calc_numbits(void)
{
    int numbits=1;
    two_mult=2;
    double max_unsigned_range=(x_max - x_min);
    chromosome_length=0;
    for(int pr=1;pr<=precision;pr++) max_unsigned_range*=10;
    while(two_mult<max_unsigned_range) {numbits++; two_mult*=2;}
    chromosome_length+=numbits;
}
```

The following section of code produces a chromosome *chro* consisting of a string of randomly generated 0s and 1s.

```
int max_chro_len; // max chromosome length.
char chro[max_chro_len];
for(int c=0;c<chromosome_length;c++) chro[c]=((rand()%2)==0)? '0' : '1';
// don't forget to finish the string with this:
chro[chromosome_length]='\0';
```

The following function codes the process of transforming a chromosome *chro* into a real value of $x$ and then returning the function value for the function already specified as $f\_x(x)$:

```
double f_chro(char *chro)
{
    char *remainderPtr;
    unsigned x_uns;
    x_uns=strtoul(chro,&remainderPtr,2); // calculate unsigned value of x.
    x=x_min+x_uns*(x_max-x_min)/(two_mult-1); // calculate double value of x.
    return f_x(x);
}
```

The genetic process can be used to minimise a function in a given range. It can also be used to maximise a function $f(x)$ by minimising the function $-f(x)$. It can also be used to solve the equation $g(x) = 0$ by minimising the function $|g(x)|$. Whatever the case, the given function is transformed into a *fitness* function, such that it is the fitness function which has to be minimised. This is done using

$$fitness(chro) \equiv \begin{cases} +f\_chro(chro) & (\text{minimise}) \\ -f\_chro(chro) & (\text{maximise}) \\ |f\_chro(chro)| & (\text{zero}) \end{cases} \tag{14.3}$$

From this point on, only the minimisation of the fitness function will be discussed.

## 14.3 The genetic operators

The three genetic operators—roulette wheel selection, crossover, and mutation, have been introduced in Sect. 14.1. One breeding cycle involves the three operators taken in that order. Of the three, the roulette wheel selection is the only one which selects the fittest chromosomes for breeding. The other two produce random changes in those selected. The operators will be discussed initially for the case of one function

$f$ which depends only on one variable $x$. The multi-function and multi-variable cases will be described in Sect. 14.4.

## 14.3.1 Stage 1: Roulette wheel selection

The case of the minimisation of a negative fitness $f(x)$ will be considered first. Then in this special case, the function $g(x) \equiv -f(x)$ is positive. In fact, for any function $f(x)$, the function $g(x)$ can be made positive by adding a sufficiently large positive constant to it. It will be shown how to do this in a consistent manner later.

Let *POP_SIZE* be the fixed population size in any generation. Imagine a straight line which is split into a number *POP_SIZE* unequal intervals, one interval for each chromosome of the population. The length of the $i$th interval has the fitness value $-f(x_i)$ of the $i$th chromosome—this is why a negative fitness is (temporarily) required. The total length of the line is

$$len \equiv - \sum_{j=1}^{POP\_SIZE} f(x_j). \qquad (14.4)$$

Next, define the *cumulative probability* for the $i$th chromosome as

$$p_{cum}(i) \equiv - \frac{\sum_{j=1}^{i} f(x_j)}{len} \qquad (14.5)$$

so that, in particular, $p_{cum}(POP\_SIZE) = 1$. Next, generate a random number *ran* in the range $0 < ran \leq 1$, and then do a sweep through the population until a chromosome $j$ is found such that

$$p_{cum}(j-1) < ran \leq p_{cum}(j). \qquad (14.6)$$

Then chromosome $j$ is selected for inclusion in the next stage. Note that the range in Equation (14.6) is relatively large if the fitness value of chromosome $j$ is small. Hence there is a larger probability that a chromosome with a smaller fitness will be selected. This selection process is done exactly *POP_SIZE* times. The resulting group of *POP_SIZE* chromosomes will consist of a subset of the original ones: some (most probably those with the lowest fitness) will be selected more than once, while others may not be selected at all. The notion of a straight line has been used to describe the process; the more usual picture is that this line is actually the circumference of a roulette wheel, and the *POP_SIZE* selections correspond to that number of separate spins of the wheel.

Since the segments of the straight line (or circumference of the roulette wheel) must be positive in order to calculate the cumulative probability, positive *working* fitness values must be obtained from the fitness values. If the fitness is negative in the given range, then the working fitness values can be taken as the negative of

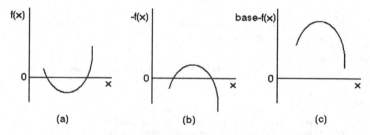

**Fig. 14.1** The steps involved in constructing a working fitness for calculating the cumulative probability: (a) the original function to be minimised, (b) the negative of the working function, and (c) the addition of *base* to produce a positive value.

the fitness values themselves. If not all of the fitness values in the given range are negative, then a sufficiently large positive constant may be added to each negative of the fitness value in order to make it positive. However, if the added constant is too large, then there will be only a small relative difference between the smallest and largest working fitness values: this will cause the selection process to operate too slowly. On the other hand, if the smallest working fitness value is close to zero, then the relative difference between the smallest and largest working fitness values will be large: this could give undue prominence to the larger values, causing the selection process to be too severe. This could cause the intermediate population to be composed almost entirely of chromosomes corresponding to the lowest fitness. On the face of it, this seems to be the situation for which we are aiming. However, this situation could cause the programme to converge too severely on a local minimum rather than on the global minimum.

This problem can be resolved in the following way. Find the minimum and maximum values $F_{min}$ and $F_{max}$ of the fitness for the intermediate population, and then add $F_{max}$ to each value of the negative of the fitness. This provides a set of positive values from which to calculate the cumulative probabilities, but note that at least one of these values will be zero, thus causing the selection process to be too severe. In order to avoid this situation, add an additional positive constant to each value which is some pre-determined proportion *fitness_factor* of the difference between $F_{min}$ and $F_{max}$. Hence for each population, subtract each fitness from a constant value *base* defined by

$$base \equiv F_{max} + fitness\_factor * (F_{max} - F_{min}). \tag{14.7}$$

A sensible choice of the quantity *fitness_factor* will make the selection process neither too severe nor too ineffective. Note that the quantity *base* is a constant to be added to the negative of the fitness of each population member of a generation, but it will vary between generations. Fig. 14.1 illustrates the steps in this process.

The following section of code shows how to calculate the cumulative probability for a negative fitness, and how to use the results of the calculation in the first stage in calculating the next generation.

```
/ ********************************************************************
double FITNESS[POP_SIZE+1];
double p_cumulative[POP_SIZE+1];
```

```
const double fitness_factor = 1.0;
/ ****************************************************************
void calc_cumulative_probability(void)
{
  double fitness_sum=0.0;
  double F_min=34567.8, F_max=-34567.8; // something big and silly.
  double base;

  // calculate base:
  for(int person=1;person<=POP_SIZE;person++) {
    if(FITNESS[person]<F_min) F_min=FITNESS[person];
    if(FITNESS[person]>F_max) F_max=FITNESS[person];
  }
  base=F_max+fitness_factor*(F_max-F_min);
  // calculate (unnormalised) cumulative probability:
  p_cumulative[0]=0.0;
  for(int person=1;person<=POP_SIZE;person++) {
    fitness_sum+=(base-FITNESS[person]);
    p_cumulative[person]=fitness_sum;
  }
  // now normalise cumulative probability:
  for(int person=1;person<=POP_SIZE;person++){
    p_cumulative[person]/=fitness_sum;
  }
}
// ****************************************************************
void breed_next_generation(void)
{
  char CHROMOSOME_temp[POP_SIZE+1][max_chro_len];
  double ran;

  // STAGE 1: roulette wheel
  // -----------------------
  calc_cumulative_probability();
  for(int person=1;person<=POP_SIZE;person++) {
    ran=0.0001*(rand()%10001);;
    for(int per=1;per<=POP_SIZE;per++) {
      if((p_cumulative[per-1]<ran)&&(p_cumulative[per]>=ran)) {
        strcpy(CHROMOSOME_temp[person],CHROMOSOME[per]);
      }
    }
  }
  for(int person=1;person<=POP_SIZE;person++){
    strcpy(CHROMOSOME[person],CHROMOSOME_temp[person]);
  }

  // STAGE 2: crossover
  // ------------------
  if(CROSSOVER) {
    // .........................................
  }
  // STAGE 3: mutation
  // -----------------
  if(MUTATION) {
    // .........................................
  }
}
```

## 14.3.2 Stage 2: Crossover

The chromosomes are numbered as CHROMOSOME[i] ($i = 1, \ldots, POP\_SIZE$), but there is no significance about the ordering: the numbering is there purely to aid the book-keeping which enables us to keep track of the chromosomes selected for crossover. A pre-determined crossover probability *p_crossover* is chosen. For each chromosome in the intermediate population selected during the roulette stage, a random number *ran* such that $0 < ran \leq 1$ is generated. If $ran < p\_crossover$, then this chromosome is selected. At the end of the process, a subset of the chromosomes has been selected. An even number of chromosomes is needed for crossover: if an odd number has been selected, then the last one selected is ignored. The resulting even number of chromosomes are paired off at random, and the unselected chromosomes go through unchanged to the mutation stage. For each pair of selected chromosomes, a random integer *pos* is generated in the range

$$1 \leq pos < chromosome\_length, \tag{14.8}$$

and the pair undergoes crossover after position *pos*.

The value of *p_crossover* is important in the efficiency of the scheme. Unfortunately, there is no foolproof method for determining its value in advance, and it is often necessary to determine its value by trial and error for each application. Michalewicz (1996) uses the value 0.250 with a population size of 20. Man et al. (1999) use a value based on the value of *POP_SIZE*: their *p_crossover* decreases linearly from a value of 0.900 when *POP_SIZE* = 30 to a value 0.600 when *POP_SIZE* = 100. Mitchell (1998) presents a discussion of the problems involved in selecting a suitable value.

The following section of code programmes the crossover routines (the mutation probability *p_mutation* needed next is also introduced at this stage):

```
// *******************************************************************
#define NO    0
#define YES   1
#define CROSSOVER           YES
const double p_crossover_min=0.600;
const double p_crossover_max=0.900;
#define MUTATION            YES
const double p_mutation_min=    0.001;
const double p_mutation_max=    0.010;
// *******************************************************************
void calc_transition_probabilities(void)
{
    double min,max;
    // Uses scheme of Man et al, with
    // min values at POP_SIZE=100, max values at POP_SIZE=30.
    min=p_mutation_min; max=p_mutation_max;
    p_mutation=(POP_SIZE>100)? min :
              (POP_SIZE<30)? max   : min+0.0142*(max-min)*(100-POP_SIZE);
    min=p_crossover_min; max=p_crossover_max;
    p_crossover=(POP_SIZE>100)? min :
              (POP_SIZE<30)? max   : min+0.0142*(max-min)*(100-POP_SIZE);
}
// *******************************************************************
void crossover(int crossover_position, char *chro1, char *chro2)
{
    char chro1_temp[max_chro_len];
    strcpy(chro1_temp,chro1);
    for(int c=0;c<=crossover_position-1;c++) {
        chro1[c]=chro2[c]; chro2[c]=chro1_temp[c];
    }
}
```

```
// ********************************************************************
void breed_next_generation(void)
{
   char CHROMOSOME_temp[POP_SIZE+1][max_chro_len];
   double ran;

   // STAGE 1: roulette wheel
   // ----------------------

   // STAGE 2: crossover
   // -----------------
   if(CROSSOVER) {
     int crossover_counter=0;
     int cross_index[POP_SIZE+1];
     int pos;
     for(int person=1;person<=POP_SIZE;person++) {
       ran=0.0001*(rand()%10001);
       if(ran<p_crossover) {
         crossover_counter++;
         cross_index[crossover_counter]=person;
       }
     }
     if((crossover_counter%2)==1) crossover_counter-=1; //get even number
     // now do crossovers:
     if(crossover_counter!=0) {
       for(int i=1;i<crossover_counter;i+=2) {
       pos=1+(rand()%(chromosome_length-1));
         crossover(pos,CHROMOSOME[cross_index[i]],
           CHROMOSOME[cross_index[i+1]]);
     }
   }
   }
   // STAGE 3: mutation
   // -----------------
   if(MUTATION) {
     // .............................
   }
}
// ********************************************************************
```

### 14.3.3 Stage 3: Mutation

In the basic implementation of the genetic scheme described here, the mutation stage is the most straightforward to implement. However, as will be seen later, there are modifications to the mutation scheme which will increase the efficiency of the genetic process.

A pre-determined mutation probability *p_mutation* is set. Then for every bit of every chromosome of the intermediate population resulting from the crossover stage, a random number *ran* in the range $0 < ran \leq 1$ is generated. If $ran \leq$ *p_mutation*, then that bit is mutated into its opposite bit. Again, the value of *p_mutation* is important in the efficiency of the scheme: but again, there is no foolproof method for determining its value, and it is often necessary to determine its value by trial and error for each application. Michalewicz (1996) uses the value 0.010 with a population size of 20 with no stated justification. Man et al. (1999) use a value based on the value of *POP_SIZE*: their *p_mutation* decreases linearly from a value of 0.010 when *POP_SIZE* = 30 to a value 0.001 when *POP_SIZE* = 100.

The following code programmes the mutation routines (the function to set the mutation probability has been described in the crossover section):

```
// *******************************************************************
void mutate(int mutation_position, char *chro)
{
  char c=chro[mutation_position-1];
  chro[mutation_position-1]=(c=='1')? '0' : '1';
}
// *******************************************************************
void breed_next_generation(void)
{
  char CHROMOSOME_temp[POP_SIZE+1][max_chro_len];
  double ran;

  // STAGE 1: roulette wheel
  // -----------------------

  // STAGE 2: crossover
  // ------------------
  if(CROSSOVER) {
    // .......................
  }
  // STAGE 3: mutation
  // -----------------
  if(MUTATION) {
    int ch;
    double mutation_prob;
    for(int person=1;person<=POP_SIZE;person++) {
      for(int c=0;c<chromosome_length;c++) {
        ran=0.0001*(rand()%10001);
        if(ran<=p_mutation) {mutate(c,CHROMOSOME[person]);}
      }
    }
  }
}
// *******************************************************************
```

Note that the same mutation probability has been applied to every bit of every chromosome. However, it is obvious that a mutation of a bit lying towards the left end of a chromosome will produce a larger change to the corresponding real value of $x$ than a mutation of a bit lying towards the right end of the chromosome. It will be seen later how a differential mutation probability can be introduced.

## 14.4 The multivariable and multifunction cases

The GA process has been described in relation to the optimisation of one function of one variable. However, it is a straightforward matter to extend to the case in which the function to be optimised is a function of several variables, or to the case in which a number of functions of several variables have to be made zero.

## 14.4.1 The multivariable case

The case of a function $f(x)$ of a single variable $x$ has been considered, and the
GA approach is easily extended to the case of finding the minimum of a function
$f(x_1, \ldots, x_{num\_vars})$ of $num\_vars$ variables $\{x_v : v = 1, \ldots, num\_vars\}$. Now, a
range and a precision must be specified for each $x_v$, and the gene length $numbits[v]$
must be calculated for each variable $v$. The chromosome population member corresponding to the vector $(x_1, \ldots, x_{num\_vars})$ is then simply the concatenation of the
genes for each variable. The total chromosome length for this population member
will be

$$chromosome\_length$$
$$\equiv \sum_{v=1}^{num\_vars} numbits[v]. \qquad (14.9)$$

The function `calc_numbits()` of Sect. 14.2 is generalised as follows:

```
// ****************************************************************
const int max_num_vars=5;  // something large enough
int prec[max_num_vars+1];
int numbits[max_num_vars+1];
unsigned two_mult[max_num_vars+1];
double x_min[max_num_vars+1],x_min[max_num_vars+1];
int chromosome_length;
// ****************************************************************
void calc_numbits(void)
{
  double max_unsigned_range;
  chromosome_length=0;
   for(int v=1;v<=num_vars;v++) {
     max_unsigned_range=(x_max[v]-x_min[v]);
     two_mult[v]=2;
     numbits[v]=1;
     for(int pr=1;pr<=prec[v];pr++) max_unsigned_range*=10;
     while(two_mult[v]<max_unsigned_range) {numbits[v]++; two_mult[v]*=2;}
     chromosome_length+=numbits[v];
   }
}
// ****************************************************************
```

When evaluating the fitness of the compound chromosome, it must be decomposed
into its $num\_vars$ separate genes: this is a straightforward task since the length
$numbits[v]$ of each gene is known. The function `f_chro()` of Sect. 14.2 is generalised as follows:

```
// ****************************************************************
double f_chro(int fn_number, char *chro)
{
  int start_c=0;
  char *remainderPtr;
  unsigned x_uns[max_num_vars+1];
  // decompose chromosome into bit strings for each x:
  for(int v=1;v<=num_vars;v++) {
     for(int c=0;c<numbits[v];c++) x_string[v][c]=chro[start_c+c];
     x_string[v][numbits[v]]='\0';
     start_c+=numbits[v];
  }
```

```
// calculate the unsigned value for each x:
for(int v=1;v<=num_vars;v++) x_uns[v]=strtoul(x_string[v],&remainderPtr,2);
// calculate double values for each x:
for(int v=1;v<=num_vars;v++)
  x[v]=x_min[v]+x_uns[v]*(x_max[v]-x_min[v])/(two_mult[v]-1);
return f_x(fn_number,x);
}
// ****************************************************************
```

The mutation and crossover operations are then applied to the compound chromo-some, and not to the sub-chromosomes (genes) separately. However, we will see in a later section that we may need to apply a differential mutation probability, with a smaller probability applying to the bits towards the left of each sub-chromosome.

## 14.4.2 The multifunction case

The case of the minimisation of a single function $f$ has already been considered. However, there may be situations in which we need to find the set of variables $x_v$ which zeros a set of functions. The extension is straightforward, and is described in the function fitness() below. Essentially, if the set of functions is $f_F$ ($F = 1, \ldots, num\_funs$), then the working fitness of a chromosome $chro$ is taken as

$$fitness[chro] \equiv \pm \sum_{F=1}^{num\_funs} |f_F(chro)| \qquad (14.10)$$

where the appropriate sign is taken depending on the optimisation task to be per-formed. Alternative forms of this fitness function are

$$\pm \sum_{F=1}^{num\_funs} |f_F(chro)|^2 \quad \text{and} \quad \pm \sum_{F=1}^{num\_funs} \sqrt{|f_F(chro)|}.$$

The following function codes this (note that the function f_chro() will need to be slightly modified):

```
// ****************************************************************
const int zero=0;
const int minimise=1;
const int maximise=3;
inline double task_index(int tsk){return (tsk==3)? -1.0 : 1.0;}
int num_funs;
// ****************************************************************
double fitness(char *chro)
{
  double sum=0.0;

  if(task==zero) { //zeros of a set of functions.
    for(int f=1;f<=num_funs;f++) {
      sum+=fabs(f_chro(f,chro));
      // or\ldots sum+=fabs(f_chro(f,chro))*fabs(f_chro(f,chro));
      // or\ldots sum+=sqrt(fabs(f_chro(f,chro)));
    }
  }
```

```
     else { //optimisation of a set of functions.
       sum=f_chro(1,chro);
     }
     return task_index(task)*sum;
   }
   // ************************************************************
```

### 14.4.3 Example: maximising a function of two variables

To illustrate the use of using several variables when maximising a function, we take the example given by Michalewicz (1996). The function to be maximised is

$$f(x_1, x_2) = 21.5 + x_1 \sin(4\pi x_1) + x_2 \sin(20\pi x_2). \tag{14.11}$$

The data for the problem is set in the following functions:

```
// ************************************************************
void set_problemdata(void)
{
  strcpy(problem_name,"Example1");  // Example of Michalewicz
  task=maximise;
  num_vars=2;   // must be <= max_num_vars
  num_funs=1;          // must be <= max_num_funs
  x_min[1]= -3.00;    x_max[1]= 12.10;    prec[1]=4;
  x_min[2]=  4.10;    x_max[2]=  5.80;    prec[2]=6;
}
// ------------------
double f_x(int fn_number, double *x)
{
  double fn_temp[max_num_funs+1];
  fn_temp[1]=21.5+x[1]*sin(4.0*PI*x[1]) + x[2]*sin(20.0*PI*x[2]);
  return fn_temp[fn_number];
}
// ************************************************************
```

A run of the programme produced the following results. The optimal fitness was achieved after 262 generations, and produced a maximum function value of 38.791084. The optimal chromosome was

$$1111011111100110011111101000110010111011$$

which decomposed into the genes

$$1111011111100110011 \quad \text{and} \quad 11110100011001011011.$$

The corresponding unsigned coordinates were 253849 and 2002107, and the corresponding coordinate values were $x_1 = 11.6222$ and $x_2 = 5.722955$. Note how the first gene is shorter than the second: this is partly because the precisions of $x_1$ and $x_2$ have been differently specified as four and six respectively in order to illustrate the process.

## 14.5 Refinements to the GA approach

The implementation of the GA routines involves the generation of random numbers. Hence there is no way of knowing if the routines will produce a sensible answer to a given problem. However, there are some methods whereby this uncertainty can be reduced. One way is to use *differential mutation* in which the probability of bit mutation depends on its position along the chromosome length. A second way is to use *contractive mapping*, which can make it more certain, but not always, that an optimal solution is being obtained. A third method is to use a preliminary solution to reduce the search range, hence speeding up the routines.

### 14.5.1 Differential mutation

A mutation on a bit which lies towards the left of a chromosome produces a much more significant change to the corresponding coordinate $x$ than a bit which lies further to the right. Consequently, when the genetic process has run a considerable way through and the optimum solution is close, we will probably not need to make a severe change to the solution already obtained. It would then be advantageous to have a smaller mutation probability acting on leftward bits, and a comparatively larger mutation probability acting on the rightward bits. A simple solution would be to specify a small value of the mutation probability at the left end of the chromosome, a larger value to apply at the right end, and a linear variation from one end to the other.

One complication in the multivariable case is that this variation (linear or otherwise) should not apply across the whole chromosome, but should apply to each of the decomposed genes separately. Again, choice of the probability values will be important, and I know of no way of producing them automatically. Implementation of this approach will be most useful when applied to the contractive mappings discussed in the next section, but the coding of this differential mutation is given in readiness. It involves a modification of the mutation stage in the function `breed_next_generation()`. Rather than specifying a minimum and maximum mutation probability, a maximum value is specified together with a factor *p_mutation_fac* $<= 1$ such that the mutation probability is reduced by this factor at the left end. If *p_mutation_fac* $= 1$, then this implies that a uniform mutation probability is taken across the chromosome. In the following section of code, the value *p_mutation_fac* $= 0.01$ has been taken. This means that the mutation factor will vary linearly from $0.01 \times p\_mutation$ at the left end of each gene to *p_mutation* at the right end. In general, there is no known method of determining what this factor should be.

```
// *****************************************************************
const double p_mutation_fac=0.010;   // (0<p<=1)
void breed_next_generation(void)
{
   double ran;
```

```
// STAGE 1: roulette wheel
// ----------------------

// STAGE 2: crossover
// -----------------
if(CROSSOVER) {
   // ..................................
}
// STAGE 3: mutation
// ----------------
if(MUTATION) {
  int ch;
   double p_m;
   for(int person=1;person<=POP_SIZE;person++) {
     for(int v=1;v<=num_vars;v++) { // split into genes
       for(int c=1;c<=numbits[v];c++) {
         ch=c;
         for(int v1=1;v1<v;v1++) {ch+=numbits[v1];}
         ran=0.0001*(rand()%10001);
         p_m=p_mutation_fac*p_mutation
                +(1.0-p_mutation_fac)*(c-1)*p_mutation/(numbits[v]-1);
         if(ran<=p_m) {mutate(ch,CHROMOSOME[person]);}
       }
     }
   }
}
// *********************************************************************
```

### 14.5.2 Contractive mapping

The GA approach would be greatly enhanced if it could be ensured that the scheme
would converge to the optimal solution. The contractive mapping scheme ensures
that there is a good chance that it will do so. It works on the basis that if a sequence
of generations can be produced such that the *average* fitness improves from one generation to the next, then the process will converge to the optimum fixed point. Only
the computer implementation of the contractive mapping scheme will be presented
in this Chapter. The theory behind the process is given in Appendix A.

Let *POP_SIZE* be the fixed population size of each generation. Let $C$ be a chromosome of a particular population $P$. Let $F(C)$ be the fitness of chromosome $C$,
and let $F_A(P)$ be the average fitness of the population:

$$F_A(P) \equiv \frac{1}{POP\_SIZE} \sum_{C \in P} F(C). \qquad (14.12)$$

Contractive mapping is implemented as follows. Starting with a newly-bred generation, breed the next generation. If the average fitness of the new generation is not
greater than that of its predecessor, then accept that generation in place of the old.
But if the fitness average of the new generation is larger than that of its predecessor,
then kill off the new generation and re-generate, if necessary repeating the kill until
a generation with lower average fitness is produced.

One obvious problem with this method is that it is possible to get into an unacceptably long loop of kills. To overcome this, a maximum number of kills must be imposed on the killing process. If this number is attained, then we must accept the last (unsatisfactory) generation in order to get out of the loop, even though it means that we have ended up with a higher fitness average than its predecessor. This process can be combined with a differential mutation probability. On the first breed from a new generation, a uniform mutation probability is used. If this new breed needs to be killed off, then we try a new breed using a differential mutation probability. If a generation is successfully bred with higher fitness average, then the uniform mutation probability is restored.

The mutation stage of the function breed_next_generation() must be modified as follows to include contractive mapping. The factor *p_mutation_factor* is changed to *p_mutation_killfactor* to reflect the fact that differential mutation is applied only when kills have been made.

```
// ******************************************************************
const double p_mutation_killfac=0.010;   // (0<p<1)
void breed_next_generation(void)
{
    char CHROMOSOME_temp[POP_SIZE+1][max_chro_len];
    double ran;

    // STAGE 1: roulette wheel
    // ----------------------

    // STAGE 2: crossover
    // -----------------
    if(CROSSOVER) {
        // .............................
    }
    // STAGE 3: mutation
    // ----------------
    if(MUTATION) {
        int ch;
        double p_m;
        for(int person=1;person<=POP_SIZE;person++) {
            for(int v=1;v<=num_vars;v++) {
                for(int c=1;c<=numbits[v];c++) {
                    ch=c;
                    for(int v1=1;v1<v;v1++) {ch+=numbits[v1];}
                    ran=0.0001*(rand()%10001);
                    p_m=(kill_ctr==0)? p_mutation:
                                       p_mutation_killfac*p_mutation
                                       +(1.0-p_mutation_killfac)*(c-1)
                                       *p_mutation/(numbits[v]-1);
                    if(ran<=p_m) {mutate(ch,CHROMOSOME[person]);}
                }
            }
        }
    }
}
// ******************************************************************
```

### 14.5.3 Range refinement

When setting up the data for a problem, a search range of values must be specified for each variable. Each range must enclose the position where we think the optimum value of the corresponding variable is likely to be. If it is possible to obtain preliminary results using these initial guesses, then it should be possible to refine each range so that subsequent runs will further pin down the optimum results. There are two main advantages of doing this. First, time is not wasted in looking at values of the variables that are too far away from the optimum. Second, the length of each chromosome depends both on the specified precisions and on the lengths of the search ranges; cutting down on the search ranges will generally produce a reduction in the required chromosome length. This will reduce the simulation time, since fewer chromosome bits will be processed.

The steps involved in restricting the range of a variable $x$ are as follows:

1. Obtain a preliminary optimal value $x_{opt}$ for the variable using the initial guess for the search range $x_{min} \le x \le x_{max}$ of the variable.
2. Obtain the *range radius* $R_{old}$ which is the minimum of the values $x_{max} - x_{opt}$ and $x_{opt} - x_{min}$:

$$R_{old} \equiv \min(x_{max} - x_{opt}, x_{opt} - x_{min}).$$

3. Choose a restriction factor with $0 < restriction\_factor \le 1$ and reduce the radius to the value $R_{new} = R_{old} \times restriction\_factor$.
4. Take a new search range $(x_{opt} - R_{new}) \le x \le (x_{opt} + R_{new})$.
5. Re-run the programme with this new range.

Therefore the new search range will be centered on the preliminary value $x_{opt}$ and have a span less than the initial range. The following section of code shows how this is programmed.

```
// *********************************************************************
inline double min(double a, double b){return (a<b)? a : b;}
const double restriction_factor = 0.10;
// *********************************************************************
void restrict_ranges(void)
{
  double fts=fitness(opt_opt_chromosome);
  double radius;
  for(int v=1;v<=num_vars;v++) {
    radius=min(x[v]-x_min[v],x_max[v]-x[v]);
    radius*=restriction_factor;
    x_min[v]=x[v]-radius; x_max[v]=x[v]+radius;
  }
}
// *********************************************************************
```

## 14.6 Simulated annealing

Simulated annealing (SA) is a method which allows the required global minimum solution to be approached in a steady, if unspectacular, manner. Just like the basic genetic approach itself, the simulated annealing method mimics natural physical processes. For example, consider the formation of a crystal from a melt. If the melt is cooled too rapidly (the process of *quenching*) then the crystal may form which is not in its lowest energy state, and will have defects in its lattice structure: the disorder of the melt becomes frozen into the crystal. On the other hand, if the temperature is lowered slowly (the process of *annealing*), then the crystal is more likely to end up in its stable state of lowest energy. Full descriptions of the SA process are given by Johnnson et al. (1987), and Laarhoven and Aarts (1987).

### 14.6.1 How simulated annealing works

An artificial temperature $T$ is introduced in the SA process for any minimisation problem. When this temperature is reduced, then the processes of mutation and crossover will be slowed down. The three main challenges of the SA scheme are

- how to quantify the slowing in the genetic processes in terms of the temperature;
- How to choose initial and final temperatures, and how to reduce the temperature smoothly as new generations are produced (the *cooling schedule*).

The first challenge is addressed by linking the mutation and crossover transition probabilities to the distributions of statistical mechanics which were introduced in Chapter 3. The precise form of this quantification will depend on the particular statistics $S$ which is used. In later sections, the three different statistics of Maxwell-Boltzmann (MB), Tsallis (TS), and Fermi-Dirac (FD) will be discussed in detail.

Using the results of statistical mechanics, the particular statistics $S$ will be used to produce an *acceptance function* $P_s(E, \Delta E, T)$, where $E$ is the energy, $\Delta E$ is an increment in the energy, and $T$ is the temperature of the system. At this stage, all that is required of the acceptance function $P_s$ is that it be positive, and that $P_s$ decreases as the temperature $T$ decreases if $\Delta E > 0$:

$$P_s(E, \Delta E, T) > 0 \tag{14.13}$$

$$\frac{\partial}{\partial T} P_s(E, \Delta E, T) > 0 \quad \text{for } \Delta E > 0. \tag{14.14}$$

Particular forms of the function $P_s(E, \Delta E, T)$ will be discussed later but, for example, it will be found that the function has the form $P_{MB}(E, \Delta E, T) = e^{-\Delta E/(k_B T)}$ when MB statistics are used.

Since the natural process of annealing involves a reduction in energy with a reduction in temperature, it will be a natural step in SA to replace the energy with the function $f$ which is being minimised, and to replace the natural temperature with an artificial temperature:

$$P_s(f, \Delta f, T) > 0 \tag{14.15}$$

$$\frac{\partial}{\partial T} P_s(f, \Delta f, T) > 0 \quad \text{for } \Delta f > 0. \tag{14.16}$$

Once the particular statistics $S$ has been decided upon, the acceptance function $P_S$ must be used in connection with the GA processes of mutation and crossover. In the basic mutation process described in Sect. 14.3, every bit of every chromosome is visited in turn. For each bit, a random number in the range $[0, 1]$ is generated, and if that number is less than the prescribed mutation probability, then that bit is flipped. This procedure is modified in the SA process as follows: for each bit of each chromosome,

1. Flip the bit, and compare the fitness $f$ of the original chromosome with the fitness $f + \Delta f$ of the new chromosome.
2. If $\Delta f \leq 0$ then accept the new chromosome, and move on to the next bit.
3. If $\Delta f > 0$ then the new chromosome may still be accepted: generate a random number in the range $[0, 1]$, and if this number is less than the acceptance function $P_s(f, \Delta f, T)$ then the new chromosome is accepted. Otherwise retain the original chromosome, and move on to the next bit.

The temperature is lowered steadily as new generations of chromosome are produced, and hence the likelihood of accepting unsuitable chromosomes is reduced.

The following section of code shows how the process of simulated annealing can be applied to mutation.

```
// ****************************************************************
anneal_Temp*=anneal_Temp_decrement_fac; //do this before each generation.
// ****************************************************************
double P_s(double f, double df, double Temp)
{
    // acceptance function for MB statistics:
    return exp(-df/Temp);
}
// ****************************************************************
void breed_next_generation(void)
{
    double ran;

    // STAGE 1: roulette wheel
    // ----------------------

    // STAGE 2: crossover
    // -----------------

    // STAGE 3: mutation
    // ----------------
    if(MUTATION) {
        char chro_new[max_chro_len];
        int ch;
        double fitness_new,p_statistics,df;
        for(int person=1;person<=POP_SIZE;person++) {
            for(int v=1;v<=num_vars;v++) {
                for(int c=1;c<=numgenes[v];c++) {
                    ch=c;
                    for(int v1=1;v1<v;v1++) {ch+=numgenes[v1];}
                    strcpy(chro_new,CHROMOSOME[person]);
                    mutate(ch,chro_new);
                    fitness_new=fitness(chro_new);
                    df=fitness_new-FITNESS[person];
                    if(df<=0.0) {
                        strcpy(CHROMOSOME[person],chro_new);
                        FITNESS[person]=fitness_new;
                    }
                    else {
```

```
            p_statistics=P_s(FITNESS[person],df,anneal_Temp);
            ran=ran_01();
            if(ran<p_statistics) {
              strcpy(CHROMOSOME[person],chro_new);
              FITNESS[person]=fitness_new;
            }
          }
        }
      }
    }
  }
}
// ******************************************************************
```

There is no point in applying simulated annealing to mutation without at the same time applying it to crossover. Select pairs for crossover as before. For each selected pair, calculate their fitnesses before and after crossover. This time, let $f_{new}$ be the average of the fitnesses of the pair after crossover, and let $f_{old}$ be the average before crossover. If $\Delta f \equiv f_{old} - f_{new} \leq 0$, then accept the new pair. But if $\Delta f > 0$, accept the pair only if a randomly generated number in the range [0, 1] is less than the acceptance function $P_s(f, \Delta f, T)$. An alternative to using the average would be to use maximum values, although this would probably make the testing condition more severe. The following section of code shows how simulated annealing can be applied to crossover:

```
// ******************************************************************
void breed_next_generation(void)
{
  double ran;

  // STAGE 1: roulette wheel
  // -----------------------

  // STAGE 2: crossover
  // ------------------
  if(CROSSOVER) {
    char chro_new1[max_chro_len],chro_new2[max_chro_len];
    double fitness_new1,fitness_new2,p_statistics,df;
    int crossover_counter=0;
    int cross_index[POP_SIZE+1];
    int pos;
    double test_new,test_old;
    for(int person=1;person<=POP_SIZE;person++) {
      ran=ran_01();
      if(ran<p_crossover) {
        crossover_counter++;
        cross_index[crossover_counter]=person;
      }
    }
    if((crossover_counter%2)==1) crossover_counter-=1;
    // now do crossovers:
    if(crossover_counter!=0) {
      for(int cc=1;cc<crossover_counter;cc+=2) {
        strcpy(chro_new1,CHROMOSOME[cross_index[cc]]);
        strcpy(chro_new2,CHROMOSOME[cross_index[cc+1]]);
        pos=ran_ab(1,chromosome_length-1);
        crossover(pos,chro_new1,chro_new2);
        fitness_new1=fitness(chro_new1);
        fitness_new2=fitness(chro_new2);
        test_new=arith_av(fitness_new1,fitness_new2);
        test_old=arith_av(FITNESS[cross_index[cc]],
          FITNESS[cross_index[cc+1]]);
        df=test_new-test_old;
```

```
        if(df<=0.0) {
          strcpy(CHROMOSOME[cross_index[cc]],chro_new1);
          strcpy(CHROMOSOME[cross_index[cc+1]],chro_new2);
          FITNESS[cross_index[cc]]=fitness_new1;
          FITNESS[cross_index[cc+1]]=fitness_new2;
        }
        else {
          p_statistics=P_s(test_old,df,anneal_Temp);
          ran=ran_01();
          if(ran<p_statistics) {
            strcpy(CHROMOSOME[cross_index[cc]],chro_new1);
            strcpy(CHROMOSOME[cross_index[cc+1]],chro_new2);
            FITNESS[cross_index[cc]]=fitness_new1;
            FITNESS[cross_index[cc+1]]=fitness_new2;
          }
        }
      }
    }
  }
  // STAGE 3: mutation
  // -----------------
}
// ***************************************************************
```

### 14.6.2 Acceptance function based on MB statistics

The standard statistical mechanics approach to SA involves the classical Maxwell-Boltzmann distribution for the mean number of particles $\langle n(E) \rangle$ associated with the single-particle energy $E$. This is derived by maximising the entropy $S$ which is based on Shannon's information content:

$$S = -k_B \sum_i p_i \ln p_i \qquad (14.17)$$

where the $p_i$ are the probabilities of finding a system with a set of microscopic states $i$, and $k_B = 1.38 \times 10^{-23}$ J K$^{-1}$ is the Boltzmann constant. The canonical distribution is obtained by maximising this entropy subject to the constraints

$$\sum_i p_i = 1, \qquad \sum_i p_i E_i = U = \text{constant}. \qquad (14.18)$$

From Equation (3.72), it follows that the mean particle occupation number at energy $E$ is

$$\langle n(E) \rangle_{MB} = C g(E) e^{-\frac{E}{k_B T}} \qquad (14.19)$$

for a system in equilibrium with energy $T$, where $C$ is independent of energy, and $g(E)$ is the degeneracy of $E$. The ratio of two such numbers associated with the energies $E$ and $E + \Delta E$ is then

$$\frac{\langle n(E + \Delta E) \rangle_{MB}}{\langle n(E) \rangle_{MB}} = e^{-\frac{\Delta E}{k_B T}} \qquad (14.20)$$

This will be taken as the form of the acceptance function $P_{MB}(f, \Delta f, T)$:

$$P_{MB}(f, \Delta f, T) \equiv e^{-\frac{\Delta f}{k_A T}}. \tag{14.21}$$

In this form, the energy has been replaced by the function $f$, and the Boltzmann constant $k_B$ has been replaced by a constant $k_A$ which is appropriate to the SA process. Further, the quantity $T$ is no longer the temperature of any physical system being considered, but is now an artificial temperature introduced into the SA process. It can be seen from Equation (14.21) that this temperature $T$ and the constant $k_A$ are not independently specified, but are linked together in the product $k_A T$. The value of $k_A$ is undetermined, and the artificial temperature scale has not yet been specified. Hence the value $k_A = 1$ can be taken, and the temperature range determined accordingly.

### 14.6.3 Acceptance function based on Tsallis statistics

The form of the entropy function in Equation (14.17) is based on the assumption that the entropy function is extensive:

$$S(AB) = S(A) + S(B)$$

for a compound system $AB$ composed of two independent systems $A$ and $B$. However, the effect of the presence of boundaries in nanosystems cannot be ignored as in the case of larger systems, and the entropy of these nanosystems will probable not be extensive. Tsallis (1988) has suggested that the entropy of the compound system of two nanosystems $A$ and $B$ could be written in the form

$$S_q(AB) = S_q(A) + S_q(B) + \frac{1}{k_B}(1 - q)S_q(A)S_q(B) \tag{14.22}$$

for some real index $q$. The entropy itself takes the form

$$S_q = k_B \frac{1 - \sum_i p_i{}^q}{q - 1}. \tag{14.23}$$

The extensive property is obtained from Equation (14.22) in the limit $q \to 1$, and it is easily verified that the standard Shannon information content is obtained in this limit:

$$\lim_{q \to 1} S_q = -k_B \sum_i p_i \ln p_i. \tag{14.24}$$

Maximisation of the entropy of Equation (14.23) subject to the constraints

$$\sum_i p_i = 1, \qquad \sum_i p_i{}^q E_i = \text{constant} \tag{14.25}$$

then leads to the distribution

$$p(E_i) \equiv p_i = \frac{1}{Z_q}\left[1 + (q-1)\frac{E_i}{k_B T}\right]^{\frac{1}{1-q}}, \tag{14.26}$$

where the normalisation factor is

$$Z_q \equiv \sum_i \left[1 + (q-1)\frac{E_i}{k_B T}\right]^{\frac{1}{1-q}}.$$

Note the power of $q$ in the second of the constraint Equations (14.25): this power has been defended by Tsallis (1995).

In common with the previous section, the acceptance function $P_{TS}(f, \Delta f, T)$ based on this Tsallis (TS) distribution can be defined as

$$P_{TS}(f, \Delta f, T) \equiv \left[\frac{p(f + \Delta f)}{p(f)}\right]^q \tag{14.27}$$

$$= \left[\frac{1 + (q-1)\frac{f}{k_A T}}{1 + (q-1)\frac{f + \Delta f}{k_A T}}\right]^{\frac{q}{q-1}} \tag{14.28}$$

where again $k_A$ is the SA equivalent of the Boltzmann constant $k_B$. Note the extra power of $q$ in Equation (14.27): this is in line with the constraint Equation (14.25), and has been shown by Andricioaei and Straub (1996) to produce an acceptance ratio which ensures that detailed balance is preserved. These authors also show that, in their particular application to conformational optimisation of a tetrapeptide, values of $q < 1$ are more likely to lead to local minima, while values of $q > 1$ are more likely to lead to the global minimum.

In order that $p(E)$ be a sensible probability, the condition

$$1 + (q-1)\frac{f}{k_A T} > 0 \tag{14.29}$$

must hold for all allowed values of the function $f$. As yet, nothing has been said about the minimum value $f_{min}$ of $f$, since this will affect the values of $q$, $k_A$ and the cooling schedule. However, it can be shown that nothing is lost by measuring $f$ from its minimum $f_{min}$. In order to see this, define the quantity

$$\alpha \equiv 1 + \frac{(q-1)f_{min}}{k_A T}. \tag{14.30}$$

Then Equation (14.28) can be written

$$P_{TS}(f, \Delta f, T) = \left[\frac{1 + (q-1)\frac{f - f_{min}}{\alpha k_A T}}{1 + (q-1)\frac{f - f_{min} + \Delta f}{\alpha k_A T}}\right]^{\frac{q}{q-1}}. \tag{14.31}$$

Hence the effect of the quantity $\alpha$ is to modify the term $k_A T$, and since the value of $k_A$ and the temperature scale has yet to be determined, we may choose $\alpha k_A = 1$ and then determine the temperature scale accordingly:

$$P_{TS}(f, \Delta f, T) = \left[ \frac{1 + (q-1)\frac{f - f_{min}}{T}}{1 + (q-1)\frac{f - f_{min} + \Delta f}{T}} \right]^{\frac{q}{q-1}}. \tag{14.32}$$

Since the value of $f_{min}$ is not known (it is this value which is to be determined), then in the implementation of the scheme based on Equation (14.32), use must instead be made of the most recently found value to replace $f_{min}$ in this equation. In the early stages of the implementation, values of $f$ will frequently be encountered which are lower than the currently found minimum, with the consequent danger that the quantity $1 + (q-1)\frac{f - f_{min}}{T}$ becomes negative. However, in these early stages, a high value of $T$ should stop this happening. Further, by arranging that $q$ starts at a suitably chosen value $q_0$, and letting $q \to 1$ as $T$ decreases, then the quantity on the left hand side should stay positive, since the currently found minimum will then be close to the actual minimum.

In the actual implementation of the scheme, the quantity $q_0$ will be taken to have a suitably high value: Andricioaei and Straub (1996) take the starting value $q_0 = 2$ to correspond to the fast SA method of Szu and Hartley (1987). Using arguments based on the equipartition of energy, Hansmann (1997) takes a constant value $q = 1 + 1/n_F$ where $n_F$ is the number of degrees of freedom in the system. In the implementation described here, $n_F$ will be taken to be the full chromosome length, and the value of $q$ will be lowered linearly with $T$ such that

$$1 + \frac{1}{n_F} \leq q \leq q_0 = 2. \tag{14.33}$$

If a lower limit of 1.0 for $q$ is used, then care must be taken to code the routines so that numerical errors do not arise when this limit is approached.

### 14.6.4 Acceptance function based on BE statistics

A third natural physical process suggests the use of Bose-Einstein (BE) statistics, through the use of an acceptance function $P_{BE}(f, \Delta f, T)$. In the SA method, the average fitness of a population is required to reduce as the temperature is reduced. This requirement bears a marked resemblance to the physical process of Bose-Einstein Condensation which was described in Chapter 4. In this process, a population of particles which obey BE statistics falls into its lowest energy state as the temperature $T$ is lowered below a certain critical temperature $T_c$, while keeping the density constant. For $T < T_c$, it has been shown that the proportion of particles in the lowest energy state is

$$P_{conds}(T) \equiv 1 - \left(\frac{T}{T_c}\right)^{\frac{3}{2}}. \tag{14.34}$$

In order for this process to occur, the chemical potential $\mu(T)$ must rise to compensate until the critical temperature $T_c$ is reached, at which point the relation $\mu(T_c) = 0$ must hold. Condensation then occurs as $T$ is reduced below $T_c$ while holding $\mu(T)$ at zero. This process has been associated with the superfluid properties of liquid helium, occurring below the $\lambda$-point of 2.1 K (Landsberg 1961; London 1938a; London 1938b; London 1939; Stearns 1974).

Consider the Grand Canonical distribution introduced in Chapter 3. This is again obtained by maximising the entropy of Equation (14.17), while now allowing the particle number $n$ to fluctuate. The constraints are

$$\sum_i p_i = 1, \qquad \sum_i p_i E_i = U = \text{constant}, \qquad \sum_i p_i n_i = N = \text{constant}.$$
$$\tag{14.35}$$

This process gives the quantum distributions for the average number of particles $\langle n(E) \rangle$ associated with the single-particle energy $E$ at temperature $T$ as

$$\langle n(E) \rangle = \frac{1}{e^{\frac{E-\mu}{k_B T}} + \alpha} \tag{14.36}$$

where $\mu$ is the chemical potential, and $\alpha$ takes the values $\alpha = +1$ for fermions obeying Fermi-Dirac statistics, and $\alpha = -1$ for bosons obeying Bose-Einstein statistics. All averaged quantities using the classical Maxwell-Boltzmann distribution in Equation (14.19) are determined formally from this quantum case in the limit $\mu/(k_B T) \to -\infty$. For large values of $E$, Equation (14.36) becomes the Maxwell-Boltzmann distribution

$$\langle n(E) \rangle_{MB} = C e^{-\frac{E}{k_B T}},$$

and all three distributions will then have the same high-energy profile.

The appropriate Bose-Einstein distribution for use in SA is then, from Equation (14.36),

$$\langle n(E) \rangle_{BE} = \frac{1}{e^{\frac{E-\mu}{k_B T}} - 1}, \tag{14.37}$$

where $E > \mu$, and the appropriate ratio of electron numbers associated with different energies is

$$\frac{\langle n(E + \Delta E) \rangle_{BE}}{\langle n(E) \rangle_{BE}} = \frac{e^{\frac{E-\mu}{k_B T}} - 1}{e^{\frac{E+\Delta E-\mu}{k_B T}} - 1}. \tag{14.38}$$

The acceptance function of this BE scheme will then be

$$P_{BE}(f, \Delta f, T) \equiv \frac{e^{\frac{f-f_{min}-\mu(T)}{T}} - 1}{e^{\frac{f+\Delta f-f_{min}-\mu(T)}{T}} - 1}. \tag{14.39}$$

As in the case of MB and TS statistics, the replacement $k_A$ for the Boltzmann constant $k_B$ has been taken as $k_A = 1$.

For this acceptance function to be of use in SA, a suitable analogy of the chemical potential function $\mu(T)$ must be found. This function is non-positive, and must increase as the temperature $T$ decreases:

$$\mu(T) \le 0 \quad \text{and} \quad \frac{d\mu}{dT} < 0. \tag{14.40}$$

It will be necessary to specify a functional form for the function $\mu(T)$ when incorporating the function into a computer simulation. The simplest way would be to write it as a linear function

$$\mu(T) = \mu_0 - \mu_1 T$$

where $\mu_0$ and $\mu_1$ are both positive constants. Using the fact that $\mu(T_c) = 0$, it follows that the function will have the form

$$\mu(T) = \mu_0 \left(1 - \frac{T}{T_c}\right) \quad \text{for } T > T_c \tag{14.41}$$

$$= 0 \quad \text{for } T \le T_c. \tag{14.42}$$

The disadvantage of this method is that there is no sensible way of choosing the value of $\mu_0$: the only restriction on its value is that it be non-negative.

Instead, it will be more useful to write Equation (14.39) in the form

$$P_{BE}(f, \Delta f, T) = \frac{e^{\frac{f-f_{min}}{T}} - M(T)}{e^{\frac{f+\Delta f - f_{min}}{T}} - M(T)} \tag{14.43}$$

where the function $M(T)$ is defined by

$$M(T) \equiv e^{\frac{\mu(T)}{T}}. \tag{14.44}$$

It follows from Equation (14.40) that this new function must satisfy

$$0 < M(T) \le 1 \quad \text{and} \quad \frac{dM}{dT} < 0. \tag{14.45}$$

Hence $M(T)$ increases as $T$ decreases. It is this new function $M(T)$ which can be written as a linear function in order to programme the SA method based on BE statistics. Let $T_{max}$ denote the temperate at which the SA process starts, and let $M_0$ be the value of $M$ at this temperature. Since the value of $M$ must reach unity at the critical temperature $T_c$, then

$$M(T_{max}) = M_0 \quad \text{and} \quad M(T_c) = 1,$$

and therefore

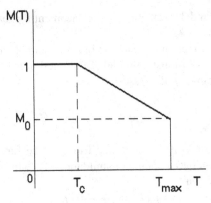

**Fig. 14.2** Plot of the piecewise linear function $M(T) \equiv \exp(\mu/T)$.

$$M(T) = \frac{T_{max} - M_0 T_c + (M_0 - 1)T}{T_{max} - T_c} \quad \text{for } T_c \leq T \leq T_{max}$$
$$= 1 \quad \text{for } T < T_c. \tag{14.46}$$

This function of $M(T)$ is plotted in Fig. 14.2. The corresponding form of $\mu(T) = T \ln M(T)$ is then given by

$$\mu(T) = T \ln \left( \frac{T_{max} - M_0 T_c + (M_0 - 1)T}{T_{max} - T_c} \right) \quad \text{for } T_c \leq T \leq T_{max}$$
$$= 0 \quad \text{for } T < T_c. \tag{14.47}$$

The advantage of this method is now that, apart from having to set the value of $T_c$, the value of $M_0$ has to be set only in the restricted range $0 \leq M_0 \leq 1$. The disadvantage of taking a piecewise linear function of $\mu(T)$ was that a value of $\mu_0$ had to be set in the unrestricted range $0 \leq \mu_0$. Consequently, it will be more convenient to base the computer implementation on Equations (14.43) and (14.46).

### 14.6.5 The cooling schedule

Once it has been decided which set of statistics to use, it is then necessary to decide on the cooling schedule. A number of cooling schedules have been described by Laarhoven and Aarts (1987). This schedule must specify the number $N_T$ of temperature steps, the initial temperature $T_{max}$, the final temperature $T_{min}$, and the condensation temperature $T_c$ if BE statistics is used. The schedule must also specify the functional form $\{T(i) : i = 0, 1, \ldots, N_T\}$ over the temperature steps, subject to the initial and final conditions

$$T(0) = T_{max} \quad \text{and} \quad T(N_T) = T_{min}. \tag{14.48}$$

This last specification, being the simplest, will be described first.

*Setting the temperature decrease function*

Suppose for the moment that the values of $N_T$, $T_{max}$ and $T_{min}$ have already been set. The simplest method is to use a *linear* decrease, in which the temperature is reduced by a constant step $(T_{max} - T_{min})/N_T$:

$$T(0) = T_{max},$$
$$T(i) = T(i-1) - \frac{T_{max} - T_{min}}{N_T}, \quad (i = 1, \ldots, N_T). \tag{14.49}$$

However, I have found that this decrease is too abrupt as $T(i)$ approaches $T_{min}$, where the optimum fitness is being approached. It can give rise to erratic behaviour in the solution at the lower temperatures. It is more sensible to approach $T_{min}$ using exponentially decreasing increments in the temperature. Instead, use the scheme

$$T(0) = T_{max},$$
$$T(i) = d.T(i-1), \quad (i = 1, \ldots, N_T) \tag{14.50}$$

where $d < 1$ is a constant. This equation has solution

$$T(i) = T_{max}d^i, \quad (i = 0, \ldots, N_T), \tag{14.51}$$

where $d$ must satisfy the relation $T_{min} = T_{max}d^{N_T}$, or

$$d = \left(\frac{T_{min}}{T_{max}}\right)^{1/N_T} = \exp\left(\frac{1}{N_T}\ln(T_{min}/T_{max})\right). \tag{14.52}$$

In this scheme, the temperature is reduced by a constant factor $d$. The decrease will be relatively large in the early stages, becoming smaller in the later stages as the minimum fitness is being approached.

*Setting the temperature limits*

The most difficult part in prescribing the cooling schedule is in setting the initial temperature $T_{max}$. One indication of how to start this process is to note that, in all three of the statistics $S$ which have been introduced, the temperature enters into the expression for $P_S(f, \Delta f, T)$ in one or several of the forms $f/T$, $\Delta f/T$, or $(f - f_{min})/T$. A very simple but crude method is to make a guess at the value $T_{max}$; this is clearly not a very satisfactory method, but can be useful when diagnostic testing. Two main methods should be adequate for most needs:

1. *Method 1.* A relatively simple schedule consists of setting $T_{max}$ as the average of the absolute value of the fitness, taken over an initial randomised population. In fact, this method is used to provide a fallback temperature if the more accurate method 2 of calculation goes wrong. An alternative way would be to take $T_{max}$ as

the maximum absolute fitness of the population, although this may give erratic results if the value is based on a single outlying random number.

2. *Method 2.* A more precise method (Johnnson et al. 1987) involves choosing a relatively high acceptance ratio $r_t$ for transitions (for example, $r_t = 0.8$), and then solving the equation $P_S(\cdots) = r_t$ to calculate the value of $T_{max}$ which produces this acceptance ratio. First, generate a random population of chromosomes, calculate the fitness of each member, and find the minimum fitness $f_m$ of this population. Ideally, the global minimum fitness $f_{min}$ should be used, but this value is as yet undetermined. Then for each chromosome, mutate each bit one at a time. For each mutated chromosome, calculate the difference $df$ between the fitness of the original unmutated chromosome and the mutated one. If $df < 0$, this mutation must be discarded, the bit left unmutated, and the next bit tried. Average all those values of $df$ which were found to be positive, to produce a value $\overline{df}$, and average the corresponding original fitnesses to produce $\overline{f}$. Then, depending on the particular statistics being used, solve for $T_{max}$ using one of the equations

$$\text{MB:} \quad r_t = e^{-\frac{\overline{df}}{T_{max}}}, \tag{14.53}$$

$$\text{BE:} \quad r_t = \frac{e^{\frac{\overline{f}-f_m}{T_{max}}} - M_0}{e^{\frac{\overline{f}+\overline{df}-f_m}{T_{max}}} - M_0}, \tag{14.54}$$

$$\text{TS:} \quad r_t = \left[ \frac{1 + (q_0 - 1)\frac{\overline{f}-f_m}{T_{max}}}{1 + (q_0 - 1)\frac{\overline{f}+\overline{df}-f_m}{T_{max}}} \right]^{\frac{q_0}{q_0-1}}. \tag{14.55}$$

In this way, it is hoped that the choice of the starting and ending values of the temperature would be relatively independent of the particular minimisation problem tackled.

It can be seen from Equations (14.53)–(14.55) that the starting temperature $T_{max}$ will depend on the particular statistics used, and their values will differ even if the same initial population is used. For the three different statistics considered, these starting temperatures can be compared for a common initial population in the following way. Let the starting temperatures in the three cases be denoted by $T_{MB}$, $T_{BE}$ and $T_{TS}$. Let $\overline{f}$ and $\overline{df}$ be as before, and suppose that the starting acceptance ratio $r_t$ is specified. Then the three starting temperatures are linked by the relations

$$r_t = e^{-\frac{\overline{df}}{T_{MB}}} = \frac{e^{\frac{\overline{f}-f_m}{T_{BE}}} - M_0}{e^{\frac{\overline{f}+\overline{df}-f_m}{T_{BE}}} - M_0} = \left[ \frac{1 + (q_0 - 1)\frac{\overline{f}-f_m}{T_{TS}}}{1 + (q_0 - 1)\frac{\overline{f}+\overline{df}-f_m}{T_{TS}}} \right]^{\frac{q_0}{q_0-1}}. \tag{14.56}$$

The first of these equalities is easily solved to give

$$T_{MB} = -\frac{\overline{df}}{\ln r_t}, \tag{14.57}$$

but there are no analytic solutions of Equation (14.56) for $T_{BE}$ and $T_{TS}$. In these two cases, the solution has to be solved iteratively. However, Equation (14.56) can be written

$$e^{\frac{\overline{df}}{T_{MB}}} = \frac{e^{\frac{\overline{f}+\overline{df}-fm}{T_{BE}}} - M_0}{e^{\frac{\overline{f}-fm}{T_{BE}}} - M_0}, \qquad (14.58)$$

from which it can be deduced that

$$T_{BE} \geq T_{MB} \qquad (14.59)$$

for a common initial population. Similarly, for a starting value $q_0 > 1$, it can be deduced from the equation

$$e^{\frac{\overline{df}}{T_{MB}}} = \left[ \frac{1 + (q_0 - 1)\frac{\overline{f}+\overline{df}-fm}{T_{TS}}}{1 + (q_0 - 1)\frac{\overline{f}-fm}{T_{TS}}} \right]^{\frac{q_0}{q_0-1}} \qquad (14.60)$$

that

$$T_{TS} \geq T_{MB}. \qquad (14.61)$$

However, there is no similar inequality linking $T_{BE}$ and $T_{TS}$, since the relative sizes will depend on the starting values $M_0$ and $q_0$. These three schemes will be applied to the problem of evaluating associated Fermi integrals in Sect. 14.7, and Tables of starting temperatures are given there for the three schemes using identical initial populations.

Once $T_{max}$ has been set, the minimum temperature $T_{min}$ can be taken as some small proportion of this, say

$$T_{min} = 10^{-4} \times T_{max}, \qquad (14.62)$$

with a long stop value of $T_{min} = 10^{-4}$ if the value calculated from Equation (14.62) is too large.

Once $T_{max}$ and $T_{min}$ have been set, it remains to set the condensation temperature $T_c$. This can be done simply by taking its value at a prescribed small proportion of the overall temperature range.

### The Markov chain

As well as reducing the temperature in small steps, a number of iterations—the *Markov chain*—must be performed at each temperature value. There seems to be no way of choosing this value with absolute certainty. However, a balance must be struck between the numbers of temperature steps and the number of Markov steps: a smaller (larger) number of temperature steps will, in general, mean a larger (smaller) number of Markov steps. The exact balance must be determined by trial and error, although some more exact approaches are detailed by Laarhoven and Aarts (1987).

## 14.7 Application: approximation to the Associated Fermi integrals

We now return to the problem of approximating the associated Fermi integrals which were introduced in Chapter 8. The problem arises in HEMT modelling because the integrals involved in the calculations of the electron density and energy density in quantum wells have non-zero lower limits which change as the simulation iterations progress. An approximation was introduced there whereby the associated integrals were written in terms of a number of parameters. These parameters were then found by minimising the average relative error $E_r$ using a somewhat crude search method. This minimisation process will now be revisited, and the minimisation will be accomplished using SA. It will be seen that the results obtained using the SA method are much more accurate than those of the original crude method. It will also present an opportunity to compare the three different statistics involved.

The associated Fermi integral $I_r(a, b)$ was written in terms of an intermediate function $L_r(a, b)$ in the form

$$I_r(a, b) = F_r(a) + L_r(a, b) \tag{14.63}$$

where

$$L_r(a, b) = F_r(a + b) - F_r(a) - \frac{1}{\Gamma(r + 1)} \int_0^b \frac{z^r}{1 + e^{z-(a+b)}} dz. \tag{14.64}$$

It was shown in Chapter 8 that the function $L_r(a, b)$ had the properties

$$L_r(a, 0) = 0, \tag{14.65}$$

$$L_0(a, b) = 0, \tag{14.66}$$

$$L_1(a, b) = b \ln(1 + e^a). \tag{14.67}$$

The evaluation of the $I$-integrals using Equation (14.63) is computationally very expensive, and clearly it would be useful to have a technique whereby these integrals may be evaluated rapidly. It would therefore be useful to have simple approximations for the functions $L_r(a, b)$ which could be evaluated rapidly.

### 14.7.1 Approximation method

An approximate function $L'_r$ is used which is a relatively simple function of the arguments $a$ and $b$, and of a number of parameters $\mathbf{x}_r$:

$$L_r(a, b) \approx L'_r(a, b; \mathbf{x}_r). \tag{14.68}$$

The optimal values of the parameters $\mathbf{x}_r$ must then be found by minimising a suitable function of them. When these parameters have been calculated, the functions $L'_r$ will be relatively easy to evaluate for any values of $a$ and $b$. As the notation implies, the values of the parameters will depend on the particular order $r$ of the Fermi integral.

Various models can be taken for the form of $L'_r$, and an investigation of them can be made to find which gives the smallest percentage error. In particular, three models which are suggested by Equations (14.65)–(14.67) are

1. *Model 1*: three parameters, $\mathbf{x}_r \equiv (p_r, \alpha_r, \beta_r)$,

$$L'_r(a, b; p_r, \alpha_r, \beta_r) = b^{p_r}(\alpha_r + \beta_r a)\ln(1 + e^a). \tag{14.69}$$

2. *Model 2*: three parameters, $\mathbf{x}_r \equiv (p_r, \alpha_r, \beta_r)$,

$$L'_r(a, b; p_r, \alpha_r, \beta_r) = b^{p_r}\alpha_r \ln(1 + \beta_r e^a). \tag{14.70}$$

3. *Model 3*: four parameters, $\mathbf{x}_r \equiv (p_r, \alpha_r, \beta_r, \gamma_r)$,

$$L'_r(a, b; p_r, \alpha_r, \beta_r) = b^{p_r}(\alpha_r + \beta_r a + \gamma_r a^2)\ln(1 + e^a). \tag{14.71}$$

For example, in model 1, it can be seen from Equation (14.67) that the minimisation process should give $p_1 \approx 1$, $\alpha_1 \approx 1$, and $\beta_1 \approx 0$. In model 2, we should get $p_1 \approx 1$, $\alpha_1 \approx 1$, and $\beta_1 \approx 1$.

The minimisation process to be carried out is as follows. Suitable ranges spanned by the variables $a$ and $b$ must be chosen. That is, the minimum and maximum values $a_{min}$, $a_{max}$, $b_{min}$ and $b_{max}$ are calculated which are likely to be encountered in the device simulation. One way of doing this is to run the simulation for several iterations without the correction function $L_r$ in order to get approximations for these minimum and maximum values. The next step is to split up the ranges $[a_{min}, a_{max}]$ and $[b_{min}, b_{max}]$ into respectively $A$ and $B$ equally spaced intervals, and to take grid values

$$a_i = a_{min} + i(a_{max} - a_{min})/A, \quad (i = 0, \ldots, A)$$
$$b_j = b_{min} + j(b_{max} - b_{min})/B, \quad (j = 0, \ldots, B)$$

with $a_0 \equiv a_{min}$, $a_A \equiv a_{max}$, $b_0 \equiv b_{min}$, and $b_B \equiv b_{max}$. The next step is to evaluate the quantities $I_r(a_i, b_j)$ at each grid point, and then to evaluate the average relative error

$$f(\mathbf{x}_r) \equiv \frac{\sum_{i=0}^{A}\sum_{j=0}^{B}|I_r(a_i, b_j) - F_r(a_i) - L'(a_i, b_j; \mathbf{x}_r)|}{\sum_{i=0}^{A}\sum_{j=0}^{B}|I_r(a_i, b_j)|}. \tag{14.72}$$

This form of the relative error is taken to give greater weight to larger values of the integral since these larger values contribute most to the values of the electron number and energy densities. The form of the function $L'_r$ will depend on which model is being used. SA techniques were used to find the values of the parameters $\mathbf{x}_r$ which minimise the function $f(\mathbf{x}_r)$.

**Table 14.3** The relative percentage errors found in Simulation 1, comparing models 1 and 2, and MB and BE statistics. The final column shows the results obtained from the earlier non-SA method.

| $r$ | Model 1 (MB) | Model 2 (MB) | Model 1 (BE) | Model 2 (BE) | non-SA |
|------|--------------|--------------|--------------|--------------|--------|
| $-1/2$ | 7.742058 | 7.688746 | 7.958087 | 7.685076 | 13.45 |
| $1/2$ | 0.616947 | 0.594567 | 0.594253 | 0.594469 | 1.93 |
| $1$ | 1.278233 | 0.013339 | 0.032610 | 0.014998 | – |
| $3/2$ | 1.405704 | 0.520529 | 0.553857 | 0.511066 | 3.48 |
| $2$ | 2.793324 | 0.743448 | 1.068047 | 0.756130 | 7.35 |
| $5/2$ | 4.018108 | 0.805126 | 0.882199 | 0.787368 | 11.41 |

**Table 14.4** The optimal parameter values obtained in Simulation 1 for model 2. The final column shows the relative percentage error, with results for no correction in brackets.

| $r$ | $p_r$ | $\alpha_r$ | $\beta_r$ | Error (uncorrected) |
|------|-------|-----------|-----------|---------------------|
| $-1/2$ | 0.142252 | $-0.153344$ | 4.999110 | 7.685076 (247.209635) |
| $1/2$ | 0.629389 | 0.510386 | 1.198911 | 0.594469 (70.442810) |
| $1$ | 1.000293 | 1.001530 | 1.001374 | 0.014998 (91.078038) |
| $3/2$ | 1.424446 | 1.129413 | 0.930935 | 0.511066 (97.265434) |
| $2$ | 1.867129 | 1.004440 | 0.885440 | 0.756130 (99.148262) |
| $5/2$ | 2.332548 | 0.726130 | 0.860104 | 0.787368 (99.732565) |

## 14.7.2 Results

A SA routine was run using data for a four-layer HEMT. A preliminary number of iterations of the device simulator without the correction factor gave the maximum and minimum values of $a$ and $b$ to be $a_{min} = -80.0$, $a_{max} = 8.0$, $b_{min} = 0.0$, and $b_{max} = 83.0$. (In order to avoid numerical problems, the value of $b_{min}$ was taken as 0.0001 rather than zero.) Three separate simulations were performed with this pre-obtained data.

*Simulation 1*

The programme was run for models 1 and 2 for several values of the Fermi index $r$, using both MB and BE statistics (no Tsallis scheme). In each case, the number of runs was 20, the population was 100, with 25 timesteps and 4 Markov steps at each timestep. A non-random seed was used for all sets of runs so that results could be compared. Results in Table 14.3 show the relative percentage errors (that is, $100 \times f(\mathbf{x}_r)$ of the optimal set of parameters. The final column labelled "non-SA" gave the corresponding results obtained by using the crude non-SA method which was described in Chapter 8. The results for the case $r = 1$ are included for completeness, and give an indication of the effectiveness of the scheme, since the exact results are known for this case.

Table 14.4 gives the actual optimal parameter values for model 2 obtained using the BE scheme. The final column shows the relative percentage errors and, for comparison, the relative percentage errors obtained for no correction ($L'_r = 0$) are given in brackets.

**Table 14.5** The relative percentage errors found Simulation 2, using a non-random seed for all runs.

| $r$ | MB | BE | TS ($q = 1.015$) | TS ($1.015 \leq q \leq 2.0$) | TS ($1.0 \leq q \leq 2.0$) |
|---|---|---|---|---|---|
| $-1/2$ | 7.683190 | 7.682270 | 7.68590 | 7.704180 | 7.692323 |
| $1/2$ | 0.596953 | 0.593808 | 0.594643 | 0.594076 | 0.604999 |
| 1 | 0.143362 | 0.0479941 | 0.0407383 | 0.012992 | 0.022650 |
| $3/2$ | 0.529227 | 0.510153 | 0.545028 | 0.516630 | 0.530384 |
| 2 | 0.744824 | 0.738380 | 0.739875 | 0.740359 | 0.772952 |
| $5/2$ | 0.765433 | 0.765723 | 0.911401 | 0.813392 | 0.802803 |

**Table 14.6** The starting temperatures $T_{max}$ for all three schemes in Simulation 2.

| $r$ | MB | BE ($M_0 = 0.5$) | BE ($M_0 = 0.25$) | TS ($q_0 = 2.0$) |
|---|---|---|---|---|
| $-1/2$ | 239451.9 | 453380.3 | 310835.5 | 452399.2 |
| $1/2$ | 4072.5 | 7711.0 | 5286.6 | 7694.3 |
| 1 | 17460.4 | 33060.1 | 22655.7 | 32988.6 |
| $3/2$ | 79106.4 | 149676.1 | 102655.0 | 149345.7 |
| 2 | 735102.8 | 1390666.4 | 953862.0 | 1387584.1 |
| $5/2$ | 19668810.4 | 37234910.6 | 25530302.8 | 37153965.0 |

Model 3 is similar to model 1, but contains an extra parameter. However, results showed that there was no advantage to be gained in using this model, in terms of the size of the relative percentage error. These results are not presented here.

It is immediately obvious that the SA method gives marked improvement over the non-SA results, which only applied to model 1. Further, model 2 produces better results than model 1 in terms of the percentage errors.

*Simulation 2*

The programme was run for model 2 only for several values of the Fermi index $r$ using the MB scheme, the BE scheme, and the Tsallis scheme for the three cases $q = 1.015$ (fixed), $1.015 \leq q \leq 2.0$, and $1.0 \leq q \leq 2.0$. The value of 1.015 was chosen as being very close to $1 + 1/n_F$. This time, the numbers of runs was 5, the population was 100, with 25 timesteps and 4 Markov steps at each timestep. The required precision was 6. Table 14.5 gives the results for the relative percentage errors when a non-random seed was used for all sets of runs. Table 14.6 shows the starting temperatures for the three schemes: note how $T_{BE}$ and $T_{TS}$ are both larger than $T_{MB}$ whatever the values of $M_0$ and $q_0$, but nothing can be said regarding the relative sizes of $T_{BE}$ and $T_{TS}$.

*Simulation 3*

The programme was run for model 2 only with the single value of the Fermi index $r = 1/2$ using the MB scheme, the BE scheme, and the Tsallis scheme for the single case $1.0 \leq q \leq 2.0$. This time, the numbers of runs was 5, the population was 100, with 100 timesteps and 1 Markov step at each timestep. The required precision was 6. The same non random seed was used for the initial populations for all three schemes. In order to compare the convergence of the three statistics, define the *relative inverse fitness* $p(T)$ by

$$p(T) \equiv \frac{f_{opt}}{a(T)} \qquad (14.73)$$

where $f_{opt}$ is the optimal value of $f$ found at temperature $T$. This function $p(T)$ satisfies the relation $p(T) \le 1$, and a value close to 1.0 will indicate that the fitnesses of a large proportion of the chromosome population are close to $f_{opt}$. It is to be hoped that $p(T) \to 1$ as $T \to T_{min}$ if the SA scheme is working correctly. An indication that this is so will be given by comparing the function $p(T)$ with that of the BE condensation function

$$p_{conds}(T) \equiv 1 - \left(\frac{T}{T_c}\right)^{\frac{3}{2}}. \qquad (14.74)$$

A comparison of the plots of the function $p(T)$ for the three distributions showed that the BE distribution came closest to the plot of the function $p_{conds}(T)$, and that this distribution corresponded to the highest proportion of chromosomes in the optimal state (Cole 2007).

# Chapter 15
# Grid generation

Many physical devices, including the MESFET and HEMT, have geometries for which the use of a rectangular grid is appropriate when modelling their behaviour. Contact edges and the interfaces between different material layers mostly lie parallel to one fixed direction.

A uniform grid, that is, one in which all of the grid lines in each direction are a constant distance apart, is the simplest to use. However, in any device, there will be regions in which the physical quantities, such as the electric field, are rapidly varying. Extra grid lines are needed in these regions. The drawback of the uniform grid approach is that an unnecessary concentration of lines will be used in regions in which the physical quantities are not varying rapidly. This slows down the iteration routines, and also gives rise to matrices which are very large. The advantage of this method is that it is very simple to code, and arrays can be used to carry the values of the physical variables at the grid points.

The alternative approach is to use a non-uniform grid. The advantage of this method is that grid lines are used only where they are needed, and iteration routines are speeded up. The disadvantage of this method is that more work has to be done in programming the routines.

As the iterations progress, the values of the physical variables will probably change markedly from their initial guesses. Hence it may be that a region requires fewer grid lines in the early stages of the iterations, but needs more in the later stages as the solution converges. On the other hand, it may be necessary to remove unnecessary grid lines from a region as the iterations progress. This allocation and de-allocation of grid lines is easily performed using a uniform grid, but not so easily achieved using a non-uniform grid. The use of vector arrays is not appropriate in the case of a non-uniform grid, and pointer allocation and de-allocation is used in this case.

This Chapter will deal only with the case of a non-uniform grid. The main challenges of this method are

- to assign an initial grid to cover the main features of the physical device,
- to allocate extra grid lines where necessary, and to de-allocate them while ensuring that the basic framework of the initial grid remains intact.

E.A.B. Cole, *Mathematical and Numerical Modelling of Heterostructure Semiconductor Devices: From Theory to Programming*, DOI 10.1007/978-1-84882-937-4_15, © Springer-Verlag London Limited 2009

## 15.1 Overview of grid generation

Some of the routines which go to make up a complete programme for the generation
of a rectangular grid will be presented. The use of these routines will enable the user
to design a non-uniform rectangular grid to fit a multi-layer device similar to a multi-
recessed HEMT. Once the functions are understood, it is a straightforward matter
to design a grid for any other similar device. Functions will also be presented for
refining the grid in regions in which a relevant physical variable is rapidly changing,
and also to de-refine the grid when the extra grid lines are no longer needed. As
described, the routines will relate to a two dimensional rectangular grid overlaying
a two-dimensional cross section of a typical HEMT structure which is shown in
Fig. 15.1. Again, once the routines are understood, it should be obvious how they
can be extended to the construction of a three dimensional grid, or indeed restricted
to a one dimensional grid.

The functions listed below are written in C++. They have been written with trans-
parency in mind: clever programming tricks are avoided, and many routines could
probably have been written more compactly. The functions are described in detail in the next section, but an overview is given here. Grid points are introduced
using dynamic memory allocation; a grid point is specified by giving its $x$- and
$y$-coordinates, and by pointers pointing to adjacent grid points. The specification
also includes the allocation of a set of physical variables to each point such as, for
example, the electrostatic potential $\psi$, the Fermi potential $\phi$, and the electron temperature $T$, but only one sample of this type is used in this programme—the variable
*typical_value*, whose value is used in the refinement or de-refinement of the grid.
Among other attributes associated with a grid point are its *positiontype* which indicates whether it is an edge, interior or exterior point, and its *durationtype* which
specifies whether or not it should remain untouched during de-refinement.

Fig. 15.1 shows the cross section of the typical device which is being modelled. It
consists of a rectangular structure containing four different material layers, although
any number may be modelled. It has four contacts—the source ($S0$–$S1$), gate ($G0$–
$G1$), fieldplate ($FP0$–$FP1$) and drain ($D0$–$D1$). As illustrated, it has two recesses,
although again it may have any number. The grid specification begins by assigning
grid points to the corners of the device, the edges of the contacts, the corners of the
recesses, and to the points where the layer boundaries meet the edges. This initial
grid is then completed by drawing grid lines from these points to the opposite edges.
Fig. 15.1 shows the state reached at this stage. The variable *durationtype* is set for
all of these points to indicate that they should not be removed if any de-refinement
is to take place. Extra grid lines are then inserted onto this initial grid, in order to
give a minimum refinement which must be specified at the start of the programme.

After the programme, in which the device equations are iterated, has been running for some time, the run must be temporarily halted to see whether any refinement
is needed. For example, if the derivative of $\psi$ in terms of the forward difference between two consecutive grid points is larger than some specified value, then an extra
grid point is placed between the two points, and a whole new grid line is constructed
through the new point. This necessitates the breaking of links between the points on

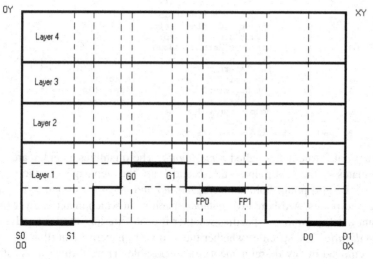

**Fig. 15.1** Cross-section of a four layer device, showing the initial allocation of permanent grid lines.

the grid lines through the two original points, and the construction of new links to the grid points on the new line. This process is illustrated by using the variable *typical_value* as the variable on which the need for refinement is tested. Similarly, at some point in the iteration process, the grid may be tested to check whether or not some grid lines have become redundant using the reverse process. Grid lines can then be disposed of, unless they pass through points which have been designated as permanent using the attribute *durationtype*.

Any complete device simulation must include extra functions for calculating physical variables, for iterating the equations, and for output of results to disc. This output could include giving the coordinates of the grid points, and the values of *typical_value* and *durationtype* at each grid point. Sufficient information should be output in order that the current grid could be reconstructed in a later run.

## 15.2 Functions of a grid generation programme

The following points show how the main functions of a grid generation programme should be constructed.

- The contact potentials *V_source*, *V_gate*, *V_drain* and *V_fplate* must be entered.
- The basic grid point specification is given as a *structure* as follows:

```
81: typedef struct something {
82:
83:             double x;
84:             double y;
85:             double h;
86:             double k;
```

```
87:                        struct something *right;
88:                        struct something *left;
89:                        struct something *up;
90:                        struct something *down;
91:                        // --------------------------------
92:                        int layernumber;      // layer containing point
93:                        int positiontype;     // exterior(0),interior(1),edge(2)
94:                        int durationtype;     // 0=disposable, 1=permanent
95:                        // --------------------------------
96:                        double typical_value;
97: } meshpoint;
98: #define sizegrid sizeof(meshpoint)
```

The whole structure is called a *meshpoint*. The quantities $x$ and $y$ are the co-ordinates of the grid point, while $h$ and $k$ are the steplengths to the right and up of the grid point. Pointers *right*, *left*, *up* and *down* then point to neighbouring grid points. Attribute *layernumber* specifies in which material layer the grid point lies, *positiontype* specifies whether it is an exterior, interior or edge point, and *durationtype* specifies whether the grid point is permanent (that is, is not to be removed in any de-refinement), or disposable. There follows a list of physical variables associated with the grid point; only one (*typical_value*) has been specified here in this sample programme, but others could be

```
double psi;
double phi;
double T;       etc...
```

For a three-dimensional grid, the extra lines

```
double z; // extra coordinate
double l; // extra steplength
struct something *forward;
struct something *backward;
```

could be added.

- The device parameters can be provided in the function *set_device_structure*(), sections of which are listed below:

```
 36: const int NO=0,no=0,YES=1, yes=1;
147: double layerdepth[max_number_layers],y_layer[max_number_layers];
158: int numberoflayers;
159: int numberofrecess;
162: double Xmin,s0,s1,g0,g1,d0,d1,Xmax;
166: double fp0=0.0,fp1=0.0;
169: double rx0[max_number_recess+1],
170:        rx1[max_number_recess+1],
171:        ry[max_number_recess+1];
213: void set_device_structure(void)
214: {
215: // Example: 4-layer device Si-GaAs Si-AlGaAs Si-GaAs GaAs
218: // Material_types: GaAs=0, AlGaAs=1, InGaAs=2.
219:    devicename="Example1";
220:    numberoflayers = 4;
221:    // contacts:
222:    Xmin = 0.000e-6;
223:    s0   = 0.000e-6;
224:    s1   = 0.300e-6;
225:    g0   = 0.850e-6;
226:    g1   = 1.850e-6;
227:    d0   = 3.400e-6;
228:    d1   = 3.600e-6;
```

```
229:    Xmax = 3.800e-6;
230:    // fieldplate structure:
231:    FIELDPLATE=YES;
232:    fp0  = 2.200e-6;
233:    fp1  = 2.700e-6;
234:    // recess structure:
235:    numberofrecess=2;
236:    rx0[1]  = 0.500e-6;
237:    rx1[1]  = 3.000e-6;
238:    ry[1]   = 0.050e-6;
239:    rx0[2]  = 0.700e-6;
240:    rx1[2]  = 2.000e-6;
241:    ry[2]   = 0.100e-6;
242:    // layer structure:
243:    layerdepth[1]=0.100e-6;   // layer 1
244:    layerdepth[2]=0.030e-6;   // layer 2
245:    layerdepth[3]=0.100e-6;   // layer 3
246:    layerdepth[4]=0.500e-6;   // layer 4
247:    // NOTES: Set trunc_ratio=1.0
248: }
```

The number of layers must be specified, and the $x$-coordinates of the end points of the source, gate, drain and fieldplate must be specified. The number of recesses is specified, and the $x$-coordinates and depth are given for each recess. The layer structure is then specified: only a skeleton specification is shown here, but a fuller specification could contain:

```
layerdepth[1]=0.100e-6;  // depth   --   layer 1
material type[1]=1;  // AlGaAs
mu[1]=0.2;  // mobility
Nd[1]=1.5e18;    // doping density
```

with similar specifications for the other layers. Very often, a device will have a final substrate layer very much thicker than the other layers, and it is wasteful of computing resources to perform iterations in this buffer layer at large distances from the other layers. This buffer layer can be truncated using the variable *trunc_ratio* so that, for example, a value of 0.1 of this variable means that the substrate is truncated to only 10% of its value. The $y$-coordinates of the layer interfaces must also be calculated from the specification of their thicknesses.

• The initial grid structure can be set from the data, or read in from previously saved data. This second method is not described here, but the first method will be described in detail. The starting point of this process is the function *set_mesh* as follows:

```
286: void set_mesh(void)
287: {
288:    cout << "\nSetting initial mesh ...";
289:    set_initial_mesh();
290:    set_positiontype();
291:    set_h_k_layernumber();
292:    calc_num_pts();
293:    cout << "... initial mesh set.";
294: }
```

– The initial outline must be set in function *set_initial_mesh*; the naming of the appropriate grid points is shown in Figs. 15.1 and 15.2. The outer rectangle OO–OY–XY–OX is set between lines 304 and 333, listed as follows:

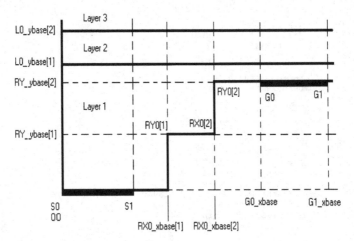

**Fig. 15.2** Naming of the boundary grid points of the device.

```
304:    // set rectangle:
305:    OO=(meshpoint*)malloc(sizegrid);
306:    OO->down=NULL;
307:    OO->left=NULL;
308:    OO->x=Xmin;
309:    OO->y=Ymin;
310:
311:    OY=(meshpoint*)malloc(sizegrid);
312:    OO->up=OY;
313:    OY->down=OO;
314:    OY->up=NULL;
315:    OY->left=NULL;
316:    OY->x=Xmin;
317:    OY->y=Y_tr;
318:
319:    XY=(meshpoint*)malloc(sizegrid);
320:    OY->right=XY;
321:    XY->left=OY;
322:    XY->up=NULL;
323:    XY->right=NULL;
324:    XY->x=Xmax;
325:    XY->y=Y_tr;
326:
327:    OX=(meshpoint*)malloc(sizegrid);
328:    XY->down=OX;
329:    OX->up=XY;
330:    OX->right=NULL;
331:    OX->down=NULL;
332:    OX->x=Xmax;
333:    OX->y=Ymin;
```

- The bases along the $x$-axis of the recesses must be set; although some of these points lie outside the profile of the device, it is useful to have a simple rectangle with which to work. Note that whenever a new grid point is introduced, it is necessary to break the links between previously adjacent grid points, and then to create new ones to the new grid point (being careful to do this in the correct order).

- The contact edges must be introduced; the source and drain edges are already on the bottom edge of the rectangle, but the base grid points of the gate and fieldplate must be introduced separately since these contacts are off the rectangle base.
- Vertical lines must be constructed through all of the points which have been introduced so far. This is done using a function *set_vertical*; the arguments of this function are the grid point through which the vertical is constructed, and the duration-type of the new points (disposable or permanent). These initial verticals are set to permanent.
- The material interfaces must be set; the bases are set along the $y$-axis, and then horizontals must be set through these base points.
- The recess base points along the $y$-axis must be set. Since the layer base points have already been set, care must be taken to interweave the recess base points with the established interface base points. Once these bases have been set in the correct order, horizontals are constructed through them.
- The rectangle is now criss-crossed by a number of grid lines; some of these are then singled out as being corners of recesses. This can be done using a function which tests the closeness of two $y$ values. In order that the iterations take place smoothly with no layer being missed out, it is necessary to ensure that each layer contains at least two grid lines. This can be done using a function which takes three arguments: (i) the grid point such that a new horizontal is inserted half way between it and the one above, (ii) the duration type of the new points on the line, and (iii) whether or not physical values on the new line should be interpolated from the corresponding values on either side. At this stage, the new points are made permanent, and no interpolation is to be made since physical quantities have yet to be assigned.
- Now that this minimal grid has been set, all of the points are made permanent (some will already have been made permanent, and some will have had no assignment).
- The initial grid is then completed by inserting sufficient numbers of horizontal and vertical grid lines. These numbers are specified in advance, but the resulting structure will contain a grid which only approximates to this specification.
- The position type of each point on the grid must then be set. The position type of each point can be edge, interior or exterior (since some grid points inside the basic rectangle fall outside the device profile).
- A function *set_h_k_layernumber()* must then be constructed to sweep the points. For each point it should calculate the distance $h$ to the point on its right, the distance $k$ to the point above, and the layer in which the point is situated.
- A function *calc_num_points()* can then be constructed to calculate the numbers of internal, external and edge points in the basic rectangle: this is mainly for diagnostic purposes, but these numbers can also be used in functions in which averages are taken over grid points.

- The next step is to input the initial values of the physical variables. In the definition of the structure *meshpoint*, the variable *typical_value* represents this set of data, although in realistic situations the data will contain the electrostatic potential $\psi$, the Fermi potential $\phi$ and the electron temperature $T$.
- In a full simulation, iterations would be made on the variables using suitable iteration processes, and these results could be stored to disc.
- After a sufficient number of iterations have taken place, the next step is to determine whether or not the original grid remains suitable. An index *refine_index* allows a limit to be placed on the number of refinement sweeps. Refinement is done on a physical value—normally the variable $\psi$, but on the variable *typical_value* in the function which is described here. The need for refinement is tested in both the $x$- and $y$-directions. Confining attention to the $x$-direction for the moment, the values of *typical_value* are tested at a grid point and the one above; if the difference in these values is greater than a specified value *x_refine_diff*, then a grid point is inserted mid way between the two, and a whole line of points is constructed along the $x$-direction. Links between existing adjacent pairs must be broken, and new ones formed. All of the new points inserted in this way are labelled as disposable, and the value of *typical_value* must be interpolated from those on either side. Similarly, sweeps must be made in the horizontal direction to see whether new lines of grid points should be inserted in the $y$-direction using a specified value *y_refine_diff*. This whole refinement sweep may be repeated a number of times. Note that in this programme, the testing has been done on differences in the value between adjacent grid points. More realistically, it will be better to test on the derivative of the value; for example, on the derivative of $\psi$ in order to put extra grid points in regions in which the electric field $-\nabla\psi$ is large.
- Once the grid has been refined, a further number of iterations of the device equations can take place. Further refinement can then be done, or de-refinement may be tested for. This will ensure that grid lines are removed from regions where they are no longer needed. Again, this testing will be done on values of *typical_value* at consecutive grid points, but the testing has to be more rigorous than for the refinement processes. For example, suppose that the difference in the values of *typical_value* at a grid point and the one to its right is less than the value *x_derefine_diff*. Before removing this point and the whole of the line in the $y$-direction on which it is situated, each pair of points above and below the points originally tested must themselves be tested, otherwise points may be removed where they are still needed. Once this de-refinement has been completed, the values of the steplengths $h$ and $k$ must be re-calculated.
- Once the grid has been refined, iterated, or de-refined as many times as necessary, the results can be put out to disc if needed. The form of the disc output is left to the user.
- The final act is to remove all the grid points which have been allocated using *malloc()*. This is accomplished using the function *kill_mesh()* whose listing is as follows:

```
885: void kill_mesh(void) // remove the mesh
886: {
```

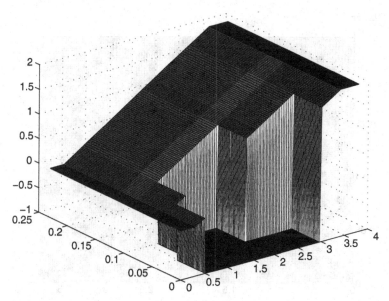

**Fig. 15.3** The initial shape of the variable *typical_value*.

```
887:    meshpoint *base,*nextbase,*cur;
888:    base=OY;
889:    cout << "\n\nRemoving mesh ...";
890:    while(base!=NULL) {
891:      nextbase=base->right; cur=base->down;
892:      while(cur!=NULL) {
893:        free(cur->up);
894:        if(cur->down!=NULL) cur=cur->down;
895:        else { free(cur);cur=NULL; }
896:      }
897:      base=nextbase;
898:    }
899:    cout << " mesh removed.";
900: }
```

## 15.3 Results from the programme

A program MESHGEN.CXX was run with an initial rough grid size of $100 \times 100$. The initial specified shape of *typical_value* is shown in Fig. 15.3; its value rises from 0.0 V at the source end to 2.0 V at the drain end, but has a constant profile in the y-direction. These profiles show the situation before refinement and de-refinement. Refinement was done using the values *x_refine_diff* = 0.01 V and *y_refine_diff* = 0.10 V. De-refinement was done using the values *x_derefine_diff* = 0.15 V and *y_derefine_diff* = 0.15 V. The truncation ratio was taken as *trunc_ratio* = 0.001. At the end of the run, the grid shown in Fig. 15.4 was produced. Fig. 15.5 plots those points for which the variable *durationtype* = 1; these are the points which remain

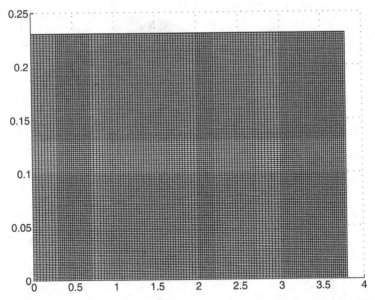

**Fig. 15.4** The final grid after refinement and de-refinement, based on an initial grid of $100 \times 100$.

after any de-refinement has taken place. The screen output of the program run is as follows:

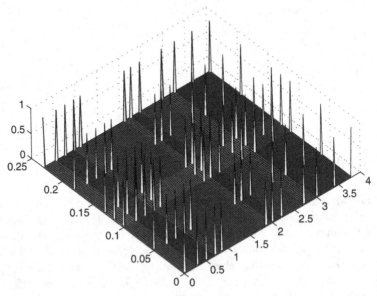

**Fig. 15.5** The final grid after refinement and de-refinement, showing the points for which *durationtype* is set to permanent.

```
-----------------------------------------------------------------
Program:   MESHGEN.CXX
Version:   1
Author:    E A B Cole
Modified: 22 September 2008
Device = Example1
Results will NOT be sent to disc.
Number of recesses = 2
Fieldplate included.
-----------------------------------------------------------------
Setting initial mesh ...... initial mesh set.
Source start            0.0000e+00
Source end              3.0000e-07
Start recess 1          5.0000e-07
Start recess 2          7.0000e-07
Gate start              8.5000e-07
Gate end                1.8500e-06
End recess   2          2.0000e-06
Fplate start            2.2000e-06
Fplate end              2.7000e-06
End recess 1            3.0000e-06
Drain start             3.4000e-06
Drain end               3.8000e-06

INITIAL MESH:
Number of points along x   =    129
Number of points along y   =    146
Number of interior points  = 14384
Number of exterior points  =  3776
Number of edge points      =    674
Total number of points     = 18834
Number of permanent points =     96

REFINED MESH:
Number of points along x   =    361
Number of points along y   =    146
Number of interior points  = 37296
Number of exterior points  = 14272
Number of edge points      =   1138
Total number of points     = 52706
Number of permanent points =     96

DE-REFINED MESH:
Number of points along x   =     20
Number of points along y   =      8
Number of interior points  =     85
Number of exterior points  =     19
Number of edge points      =     56
Total number of points     =    160
Number of permanent points =     96

Removing mesh ... mesh removed.
```

Notice that no refinement has taken place in the $y$-direction since the initial profile of *typical_value* was constant in that direction.

# Appendix A
# The theory of contractive mapping

This outline of the theory behind contractive mapping is based on the explanation of Michalewicz (1996), and involves an application of the *Banach theorem*. We first need to introduce the formal ideas of a *contractive mapping* and a *complete metric space* (Sutherland 1975).

**Definition A.1.** A *metric space* $\langle S, d \rangle$ consists of a set $S$ together with a mapping $d: S \times S \to R$ such that for all $p, q, r \in S$,

$$
\begin{array}{ll}
\text{(i)} & d(p,q) \geq 0, \\
\text{(ii)} & d(p,q) = 0 \quad \text{if and only if } p = q, \\
\text{(iii)} & d(p,q) = d(q,p), \\
\text{(iv)} & d(p,r) \leq d(p,q) + d(q,r).
\end{array}
\tag{A.1}
$$

**Definition A.2.** A mapping $\Phi: S \to S$ has a *fixed point* if there exists $p \in S$ such that $\Phi(p) = p$.

**Definition A.3.** If $\langle S, d \rangle$ is a metric space, a mapping $\Phi: S \to S$ is a *contractive mapping* if and only if there exists $\epsilon$ in the range $0 \leq \epsilon < 1$ such that, for all $p, q \in S$,

$$
d(\Phi(p), \Phi(q)) \leq \epsilon d(p, q).
\tag{A.2}
$$

**Definition A.4.** A *Cauchy sequence* is a sequence of elements $p_0, p_1, p_2, \ldots \in S$ of a metric space $\langle S, d \rangle$ if and only if for any $\epsilon \geq 0$ there exists $k$ such that $d(p_m, p_n) < \epsilon$ for all $m, n > k$.

**Definition A.5.** A metric space $\langle S, d \rangle$ is *complete* if any Cauchy sequence $\{p_n\}$ has a limit

$$
p \equiv \lim_{n \to \infty} p_n.
$$

**Theorem A.1.** *The Banach theorem: Suppose that $\langle S, d \rangle$ is a complete metric space. Let $\Phi: S \to S$ be a contractive mapping. Then $\Phi$ has an unique fixed point $p \in S$ such that for any $p_0 \in S$,*

E.A.B. Cole, *Mathematical and Numerical Modelling of Heterostructure Semiconductor Devices: From Theory to Programming*, DOI 10.1007/978-1-84882-937-4, © Springer-Verlag London Limited 2009

$$p = \lim_{i \to \infty} \Phi^i(p_0)$$

*where* $\Phi^0(p_0) \equiv p_0$ *and* $\Phi^{i+1}(p_0) \equiv \Phi(\Phi^i(p_0))$.

See the work of Sutherland (1975) for a proof of this theorem.

We are now in a position to apply the Banach theorem to our minimisation problem. In effect, we need to define an appropriate complete metric space and mapping $\Phi$ for our problem.

For a given minimisation problem with given variable ranges and precision, let $P$ be a population of fixed size POP_SIZE. Let $S$ be the set of all such populations:

$$S \equiv \{P : P \text{ has population size POP\_SIZE}\}.$$

Let $F(C)$ be the fitness of the chromosome $C \in P$, let $F_A(P)$ be the average fitness of the population $P$, and let $F_M$ be the minimum possible fitness for the problem. Then

$$F(C) \geq F_M \quad \text{for all } C \in P \text{ and all } P \in S,$$
$$F_A(P) \geq F_M \quad \text{for all } P \in S.$$

For any $P_1, P_2 \in S$, define the function $d(P_1, P_2)$ such that

$$d(P_1, P_2) = \begin{cases} 0 & \text{if } P_1 = P_2 \\ |1 + F_A(P_1) - F_M| + |1 + F_A(P_2) - F_M| & \text{otherwise.} \end{cases} \quad (A.3)$$

Then it is easily verified that properties (i) to (iv) of Equation (A.1) are satisfied. Hence $\langle S, d \rangle$ is a metric space. Note also that

$$d(P_1, P_2) \geq 2 \quad \text{if} \quad P_1 \neq P_2. \quad (A.4)$$

Hence this metric space is also complete since, for any $0 < \epsilon < 2$, $d(P_m, P_n) < \epsilon \Rightarrow P_m = P_n$. Completeness also follows automatically since $S$ is a finite set. Further, suppose that $P_1, P_2, P_1', P_2' \in S$, with $P_1' \neq P_2'$ and $F_A(P_i) > F_A(P_i')$ ($i = 1, 2$). Then either $d(P_1, P_2) = 0$ or

$$d(P_1', P_2') = |1 + F_A(P_1') - F_M| + |1 + F_A(P_2') - F_M|$$
$$< |1 + F_A(P_1) - F_M| + |1 + F_A(P_2) - F_M|$$
$$= d(P_1, P_2).$$

Hence

$$\{F_A(P_i') < F_A(P_i) \, (i = 1, 2) \text{ and } P_1' \neq P_2'\} \quad \Rightarrow \quad d(P_1', P_2') < d(P_1, P_2). \quad (A.5)$$

Now let $\Phi$ be the mapping which breeds a new population $\Phi(P)$ from population $P$ such that

$$F_A(\Phi(P)) < F_A(P). \quad (A.6)$$

Then for any two populations $P_1$ and $P_2$, Equation (A.5) implies that

$$d(\Phi(P_1), \Phi(P_2)) < d(P_1, P_2). \tag{A.7}$$

Hence $\Phi$ is a contractive mapping by Equation (A.2). It follows from the Banach theorem that there exists a unique fixed point population $P_M$ such that

$$P_M = \lim_{i \to \infty} \Phi^i(P_0)$$

for any initial $P_0 \in S$.

Hence to achieve this outcome in the optimisation problem, a breeding strategy must be chosen which produces Equation (A.6). In one breeding cycle, this will involve the production of offspring, and perhaps their destruction, until a population with a better fitness average is obtained.

# References

Adachi S (1985) J Appl Phys 58:R1

Ando Y, Itoh T (1988) Analysis of charge control in pseudomorphic two dimensional electron gas field effect transistors. IEEE Trans Electron Dev 35:2295–2301

Andricioaei I, Straub JE (1996) Generalised simulated annealing algorithms using Tsallis statistics: Application to conformational optimization of a tetrapeptide. Phys Rev E 53:R3055–R3058

Arpigny C (1963) Astrophys J 138:607–609

Asenov A, Reid D, Barker J, Cameron N, Beaumont S (1993) Application of quadrilateral finite elements for simulation of T-gate MESFETs and HEMTs. In: Proc of the int workshop on computational electronics, University of Leeds

Aymerich-Humet X, Serra-Mestres F, Millan J (1981) Solid State Electron 24:981

Barton TM (1989) Computer simulations. In: Snowden CM (ed) Semiconductor device modelling. Springer, London, pp 227–247

Bastard G (1981) Superlattice band structure in the envelope function approximation. Phys Rev B 24:5693–5697

Battocletti FE (1965) Polynomial approximation of the Fermi integral. Proc IEEE 53:2162–2163

Bednarczyk D, Bednarczyk J (1978) The approximation of the Fermi-Dirac integral $F(1/2)$. Phys Lett A 64:409

Bertoni A, Bordone P, Brunetti R, Jacoboni C, Reggiani S (2001) Numerical simulation of quantum logic gates based on quantum wires. VLSI Des 13:97–102

Blakemore JS (1962) Semiconductor statistics. Pergamon, Elmsford

Blakemore JS (1982) Approximation for the Fermi-Dirac integrals, especially the function $F(1/2)$ used to describe electron density in a semiconductor. Solid State Electron 25:1067–1076

Blanchard P, Devaney RL, Hall G (2002) Differential equations. Brooks/Cole, Stamford

Blotekjaer K (1970) Transport equations for electrons in two-valley semiconductors. IEEE Electron Dev Lett ED-17:38–47

E.A.B. Cole, *Mathematical and Numerical Modelling of*        393
*Heterostructure Semiconductor Devices: From Theory to Programming*,
DOI 10.1007/978-1-84882-937-4, © Springer-Verlag London Limited 2009

Bodine F, Holst M, Kerkhoven T (1993) Computation of the three-dimensional depletion approximation by Newton's method and multigrid. In: Proc of the int workshop on computational electronics, University of Leeds

Bordone P, Bertoni A, Brunetti R, Jacoboni C (2001) Wigner paths method in quantum transport with dissipation. VLSI Des 13:211–220

Bramble JH (1993) Multigrid methods. Longman, Harlow

Brandt A (1977) Math Comput 31:333–390

Brandt A, McCormick SF, Ruge J (1983) Multigrid methods for differential eigenproblems. SIAM J Sci Stat Comput 4:244–260

Brenner SC, Scott LR (1994) The mathematical theory of finite element method. Springer, Berlin

Briggs WL, McCormick SF (2000) A multigrid tutorial, 2nd edn. SIAM, Philadelphia

Capper DM (2001) C++ for scientists, engineers and mathematicians. Springer, London

Chapman S, Cowling TG (1970) The mathematical theory of non-uniform gases. Cambridge University Press, Cambridge

Cheng M-C, Chennupati R (1995) Evolution of non-equilibrium electron distribution functions at hydrodynamic scales in multi-valley semiconductors. J Phys D Appl Phys 28:160–173

Cody WJ, Thacher HC (1967) Rational Chebyshev approximations for Fermi-Dirac integrals. Math Comput 21:30–40

Cole EAB (2001) Integral evaluation in the mathematical and numerical modelling of high-electron-mobility transistors. J Phys, Condens Matter 13:515–524

Cole EAB (2004) The phase plane method for the solution of equations applied to semiconductor device modelling. Int J Numer Model Electron Netw Dev Fields 17:335–352

Cole EAB (2007) Integral evaluation in semiconductor device modelling using simulated annealing with Bose-Einstein condensation. Int J Numer Model Electron Netw Dev Fields 20:197–215

Cole EAB, Snowden CM (2000) The mathematical and computer modelling of microwave semiconductor devices. Math Comput Model 31:15–34

Cole EAB, Snowden CM, Boettcher T (1997) Solution of the coupled Poisson-Schrödinger equations using the multigrid method. Int J Numer Model Electron Netw, Dev Fields 10:121–136

Cole EAB, Boettcher T, Snowden CM (1998) Two-dimensional modelling of HEMTs using multigrids with quantum correction. VLSI Des 8:29–34

Cook RK, Frey J (1982) An efficient technique for two dimensional simulation of velocity overshoot effects in Si and GaAs devices. COMPEL—Int J Comput Math Electr Electron Eng 1:65–87

Das Sarma S, Kalia M, Nakayama M, Quinn JJ (1979) Stress and temperature dependence of subband structure in silicon inversion layers. Phys Rev B 19(12):6397–6406

Davies JH (1998) The physics of low-dimensional semiconductors—an introduction. Cambridge University Press, Cambridge de Graaff HC, Klaassen FM (1990) Compact transistor modelling for circuit design. Springer, Berlin

Deitel HM, Deitel PJ (2001) C++ how to program. Prentice Hall, New Jersey

Dekker AJ (1963) Solid state physics. MacMillan & Co, London

Dick E, Riemslagh K, Vierendeels J (1999) Multigrid methods VI. Springer, Berlin

Drury R, Snowden CM (1995) A quasi two-dimensional HEMT model for microwave CAD applications. IEEE Trans Electron Dev 42:1026–1031

Dugdale JS (1966) Entropy and low temperature physics. Hutchinson, London

Dunning Davies J (1996) Concise thermodynamics. Albion Press, Cape Torin

Ezaki T, Mori N, Hamaguchi C (1998) Electron-LA phonon interaction in a quantum dot. VLSI Des 8:225–230

Feng Y-K, Hintz A (1988) Simulation of submicrometer GaAs MESFETs using a full dynamic transport model. IEEE Trans Electron Dev ED-35:1419

Ferry DK (2001) Simulation at the start of the new millennium: crossing the quantum-classical threshold. VLSI Des 13:155–161

Ferry DK, Goodnick SM (1997) Transport in nanostructures. Cambridge University Press, Cambridge

Fish J (2008) A first course in finite elements. Wiley-Blackwell, New York

Franz AF, Franz GA, Selberherr S, Ringhofer C, Markowich P (1983) Finite Boxes—a generalization of the finite-difference method suitable for semiconductor device simulation. IEEE Trans Electron Dev ED-30:1070–1082

Gardner C (1991) Numerical simulation of steady state electron shock wave in a submicrometer semiconductor device. IEEE Trans Electron Dev 38:392–398

Gardner C (1994) The quantum hydrodynamic model for semiconductor devices. SIAM J Appl Math 54:409–427

Gol'dman II, Krivchenkov VD (1961) Problems in quantum mechanics. Pergamon, Oxford

Gora T, Williams F (1969) Theory of electronic states and transport in graded mixed semiconductors. Phys Rev 177:1179

Grinstein FF, Rabitz H, Askar A (1983) The multigrid method for accelerated solution of the discretized Schrödinger equation. J Comput Phys 51:423–443

Grubin HL, Buggeln RC (2001) Wigner function methods in modeling of switching in resonant tunneling devices. VLSI Des 13:221–228

Gummel HK (1964) A self-consistent interactive scheme for one-dimensional steady-state transistor calculations. IEEE Trans Electron Dev ED-11:455–465

Hansmann UHE (1997) Simulated annealing with Tsallis weights: a numerical comparison. arXiv:cond-mat/9710190v1

Hedin L, Lundqvist BI (1971) J Phys C 4:2064

Horenstein MN (1995) Microelectronic circuits and devices. Prentice Hall, New Jersey

Hussain S, Cole EAB, Snowden CM (2003) Hot electron numerical modelling of short gate length pHEMTs applied to novel field plate structures. Int J Numer Model Electron Netw Dev Fields 16:15–28

Ingham DB (1989) Numerical techniques—finite difference and boundary element method. In: Snowden CM (ed) Semiconductor device modelling. Springer, London, pp 34–48

Johnnson DS, Aragon CR, McGeoch LA, Schevon C (1987) Optimization by simulated annealing: an experimental evaluation, parts I and II. AT & T Bell, Laboratories, preprint

Jones EL (1966) Rational Chebyshev approximation of the Fermi-Dirac integrals. Proc IEEE 54:708

Jordan DW, Smith P (1979) Nonlinear ordinary differential equations. Clarendon, Oxford

Kalos MH, Whitlock PA (2008) Monte Carlo methods. Wiley-Blackwell, New York

Karmalkar S, Mishra UK (2001) Enhancement of breakdown voltage in AlGaN/GaN high electron mobility transistors using a field plate. IEEE Trans Electron Dev 48:1515–1521

Kelmanson M (2000) Rapid zonal algorithm for polyelliptic PDEs in domains with high aspect ratio. Math Comput Model 31:45–60

Kelsall RW (1998) Monte Carlo simulations of intersubband hole relaxation in a GaAs/AlAs quantum well. VLSI Des 8:367–373

King AC, Billingham J, Otto SR (2008) Differential equations: linear, nonlinear, ordinary, partial. Cambridge University Press, Cambridge

Korteweg DJ, de Vries F (1895) On the change of form of long waves advancing in a rectangular canal, and on a new type of long stationary waves. Philos Mag 39:422–443

Kreskovski JP (1987) A hybrid central difference scheme for solid state device simulation. IEEE Trans Electron Dev ED-34:1128–1133

Kumar A, Laux SE, Stern F (1990) Electron states in a GaAs quantum dot in a magnetic field. Phys Rev B 42:5166–5175

Landau D, Binder K (2005) A guide to Monte Carlo simulations in statistical physics. Cambridge University Press, Cambridge

Landau LI, Lifshitz EM (1981) Quantum mechanics (non-relativistic theory). Course of theoretical physics, vol 3. Butterworth-Heinemann, Stoneham

Landsberg PT (1961) Thermodynamics. Interscience, New York

Landsberg PT (1969) Solid state theory—methods and applications. Wiley-Interscience, London

Lax P (1968) Integrals of non linear equations of evolution and solitary waves. Commun Pure Appl Math 21:467–490

Lehmann D, Jasiukiewicz CZ (2002) About the shortcomings of using the Fang-Howard electron wave functions for phonon emission rate calculations in single heterostructures. Phys B, Condens Matter 316–317:226–229

Liou JJ (1992) Semiconductor device physics and modelling, Parts 1 and 2. IEEE Proc G 139:646–660

London F (1938a) Nature (Lond) 141:643

London F (1938b) Phys Rev 54:947

London F (1939) J Phys Chem 43:49

Lugli P (1993) Monte Carlo models and simulations. In: Snowden CM, Miles RE (eds) Compound semiconductor device modeling. Springer, Berlin, pp 210–231

Man KF, Tang KS, Kwong S (1999) Genetic algorithms. Springer, Berlin

McAndrew CC, Singhal K, Heasell EL (1985) IEEE Electron Dev Lett EDL-6:446

Meirav U, Kastner MA, Wind SJ (1990) Single-electron charging and periodic conductance resonance in GaAs Nanostructures. Phys Rev Lett 65:771–774

Michalewicz Z (1996) Genetic algorithms + data structures = evolution programs. Springer, Berlin

Mietzner T, Jakumeit J, Ravaioli U (2001) Influence of electron-electron interaction on electron distributions in short Si-MOSFETs analysed using the Local Iterative Monte Carlo technique. VLSI Des 13:175–178

Miles RE (1989) Review of semiconductor device physics. In: Snowden CM (ed) Semiconductor device modelling. Springer, London, pp 2–15

Mitchell M (1998) An introduction to genetic algorithms. MIT Press, Cambridge

Mobbs SD (1989) Numerical techniques—the finite element method. In: Snowden CM (ed) Semiconductor device modelling. Springer, London, pp 49–59

Morrow RA, Brownstein KR (1984) Model effective mass hamiltonians for abrupt heterojunctions and the associated wave function matching conditions. Phys Rev B 30:678

Morton KW, Suli E (1990) Finite volume methods and their analysis. Oxford University Computing Laboratory, Report 90/14, pp 3–25

Nakazato K, Blaikie RJ (1994) Single-electron memory. J Appl Phys 75:5123–5134

Ng S-H, Khoie R (1991) A two-dimensional self-consistent numerical model for high electron mobility transistor. IEEE Trans Electron Dev 38(4):852–861

Norris GB, Look DC, Kopp W, Klem J, Morkoc H (1985) Theoretical and experimental capacitance-voltage behavior of $Al_{0.3}Ga_{0.7}As$/GaAs modulation-doped heterojunctions: Relation of conduction band discontinuity to donor energy. Appl Phys Lett 47(4):423–425

Pothier H, Weis J, Haug RJ, Klitzing Kv (1993) Realization of an in-plane-gate single-electron transistor. Appl Phys Lett 62:3174–3176

Press WH, Flannery BP, Teukolsky SA, Vetterling WT (2002) Numerical recipes in C++. Cambridge University Press, Cambridge

Sakura N, Matsunaga K, Ishikura K, Takenaka I, Asano K, Iwata N, Kanamori M, Kuzuhara M (2000) In: 2000 IEEE MTT-S int microwave symp dig, pp 1715–1718

Sandborn PA, Rao A, Blakey PA (1989) An assessment of approximate nonstationary charge transport models used for GaAs device modeling. IEEE Trans Electron Dev 36:1244–1253

Scharfetter DL, Gummel HK (1968) Large signal analysis of a silicon read diode oscillator. IEEE Trans Electron Dev ED-16:64

Schiff LI (1955) Quantum mechanics. McGraw-Hill, New York

Scholze A, Wettstein A, Schenk A, Fichtner W (1998) Self-consistent calculations of the ground state and the capacitance of a 3D $Si/SiO_2$ quantum dot. VLSI Des 8:231–236

Scratton RE (1987) Further numerical methods. Edward Arnold, London

Selberherr S (1984) Analysis and simulation of semiconductor devices. Springer, New York

Smith GD (1985) Numerical solution of partial differential equations: finite difference methods. Clarendon, Oxford

Sneddon IN (1961) Special functions of mathematical physics and chemistry. Oliver & Boyd, Edinburgh

Snowden CM (1988) Semiconductor device modelling. Peter Peregrinus, London

Snowden CM, Loret D (1987) Two-dimensional hot-electron models for shortgate-length GaAs MESFETs. IEEE Trans Electron Dev ED-34:212

Stearns RL (1974) Bose-Einstein statistics and bosons. In: Besancon RM (ed) The encyclopedia of physics, 2nd edn. Van Nostrand-Reinhold, New York

Stern F (1970) Iteration methods for calculating self-consistent fields in semiconductor inversion layers. J Comput Phys 6:56–67

Stern F, Das Sarma S (1984) Electron energy levels in GaAs-Ga$_{1-x}$Al$_x$As heterojunctions. Phys Rev B 30:840–848

Stoer J, Bulirsch R (1980) Introduction to numerical analysis. Springer, New York

Streetman BG (1990) Solid state electronic devices. Prentice-Hall, London

Sutherland WA (1975) Introduction to metric and topological spaces. Clarendon, Oxford

Szu H, Hartley R (1987) Fast simulated annealing. Phys Lett A 122:157

Tang T-W (1984) Extension of the Scharfetter-Gummel algorithm to the energy balance equation. IEEE Trans Electron Dev ED-31:1912

Tang T-W, Ieong M-K (1995) Discretisation of flux densities in device simulations using optimum artificial diffusivity. IEE Trans Comput-Aided Des Integr Circuits Syst 14:1309–1315

Trellakis A, Ravaioli U (2001) Three-dimensional spectral solution of Schrödinger equation. VLSI Des 13:341–347

Tsai H-P, Coccioli R, Itoh T (2002) Time domain global modelling of EM propagation in semiconductor using irregular grids. Int J Num Modell Electron Netw Dev Fields 15:355–370

Tsallis C (1988) J Stat Phys 52:479

Tsallis C (1995) Non-extensive thermostatistics: brief review and comments. Phys A 221:277–290

Valadares EC, Sheard FW (1993) Self-consistent calculations of emitter accumulation layers in biased GaAs-Al$_x$Ga$_{1-x}$ heterostructures. Semicond Sci Technol 8:1746–1749

van Laarhoven PJM, Aarts EHL (1987) Simulated annealing: theory and applications. Kluwer Academic, Dordrecht

Varga RS (1962) Matrix iterative algebra. Prentice Hall, New Jersey

von Roos O (1983) Phys Rev B 27:7547

Wakejima A, Ota K, Matsunaga K, Contrata W, Kuzuhara M (2001) Field-modulating plate (FP) InGaP MESFET with high breakdown voltage and low distortion. In: MON4C-4 2001 IEEE radio frequency integrated circuits symposium

Wasshuber C, Kosina H, Selberherr S (1998) Singel-electron memories. VLSI Des 8:219–224

Wigner E (1932) On the quantum correction for thermodynamic equilibrium. Phys Rev 40:749–759

Wilkinson JH (1965) The algebraic eigenvalue problem. Clarendon, Oxford

Wilkinson JH, Reinsch C (1971) Handbook for automatic computation, vol 2. Springer, Berlin

Young DM (1971) Iterative solutions of large linear systems. Academic Press, New York

Yousef S (1996) Iterative methods for sparse linear systems. PWS-Kent, Boston

Zhu Q-G, Kroemer H (1983) Phys Rev B 27:3519

Zhu Y, Cangellaris A (2006) Multigrid finite element methods for electromagnetic field modelling. Wiley, Hoboken

# Index

E.A.B. Cole, *Mathematical and Numerical Modelling of*
*Heterostructure Semiconductor Devices: From Theory to Programming*,
DOI 10.1007/978-1-84882-937-4, © Springer-Verlag London Limited 2009